Guidelines for Vapor Cloud Explosion, Pressure Vessel Burst, BLEVE, and Flash Fire Hazards

Guidelines for Vapor Cloud Explosion, Pressure Vessel Burst, BLEVE, and Flash Fire Hazards

Second Edition

Center for Chemical Process Safety
New York, New York

A JOHN WILEY & SONS, INC., PUBLICATION

A Joint Publication of the Center for Chemical Process Safety of the American Institute of Chemical Engineers and John Wiley & Sons, Inc.

Published by John Wiley & Sons, Inc., Hoboken, New Jersey.
Published simultaneously in Canada.

For general information on our other products and services or for technical support, please contact our Customer Care Department within the United States at (800) 762-2974, outside the United States at (317) 572-3993 or fax (317) 572-4002.

Wiley also publishes its books in a variety of electronic formats. Some content that appears in print may not be available in electronic format. For information about Wiley products, visit our web site at www.wiley.com.

Library of Congress Cataloging-in-Publication Data:

Guidelines for vapor cloud explosion, pressure vessel burst, BLEVE, and flash fire hazards. — 2nd ed.
   p. cm.
  "Center for Chemical Process Safety."
  Includes index.
  ISBN 978-0-470-25147-8 (cloth)
  1. Chemical plants—Fires and fire prevention. 2. Chemical plants—Safety measures. 3. Pressure vessels—Safety measures. 4. Chemicals—Fires and fire prevention. 5. Explosions—Prevention. I. American Institute of Chemical Engineers. Center for Chemical Process Safety.
  TH9445.C47G86 2010
  660'.2804—dc22
                                        2010003430

10 9 8 7 6 5 4 3 2 1

CONTENTS

List of Tables .... *xi*

List of Figures .... *xiii*

Glossary .... *xxi*

Acknowledgements .... *xxv*

1. INTRODUCTION **1**

2. MANAGEMENT OVERVIEW **3**

2.1. Flash Fires 4

2.2. Vapor Cloud Explosions 4

2.3. Pressure Vessel Bursts 5

2.4. BLEVEs 5

2.5. Prediction methodologies 6

3. CASE HISTORIES **7**

3.1. Historical experience 7

3.2. Flash fires 7

3.2.1. Donnellson, Iowa, USA: Propane Fire 7

3.2.2. Lynchburg, Virginia, USA: Propane Fire 8

3.2.3. Quantum Chemicals, Morris, Illinois, USA: Olefins Unit Flash Fire 11

3.3. Vapor Cloud Explosions 13

3.3.1. Flixborough, UK: Vapor Cloud Explosion in Chemical Plant 13

3.3.2. Port Hudson, Missouri, USA: Vapor Cloud Explosion after

Propane Pipeline Failure 19

3.3.3.   Jackass Flats, Nevada, USA: Hydrogen-Air Explosion during Experiment 21

3.3.4.   Ufa, West-Siberia, USSR: Pipeline Rupture Resulting In a VCE 23

3.3.5.   Phillips, Pasadena, Texas USA: Propylene HDPE Unit VCE and BLEVEs 26

3.3.6.   BP, Texas City, Texas USA: Discharge from Atmospheric Vent Resulting in a VCE 29

3.4.   Pressure Vessel Burst 32

3.4.1.   Kaiser Aluminum, Gramercy, Louisiana USA: Alumina Process Pressure Vessel Burst 32

3.4.2.   Union Carbide Seadrift, Texas USA: Ethylene Oxide Distillation Column Pressure Vessel Burst 35

3.4.3.   Dana Corporation, Paris, Tennessee USA: Boiler Pressure Vessel Burst 36

3.5.   BLEVE 40

3.5.1.   Procter and Gamble, Worms, Germany: Liquid CO_2 Storage Vessel Explosion 40

3.5.2.   San Juan Ixhuatepec, Mexico City, Mexico: Series of BLEVEs at LPG Storage Facility 41

3.5.3.   San Carlos de la Rapita, Spain: Propylene Tank Truck Failure 44

3.5.4.   Crescent City, Illinois, USA: LPG Rail Car Derailment 45

3.5.5.   Kingman, Arizona USA: LPG Railroad Tank Car BLEVE 48

4.   BASIC CONCEPTS 51

4.1.   Atmospheric Vapor Cloud Dispersion 51

4.2.   Ignition 54

4.3.   Thermal Radiation 55

4.3.1.   Point-Source Model 56

4.3.2.   Solid-Flame Model 57

4.4.   Explosions — VCE 64

4.4.1. Deflagration 64

4.4.2. Detonation.. 66

4.5.   Blast Effects.. 70

4.5.1. Manifestation 70

4.5.2. Blast Loading.. 71

4.5.3. Ground Reflection... 73

4.5.4. Blast Scaling 74

5.   FLASH FIRES...77

5.1.   Overview of Experimental Research 81

5.1.1. China Lake and Frenchmen Flats cryogenic liquid tests.... 81

5.1.2. Maplin Sands Tests... 82

5.1.3. Musselbanks Propane Tests 83

5.1.4. HSE LPG Tests of Flash Fires and Jet Fires 84

5.2.   Flash-Fire Radiation Models................................. 86

5.3.   Sample Calculations 92

6.   VAPOR CLOUD EXPLOSIONS.................................97

6.1.   Introduction.. 97

6.1.1. Organization of Chapter....................................... 97

6.1.2. VCE Phenomena... 97

6.1.3. Definition of VCE... 99

6.1.4. Confinement and Congestion.............................. 101

6.2.   Vapor Cloud Deflagration Theory and Research............ 104

6.2.1. Laminar Burning Velocity and Flame Speed.................. 104

6.2.2. Mechanisms of Flame Acceleration................................ 107

6.2.3. Effect of Fuel Reactivity 111

6.2.4. Effect of Confinement... 113

6.2.6. Effects of Other Factors 140

6.2.7. University of Leeds Correlation............................ 141

6.2.8. TNO GAME Correlation 142

6.2.9. Shell CAM Correlation 143

6.3.   Vapor Cloud Detonation Theory and Research 152

6.3.1.   Direct Initiation of Vapor Cloud Detonations................. 152

6.3.2.   Detonability of Commonly Used Fuels 153

6.3.3.   Deflagration-to-Detonation Transition (DDT)................. 156

6.3.4.   Blast Effects Produced by Vapor Cloud Detonations 159

6.4.   VCE Prediction Methods 166

6.4.1.   TNT Equivalency Method 168

6.4.2.   VCE Blast Curve Methods.. 174

6.4.3.   TNO Multi-Energy Method 176

6.4.4.   Baker-Strehlow-Tang (BST) Method 188

6.4.5.   Congestion Assessment Method 201

6.4.6.   Numerical Methods... 207

6.5.   Sample problems.. 218

6.5.1.   Sample Problem – TNT Equivalence Method 218

6.5.2.   Sample Problem - Multi-Energy Method........................ 225

6.5.3.   BST Sample Problem... 233

6.5.4.   CAM Example Problem.. 236

7.   PRESSURE VESSEL BURSTS...................................241

7.1.   Mechanism of a PVB.. 242

7.1.1.   Accident Scenarios 242

7.1.2.   Damage Factors 244

7.1.3.   Phenomena... 244

7.1.4.   Factors that Reduce Available Explosion Energy........... 245

7.2.   Scaling Laws Used in PVB Analyses 246

7.3.   Blast Eeffects of Pressure-Vessel Bursts 247

7.3.1.   Free-Air Bursts of Gas-Filled, Massless, Spherical Pressure
          Vessels 248

7.3.2.   Effects Due to Surface Bursts 254

7.3.3.   Effects Due to Nonspherical Bursts 255

7.4.   Methods for Predicting Blast Effects from Vessel Bursts 260

7.4.1.   Development of Blast Curves 261

7.4.2.   Factors Influencing Blast Effects from Vessel Bursts 267

7.4.3.   Procedure for Calculating Blast Effects 267

7.4.4.   Adjustments for Vessel Temperature and Geometry 270

7.4.5.   Sample Problem: Airblast from a Spherical Vessel 274

7.5.   Fragments from a PVB 277

7.5.1.   Generation of Fragments from PVBs 277

7.5.2.   Initial Fragment Velocity for Ideal-Gas-Filled Vessels ... 279

7.5.3.   Ranges for Free Flying Fragments 288

7.5.4.   Ranges for Rocketing Fragments 293

7.5.5.   Statistical   Analysis   of   Fragments   from   Accidental
          Explosions 293

7.6.   Predicting Fragment Effects from Vessel Bursts 298

7.6.1.   Analytical Analysis 298

7.6.2.   Example Problem - Failure during Testing 306

8.   BASIC PRINCIPLES OF BLEVEs 311

8.1.   Introduction 311

8.2.   Definition of a BLEVE 311

8.3.   Theory 312

8.3.1.   Thermodynamics of Boiling 312

8.3.2.   Mechanics of Vessel Failure 313

8.3.3.   Description of a "Typical" BLEVE 317

8.4.   BLEVE Consequences 320

8.4.1.   Airblast 320

8.4.2.   Thermal Hazards 336

8.4.3.   Fragment and Debris Throw 342

8.4.4.   Ranges for Rocketing Fragments 344

8.5.   Analytical Models 349

8.6.   Sample Problems 349

8.6.1.   Sample Problem #1:   Calculation of Air Blast from
          BLEVEs 349

8.6.2.   Sample Problem #2:   Calculation of Fragments from

         BLEVEs...355

8.6.3.   Sample Problem #3: Thermal Radiation from a BLEVE 359

9.   REFERENCES 361

APPENDIX A – VIEW FACTORS FOR SELECTED CONFIGURATIONS..407

APPENDIX B – TABULATION OF SOME GAS PROPERTIES IN METRIC UNITS 409

APPENDIX C – CONVERSION FACTORS TO SI FOR SELECTED QUANTITIES...411

LIST OF TABLES

Table 3.1. Pressure at time of explosion in the digestion area equipment......... 33

Table 4.1. Explosion Properties of Flammable Gases and Vapors in Air at Standard Pressure [a] 53

Table 4.2. Initiation Energies for Deflagration and Detonation for Some Fuel-Air Mixtures[a] 55

Table 4.3. Characteristic detonation cell sizes for some stoichiometric fuel-air mixtures[a] 70

Table 5.1. Experimental Conditions and Flame Speeds for HSL LPG Tests.... 85

Table 5.2. Wide-Gauge Radiometer Measurements of Surface Emissive Power for Flash and Pool Fires 90

Table 5.3. Results of calculations 95

Table 6.1. Test Results of VCE Deflagration in Tubes.................................. 114

Table 6.2. Maximum flame speeds for various fuels and configurations
(Van Wingerden and Zeeuwen, 1983) 117

Table 6.3. Small scale test results on VCE deflagration in 2-D configuration 118

Table 6.4. Large scale test results on VCE deflagration in 2-D configuration 123

Table 6.5. Results of experiments under unconfined conditions without obstacles
.... 126

Table 6.6. Experimental results on VCE deflagration under unconfined conditions with obstacles (low congestion) 129

Table 6.7. Flame speed and overpressure from 3-D configurations................ 131

Table 6.8. Effect of blockage ratio (Mercx, 1992)... 139

Table 6.9. CAM Coefficients 144

Table 6.10. Critical initiation energy and detonability for hydrocarbon fuels (Matsui and Lee, 1978) 155

Table 6.11. DDT in ethylene/air mixtures.. 158

Table 6.12. Detonation properties for some stoichiometric fuel-air mixtures (McBride, 1996).. 160

Table 6.13. BST flame speed correlations (flame speed Mach no. M_f) (Pierorazio et al. 2004) 192

Table 6.14. Congestion description for the BST method................................. 195

Table 6.15. CAM Fuel Factor F and Expansion Ratio E for Common Fuels . 202

Table 6.16. CFD codes used to predict VCE blast loads 212

Table 6.17. Side-on peak overpressure for several distances from charge...... 221

Table 6.18. Side-On peak overpressure for several distances from charge expressing explosion severity of the Flixborough vapor cloud explosion. 224

Table 6.19. Characteristics and locations of fuel-air charges expressing potential explosion severity of the Flixborough vapor cloud... 226

Table 6.20. Nondimensionalized blast parameters at 1,000m distance from two charges, read from charts in Figure 6.40... 227

Table 6.21. Side-on peak overpressure and positive-phase duration of blast produced by Charge I   (E = 175,000 MJ, strength number 10)....................... 229

Table 6.22. Side-on peak overpressure and positive-phase duration of blast produced by Charge II   (E = 1,412,800 MJ, strength number 2) 229

Table 6.23. Constants used in the BST sample problem................................. 233

Table 6.24. Blast overpressure and impulse for different standoff distances using the BST method 235

Table 6.25. Predicted blast loads using the CAM method 238

Table 7.1. Drag coefficients (Baker et al. 1983) 292

Table 7.2. Groups of like PVB events used in fragmentation statistical analysis 294

Table 7.3. Ranges for various initial trajectory angles.................................... 309

Table 8.1. Empirical relationships for fireball durations and diameters 338

Table 8.2. Analytical relationships for fireball durations and diameters......... 339

Table 8.3. Thermodynamic data for propane 350

Table 8.4. Results of sample problem #1 355

Table 8.5. Results of sample problem #2 358

Table 8.6. Results of sample problem #3 360

LIST OF FIGURES

Figure 3.1. Details of Lynchburg, VA accident site..10

Figure 3.2. Damage resulting from the Morris, Illinois flash fire. 12

Figure 3.3 Flixborough works prior to the explosion.......................................13

Figure 3.4 Flixborough cyclohexane oxidation plant (six reactors on left).......14

Figure 3.5. Area of spill showing removed reactor..15

Figure 3.6. Bypass on cyclohexane reactors at Flixborough............................15

Figure 3.7 Aerial view of damage to the Flixborough works............................16

Figure 3.8. Damage to the Office Block and Process Areas at the Flixborough works..17

Figure 3.9. Blast-distance relationship outside the cloud area of the Flixborough explosion...18

Figure 3.10. Damage to a farm 600 m (2,000 ft) from explosion center...........20

Figure 3.11. Damage to a home 450 m (1,500 ft) from the blast center............20

Figure 3.12. Test-cell layout at Jackass Flats, NV...22

Figure 3.13 Ufa accident: (a) topographical sketch of demolished area with directions trees fell shown by arrows; (b) terrain profile (not to scale). (Makhviladze, 2002) 24

Figure 3.14. Aerial views of Ufa accident site: (upper) broad view of the forest and rail line (Makhviladze, 2002); (lower) closer view of the area where the trains passed. (Lewis, 1989)..25

Figure 3.15. Phillips Pasadena plant prior to the incident................................26

Figure 3.16 BLEVE at the Phillips Pasadena site..27

Figure 3.17 Phillips Pasadena process area damage...28

Figure 3.18. Explosion locations at Phillips Pasadena site...............................29

Figure 3.19. Aerial view of the ISOM unit after the explosion. (CSB, 2007)...30

Figure 3.20. Destroyed trailers west of the blowdown drum. (Arrow in upper left of the figure) 31

Figure 3.21. Kaiser slurry digester area flow schematic...................................32

Figure 3.22. Kaiser aluminum digester area before and after the explosion.

(MSHA, 1999) 34

Figure 3.23. Remaining No. 1 ORS base section and skirt with the attached vertical thermosyphon reboiler. 36

Figure 3.24. Final location of the boiler after explosion. 37

Figure 3.25. Hole created by boiler through roll-up door wall (west wall)....... 38

Figure 3.26. Damaged exterior wall viewed from inside boiler room (east wall). 39

Figure 3.27. View of east wall from outside plant (note rear boiler door in ditch). 39

Figure 3.28. Interior wall of boiler room (south wall). 40

Figure 3.29. Installation layout at San Juan Ixhuatepec, Mexico.................... 42

Figure 3.30. Area of damage at San Juan Ixhuatepec, Mexico. 42

Figure 3.31. Directional preference of projected cylinder fragments of cylindrical shape. 43

Figure 3.32. Reconstruction of scene of the San Carlos de la Rapita campsite disaster. 44

Figure 3.33. Derailment configuration. 46

Figure 3.34. Trajectories of tank car fragments. 47

Figure 3.35. Kingman explosion fireball. (Sherry, 1974) 49

Figure 4.1. Configuration for radiative exchange between two differential elements. 62

Figure 4.2. View factor of a fireball.. 63

Figure 4.3. Temperature distribution across a laminar flame........................... 65

Figure 4.4. Positive feedback, the basic mechanism of a gas explosion. 66

Figure 4.5. The CJ-model. 67

Figure 4.6. The ZND-model. 68

Figure 4.7. Instability of ZND-concept of a detonation wave.......................... 69

Figure 4.8. Cellular structure of a detonation. 69

Figure 4.9. Blast wave shapes... 71

Figure 4.10. Interaction of a blast wave with a rigid structure (Baker, 1973)... 72

Figure 4.11. Blast-wave scaling. (Baker, 1973).. 75

Figure 5.1. Illustration of idealized flame fronts for a flash fire. 78

Figure 5.2. Flame front progression in LPG vapor cloud (2.0 m/s wind, 2.6 kg/s discharge for 51 s, ignition 25 m from source, [HSL, 2001)........................... 79

Figure 5.3. Moment of ignition in a propane-air cloud. (Zeeuwen et al., 1983)84

Figure 5.4. Radiant heat flux from HSL LPG flash fire test 14. 86

Figure 5.5. Schematic representation of unconfined flash fire........................ 87

Figure 5.6. Flame shape assumptions. (*= ignition source)........................... 91

Figure 5.7. Definition of view factors for a vertical, flat radiator. 94

Figure 5.8. Graphical presentation for sample problem of the radiation heat flux as a function of time.. 96

Figure 6.1. Three dimensional (3-D) flame expansion geometry.................. 101

Figure 6.2. Two dimensional (2-D) flame expansion geometry.................... 102

Figure 6.3. One dimensional (1-D) flame expansion geometry 103

Figure 6.4. Flame speed S_s, gas flow velocity S_g, and laminar burning velocity, S_u, for various methane-air mixture equivalence ratios at 1 atm and 298° K (Andrews, 1997). 105

Figure 6.5. Overpressure as a function of flame speed for three geometries. (Tube-like geometry is 1-D; double plane is 2-D, and dense obstacle environment is 3-D confinement). (Kuhl et al. 1973) 108

Figure 6.6. Flame propagation in 1-D (channel) and 2-D (sector) geometries. (Stock et al. 1989)... 109

Figure 6.7. Flow visualization image sequence of flame propagation over rectangular, square and cylindrical obstacles with stoichiometric fuel-air mixtures. Left frame, t=32ms after ignition; time between frames is 1.66 ms. (Hargrave, 2002) 110

Figure 6.8. Flame speeds versus distance, non-dimensionalized with respective laminar flame speeds (fixed test conditions P = 6D, ABR = 0.5, H = 2D). 112

Figure 6.9. Flame speed versus distance for three different fuels. (Mercx 1992) 113

Figure 6.10. Experimental set-up for TNO small scale tests.......................... 116

Figure 6.11. Experimental set-up for TNO tests with horizontal obstacles. (van Wingerden, 1989)... 118

Figure 6.12. Blast produced from double plate configuration with variable heights. 119

Figure 6.13. Experimental set-up for 2-D configuration (van Wingerden, 1989).

.... 120

Figure 6.14. Flame speeds versus distance for various porosities. (van Wingerden 1989).. 120

Figure 6.15. Experimental setup to study flame propagation in a cylindrical geometry. (Moen, 1980b) 121

Figure 6.16. Flame speed-distance relationship of methane-air flames in a double plate geometry (2.5 × 2.5 m), by Moen et al. (1980b). (a) $H/D = 0.34$; (b) $H/D = 0.25$; (c) $H/D = 0.13$... 122

Figure 6.17. Large-scale test setup for investigation of flame propagation in a cylindrical geometry. Dimensions: 25 m long; 12.5 m wide; and 1 m high. Obstacle diameter 0.5 m. 123

Figure 6.18. Experimental apparatus for investigation of effects of pipe racks on flame propagation. (Harrison and Eyre, 1986 and 1987)............................... 128

Figure 6.19. Flame speed-distance graph showing transition to detonation in a cyclohexane-air experiment. (Harris and Wickens 1989).............................. 132

Figure 6.20. Effect of obstacle pitch on flame speed (dimensional distance). (van Wingerden 1989) 134

Figure 6.21. Effect of obstacle pitch on flame speed (non-dimensional distance). (van Wingerden 1989) 135

Figure 6.22. Flame speed versus distance for different pitches (Mercx, 1992).136

Figure 6.23. Flame speed versus dimensionless distance R/P. (Mercx, 1992)137

Figure 6.24. Effect of ABR on flame speed. (van Wingerden and Hjertager, 1991) 138

Figure 6.25. Effect of obstacle shape on pressure. (Hjertager 1984) 139

Figure 6.26. Flame velocity, peak overpressure, and overpressure duration in gas cloud explosions following vessels bursts. (Giesbrecht et al. 1981)................ 145

Figure 6.27. Maximum overpressure in vapor cloud explosions after critical-flow propane jet release dependent on orifice diameter: (a) undisturbed jet; (b) jet into obstacles and confinement. 146

Figure 6.28. Experimental apparatus for investigating jet ignition of ethylene-air and hydrogen-air mixtures. (Schildknecht et al., 1984) 147

Figure 6.29. Effect of the gap between two congested areas on flame speeds. (van Wingerden, 1989) 149

Figure 6.30. Comparison of flame propagation between two adjacent arrays in medium and large scale tests (Mercx 1992). 150

Figure 6.31 Flame speed/distance showing acceleration in the region of repeated obstacles and deceleration on emerging into the unobstructed region , cyclohexane-air experiment. (Harris and Wickens, 1989) 151

Figure 6.32. Flame speed/distance showing rapid deceleration on exit from a region containing repeated obstacles into an unobstructed region, natural gas-air. (Harris and Wickens, 1989) 151

Figure 6.33. Critical initiation energies of some fuel-air mixtures. (Bull et al. 1978) 153

Figure 6.34. Positive phase characteristics from VCE detonations. (Brossard et al. 1983) 162

Figure 6.35. Total amplitude of characteristics from VCE detonations. (Brossard et al. 1983) 162

Figure 6.36. Positive overpressure versus distance for gaseous detonations. . 164

Figure 6.37. Positive impulse versus distance (c_0 is the same as a_0) (Dorofeev, 1995) 164

Figure 6.38. Positive overpressure versus distance for heterogeneous detonations. (Dorofeev, 1995) 165

Figure 6.39. Positive impulse versus distance for heterogeneous detonations. (Dorofeev, 1995) 166

Figure 6.40. Side-on blast parameters for a TNT hemispherical surface burst. (Lees, 1996 after Kingery and Bulmash, 1984) 170

Figure 6.41. Multi-energy method positive-phase side-on blast overpressure and duration curves. 178

Figure 6.42. Observed overpressures from three datasets correlated to the parameter combination in the GAME relation. (Mercx, 2000) 188

Figure 6.43. BST positive overpressure vs. distance for various flame speeds.190

Figure 6.44. BST positive impulse vs. distance for various flame speeds. 190

Figure 6.45. BST negative overpressure vs. distance for various flame speeds.191

Figure 6.46. BST negative impulse vs. distance for various flame speeds. 191

Figure 6.47. Quasi two dimensional (2.5-D) flame expansion geometry........ 193

Figure 6.48. Scaled source overpressure as a function of Scaled Severity Index.
.... 205

Figure 6.49. CAMS pressure decay as a function of distance $(R_0+r)/R_0$ for P_{max} =
0.2, 0.5, 1, 2 4 and 8 bar (contours bottom to top). 206

Figure 6.50. BFETS FLACS model and target distribution. 214

Figure 6.51. Comparison of FLACS results and experimental data (internal)
pressure histories). 214

Figure 6.52. Comparison of FLACS results and experimental data (external).215

Figure 6.53. FLACS model of an onshore installation. 216

Figure 6.54. Flame front contour. 216

Figure 6.55. Pressure contours at selected times (northeast view). 217

Figure 6.56. (a) View of a storage tank farm for liquefied hydrocarbons. (b) Plot
plan of the tank farm. 219

Figure 6.57. Plot plan of Nypro Ltd. plant at Flixborough, UK. 222

Figure 7.1. Pressure-time history of a blast wave from a PVB (Esparza and Baker
1977a). 247

Figure 7.2. Pressure contours of a blast field for a cylindrical burst (X and Y axes
are scaled distances based on characteristic distance r_o). (Geng, 2009).......... 257

Figure 7.3. Pressure contours of a blast field for an elevated spherical burst (X and
Y axes are scaled distances based on characteristic distance r_o). (Geng, 2009)258

Figure 7.4. Surface burst scaled side-on overpressure generated by a cylindrical
burst at angles of 0, 45 and 90° compared to a bursting sphere. (Geng, 2009) 259

Figure 7.5. Surface burst scaled side-on impulse generated by a cylindrical burst
at angles of 0, 45 and 90° compared to a bursting sphere. (Geng, 2009)......... 260

Figure 7.6. Positive overpressure curves for various vessel pressures. (Tang, et al.
1996) 263

Figure 7.7. Negative pressure curves for various vessel pressures. 264

Figure 7.8. Positive impulse curves for various vessel pressures. (Tang, et al.
1996) 265

Figure 7.9. Negative impulse curves for various vessel pressures. (Tang, et al.
1996) 266

Figure 7.10. Adjustment factors for cylindrical free air PVBs compared to a spherical free air burst. (Geng, 2009)..272

Figure 7.11. Adjustment factors for elevated spherical PVBs compared to a hemispherical surface burst. (Geng, 2009)..273

Figure 7.12. Equivalent surface burst cylindrical PVB geometries to a free air burst 274

Figure 7.13. Fragment velocity versus scaled pressure. (Baker, 1983)...........281

Figure 7.14. Adjustment factor for unequal mass fragments (Baker et al. 1983)283

Figure 7.15. Calculated fragment velocities for a gas-filled sphere with $\gamma = 1.4$ (taken from Baum 1984; results of Baker et al. 1978a were added). 286

Figure 7.16. Scaled curves for fragment range predictions (taken from Baker et al. 1983) (– – –): neglecting fluid dynamic forces. 289

Figure 7.17. Fragment range distribution for event groups 1 and 2 (Baker et al. 1978b). 295

Figure 7.18. Fragment range distribution for event groups 3, 4, 5, and 6 (Baker et al. 1978b). 296

Figure 7.19. Fragment-mass distribution for event groups 2 and 3 (Baker et al. 1978b). 297

Figure 7.20. Fragment-mass distribution for event group 6 (Baker et al. 1978b). 297

Figure 8.1. 500-gallon (1.9 m³) pressure vessel opened and flattened on the ground after a fire-induced BLEVE. (Birk et al., 2003)....................................315

Figure 8.2. Fire test of 500-gallon (1.9 m³) propane pressure vessel resulting in massive jet release (not a BLEVE). (Birk et al., 2003) 315

Figure 8.3. Sample of high temperature stress rupture data for two pressure vessel steels. (Birk and Yoon, 2006)..317

Figure 8.4. Comparison between energy definitions: Eex, wo/Eex, Br. 323

Figure 8.5. Overpressure Decay Curve for Propane Tank BLEVE. (Birk et al., 2007)..329

Figure 8.6. Calculation of energy of flashing liquids and pressure vessel bursts filled with vapor or nonideal gas...330

Figure 8.7. Measured first peak overpressures vs scaled distance (based on vapor energy) from 2000-liter propane tank BLEVEs. (Birk et al., 2007)................335

Figure 8.8. Measured first peak overpressures vs scaled distance (based on liquid

GLOSSARY

Blast: A transient change in the gas density, pressure, and velocity of the air surrounding an explosion point. The initial change can be either discontinuous or gradual. A discontinuous change is referred to as a shock wave, and a gradual change is known as a pressure wave.

BLEVE (Boiling Liquid, Expanding Vapor Explosion): The explosively rapid vaporization and corresponding release of energy of a liquid, flammable or otherwise, upon its sudden release from containment under greater-than-atmospheric pressure at a temperature above its atmospheric boiling point. A BLEVE is often accompanied by a fireball if the suddenly depressurized liquid is flammable and its release results from vessel failure caused by an external fire. The energy released during flashing vaporization may contribute to a shock wave.

Burning velocity: The velocity of propagation of a flame burning through a flammable gas-air mixture. This velocity is measured relative to the unburned gases immediately ahead of the flame front. Laminar burning velocity is a fundamental property of a gas-air mixture.

Deflagration: A propagating chemical reaction of a substance in which the reaction front advances into the unreacted substance rapidly but at less than sonic velocity in the unreacted material.

Detonation: A propagating chemical reaction of a substance in which the reaction front advances into the unreacted substance at or greater than sonic velocity in the unreacted material.

Emissivity: The ratio of radiant energy emitted by a surface to that emitted by a black body of the same temperature.

Emissive power: The total radiative power discharged from the surface of a fire per unit area (also referred to as surface-emissive power).

Explosion: A release of energy that causes a blast.

Fireball: A burning fuel-air cloud whose energy is emitted primarily in the form of radiant heat. The inner core of the cloud consists almost completely of fuel, whereas the outer layer (where ignition first occurs) consists of a flammable fuel-air mixture. As the buoyancy forces of hot gases increase, the burning cloud tends to rise, expand, and assume a spherical shape.

Flame speed: The speed of a flame burning through a flammable mixture of gas and air measured relative to a fixed observer, that is, the sum of the burning and translational velocities of the unburned gases.

Flammable limits: The minimum and maximum concentrations of combustible material in a homogeneous mixture with a gaseous oxidizer that will propagate a flame.

Flash vaporization: The instantaneous vaporization of some or all a liquid whose temperature is above its atmospheric boiling point when its pressure is suddenly reduced to atmospheric.

Flash fire: The combustion of a flammable gas or vapor and air mixture in which the flame propagates through that mixture in a manner such that negligible or no damaging overpressure is generated.

Impulse: A measure that can be used to define the ability of a blast wave to do damage. It is calculated by the integration of the pressure-time curve.

Jet: A discharge of liquid, vapor, or gas into free space from an orifice, the momentum of which induces the surrounding atmosphere to mix with the discharged material.

Lean mixture: A mixture of flammable gas or vapor and air in which the fuel concentration is below the fuel's lower limit of flammability (LFL).

Negative phase: That portion of a blast wave whose pressure is below ambient.

Overpressure: Any pressure above atmospheric caused by a blast.

Positive phase: That portion of a blast wave whose pressure is above ambient.

Pressure wave: See Blast.

Reflected pressure: Impulse or pressure experienced by an object facing a blast.

Rich mixture: A mixture of flammable gas or vapor and air in which the fuel concentration is above the fuel's upper limit of flammability (UFL).

Shock wave: See Blast.

Side-on pressure: The impulse or pressure experienced by an object as a blast wave passes by it.

Stoichiometric ratio: The precise ratio of air (or oxygen) and flammable material which would allow all oxygen present to combine with all flammable material present to produce fully oxidized products.

Superheat limit temperature: The temperature of a liquid above which flash vaporization can proceed explosively.

Surface-emissive power: See Emissive power.

Transmissivity: The fraction of radiant energy transmitted from a radiating object through the atmosphere to a target after reduction by atmospheric absorption and scattering.

TNT equivalence: The amount of TNT (trinitrotoluene) that would produce observed damage effects similar to those of the explosion under consideration. For non-dense phase explosions, the equivalence has meaning only at a considerable distance from the explosion source, where the nature of the blast wave arising is more or less comparable with that of TNT.

Turbulence: A random-flow motion of a fluid superimposed on its mean flow.

Vapor cloud explosion: The explosion resulting from the ignition of a cloud of flammable vapor, gas, or mist in which flame speeds accelerate to sufficiently high velocities to produce significant overpressure.

View factor: The ratio of the incident radiation received by a surface to the emissive power from the emitting surface per unit area.

ACKNOWLEDGMENTS

This *Guideline* book was developed as a result of two projects sponsored by The Center for Chemical Process Safety of the American Institute of Chemical Engineers. The second edition of the *Guideline* was prepared under the direction of the Vapor Cloud Explosion subcommittee comprised of the following engineers and scientists:

Larry J. Moore (FM Global), chair

Chris R. Buchwald (ExxonMobil)

Gary A. Fitzgerald (ABS Consulting)

Steve Hall (BP plc)

Randy Hawkins (RRS Engineering)

David D. Herrmann (DuPont)

Phil Partridge (The Dow Chemical Company)

Steve Gill Sigmon (Honeywell – Specialty Materials)

James Slaugh (LyondellBasell)

Jan C. Windhorst (NOVA Chemical, emeritus)

The second edition was authored by the Blast Effects group at Baker Engineering and Risk Consultants, Inc. The authors were:

Quentin A. Baker

Ming Jun Tang

Adrian J. Pierorazio

A. M. Birk (Queen's University)

John L. Woodward

Ernesto Salzano (CNR – Institute of Research on Combustion)

Jihui Geng

Donald E. Ketchum

Philip J. Parsons

J. Kelly Thomas

Benjamin Daudonnet

The authors and the subcommittee were well supported during the project by John Davenport, who served as the CCPS staff representative.

The efforts of the document editors at BakerRisk are gratefully acknowledged for their contributions in editing, layout and assembly of the book. They are Moira Woodhouse and Phyllis Whiteaker.

CCPS also gratefully acknowledges the comments submitted by the following peer reviews:

Eric Lenior (AIU Holding)

Fred Henselwood (NOVA Chemicals)

John Alderman (RRS Engineering)

Lisa Morrison (BP International Limited)

Mark Whitney (ABS Consulting)

William Vogtman (SIS-TECH Solutions)

David Clark (DuPont, emeritus)

A NOTE ON NOMENCLATURE AND UNITS

The equations in this volume are from a number of reference sources, not all of which use consistent nomenclature (symbols) and units. In order to facilitate comparisons within sources, the conventions of each source were presented unchanged.

Nomenclature and units are given after each equation (or set of equations) in the text. Readers should ensure that they use the proper values when applying these equations to their problems.

1. INTRODUCTION

The American Institute of Chemical Engineers (AIChE) has been involved with process safety and loss control for chemical and petrochemical plants for more than forty years. Through its strong ties with process designers, builders, operators, safety professionals, and academia, AIChE has enhanced communication and fostered improvements in the safety standards of the industry. Its publications and symposia on causes of accidents and methods of prevention have become information resources for the chemical engineering profession.

Early in 1985, AIChE established the Center for Chemical Process Safety (CCPS) to serve as a focus for a continuing program for process safety. The first CCPS project was the publication of a document entitled *Guidelines for Hazard Evaluation Procedures*. In 1987, *Guidelines for Use of Vapor Cloud Dispersion Models* was published, and in 1989, *Guidelines for Chemical Process Quantitative Risk Analysis* and *Guidelines for Technical Management of Chemical Process Safety* were published.

The first edition of this book was published in 1994, and it remains the most in-depth technical material produced in a CCPS project.

This current edition is intended to provide an overview of methods for practicing engineers to estimate the characteristics of a flash fire, vapor cloud explosion (VCE), pressure vessel burst (PVB), and boiling-liquid-expanding-vapor explosion (BLEVEs). This edition summarizes and evaluates these methods, identifies areas in which information is lacking, and provides an overview of ongoing work in the field. The arrangement of this book is considerably different from previous editions, including separating pressure vessel bursts into its own chapter.

For a person new to the field of explosion and flash fire hazard evaluation this book provides a starting point for understanding the phenomena covered and presents methods for calculating the possible consequences of incidents. It provides an overview of research in the field and numerous references for readers with more experience. Managers will be able to utilize this book to develop a basic understanding of the governing phenomena, the calculational methods to estimate consequences, and the limitations of each method.

Chapter 2 of this book was written for managers, and it contains an overview of the hazards associated with flash fires, vapor cloud explosions (VCEs), pressure vessel bursts (PVBs), and boiling liquid expanding vapor explosions (BLEVEs). Chapter 3 provides a review of case histories involving these hazards. These case histories illustrate the conditions present at the time of the event, highlighting the serious consequences of such events and the need for evaluation of the hazards.

Chapter 4 provides an overview of the basic concepts associated with flash fires, VCEs, PVBs and BLEVEs. This chapter includes a discussion of dispersion, ignition, fires, thermal radiation, VCEs, and blast waves.

Chapters 5 through 8 separately address the phenomena of each type of hazard (i.e., flash fires, VCEs, PVBs and BLEVEs). These chapters include a description of the relevant phenomena, an overview of the related past and present experimental work and theoretical research, and selected consequence estimation methodologies. Each chapter includes sample problems to illustrate application of the methodologies presented. References are provided in Chapter 9.

The goal of this book is to provide the reader with an adequate understanding of the basic physical principles of flash fires and explosions and the current state of the art in hazard estimation methodologies. It is not the goal of this book to provide a comprehensive discussion of all of the experimental work and theoretical research that has been performed in the field of flash fire and explosion evaluation.

This book does not address subjects such as toxic effects, confined explosions (e.g., an explosion within a building), dust explosions, runaway reactions, condensed-phase explosions, pool fires, jet flames, or structural responses of buildings. Furthermore, no attempt is made to address frequency or likelihood of accident scenarios. References to other works related to these topics are provided for the interested reader.

2. MANAGEMENT OVERVIEW

Accidents involving fires and explosions have occurred since flammable liquids or gases began to be used broadly as fuels for industrial and consumer purposes. Summaries of such accidents are given by Davenport (1977), Strehlow and Baker (1976), Lees (1980), and Lenoir and Davenport (1993). Among the types of accidents that can occur with flammable gases or liquids are a BLEVE, flash fire, and VCE, depending on the circumstances .

Industrial fires and explosions are neither infrequent nor inconsequential. According to Marsh (2007), twenty-three major industrial explosion and fire accidents were reported worldwide in 2006. These explosions directly resulted in over 67 fatalities and 394 injuries. Of these, chemical plants accounted for 24 fatalities and 56 injuries, with 22 fatalities and 29 injuries occurring in a single accident in China. In addition, vandalism of fuel pipelines accounted for another 336 fatalities and 124 injuries. Combined accident and vandalism property damage losses totaled $259 million.

This book explores the consequences of accidental releases of flammable materials and provides practical means of estimating the consequences of fire and explosion hazards, knowledge that is essential for proper process safety management of an industrial facility. Ignition of flammable materials can produce thermal and blast overpressure hazards, the strength of which increases with the combustion energy of the material involved and how quickly that energy is released and dissipated. This book also explores explosion hazards not associated with an accidental release of flammable materials, such as failure of a pressure vessel, with and without liquid content that flashes to vapor. With a clear understanding of the threats posed by these hazards, personnel can be located and buildings can be designed to provide an appropriate level of protection. For example, overpredicting the potential loads on an occupied building may prompt unnecessary and costly structural upgrades, while underpredicting these loads may leave buildings and persons inside them vulnerable.

The likelihood of such occurrences can be reduced by appropriate process design and reliability engineering that meets or exceeds established industry standards and practices. These practices include well-designed pressure relief and blowdown systems, adequate maintenance and inspection programs, and management of human factors in system design. In addition, and perhaps most important to the success of risk management efforts, the full support of responsible management is required.

Mathematical models for calculating the consequences of such events can be employed to support mitigation efforts. Mitigating measures may include reduction of inventory; reduction of vessel volumes; isolation and depressurization systems, modification of plant siting and layout, including location and reinforcement of control rooms; strengthening of vessels; and improved mechanical integrity.

Knowledge of the consequences of flash fires, VCEs, PVBs, and BLEVEs has grown significantly in recent years as a result of international study and research efforts, and continuing incidents. Insights gained regarding the generation of overpressure, radiation, and fragmentation has resulted in the development of reasonably descriptive models for calculating the effects of these phenomena.

The remainder of this chapter provides brief descriptions of flash fires, VCEs, PVBs, and BLEVEs. Several examples of flash fires and explosions are provided that illustrate how these events occur under relevant conditions, highlighting the serious consequences of such events and the need for predicting their consequences. Chapter 3 of this book provides more detailed review of case histories for all of these event types.

2.1.   FLASH FIRES

A flash fire is the combustion of a flammable gas/air mixture that produces relatively short term thermal hazards with negligible overpressure (blast wave). As an example, consider the real case of a tractor-semitrailer carrying liquid propane that overturned near Lynchburg, Virginia, in 1972. That accident caused the tank to fail, allowing approximately 15 m^3 (4,000 gallons) of liquid propane to escape. The resulting propane vapor cloud extended at least 120 m (400 ft) from the truck prior to ignition. Upon ignition, a flash fire occurred followed by a fireball. The fireball engulfed and killed the truck driver and others outside of the fireball received serious burns. This case history is described in more detail in Section 3.2.2.

2.2.   VAPOR CLOUD EXPLOSIONS

A VCE is the combustion of a flammable gas/air mixture at a more rapid rate than in a flash fire (often due to interaction of the flame with congestion and confinement), resulting in the development of overpressure (i.e., a blast wave).

One of the most well-known large VCEs occurred at the Flixborough Works in the UK in 1974. Approximately 30,000 kg (66,000 lb) of cyclohexane was released from a cyclohexane oxidation plant reactor and formed a large vapor cloud. The vapor cloud was ignited roughly one minute after the release. The flame accelerated due to the presence of significant congestion and confinement associated with the process plant equipment and structure in the flammable vapor cloud. The blast waves resulting from the VCE caused the main office block and the control room to collapse. There were 28 fatalities as a result of this event, of which 18 were in the control room. Approximately 2,000 homes in the surrounding community were damaged. This case history is described in more detail in Section 3.3.1.

2.3.   PRESSURE VESSEL BURSTS

In a pressure vessel burst (PVB), the sudden expansion of a compressed gas generates a blast wave that propagates outward from the source, along with hazardous debris. The explosion at the Union Carbide chemical plant in Seadrift, Texas in 1991 is illustrative of a PVB. The No. 1 Ethylene Oxide Redistillation Still (ORS) was shut down for maintenance and repair several days before the incident. The No. 1 ORS distillation column was designed for a maximum allowable working pressure (MAWP) of 6 bars (90 psig). About one hour after startup, it exploded. The explosion resulted from the autodecomposition of the ethylene oxide. The pressure buildup in the No. 1 ORS reached four times the design MAWP, causing a ductile failure. The column shell fragmented over the upper 2/3 of its height. This case history is described in more detail in Section 3.4.2.

2.4.   BLEVEs

A Boiling Liquid Expanding Vapor Explosion (BLEVE) is associated with the bursting of a pressurized vessel containing liquid above its atmospheric boiling point. The liquid in the vessel may be flammable or non-flammable, such as in a hot water boiler. About one-fifth of all BLEVEs occur with non-flammable pressure-liquefied gas (Abbasi, 2007). If non-flammable, the hazard will be primarily an overpressure event with possible vessel fragmentation. If a flammable material is ignited, it will usually produce a fireball; a secondary effect will be a pressure wave due to the explosively rapid vaporization of the liquid.

The effects of a BLEVE are illustrated by the explosion resulting from a train derailment that occurred in Crescent City, Illinois in 1970. The train included nine cars carrying liquefied petroleum gas (LPG). One of the LPG cars was punctured in the derailment. Five of the LPG cars underwent BLEVEs within four hours due to the resulting fire, with the first BLEVE occurring approximately one hour after the derailment. Sections of the cars were propelled from the derailment site as a result of these explosions, with one car section being thrown over 480 m (1600 ft). Nearby buildings sustained severe damage. No fatalities occurred, although sixty-six injuries were reported. This case history is described in more detail in Section 3.5.3.

2.5.   PREDICTION METHODOLOGIES

A variety of prediction methodologies are available for each of the hazard types addressed in this book. They range from simplified methods that require relatively few calculations to complex numerical models involving millions of calculations performed on large computers. Of course, there are tradeoffs among the various methods that can be employed. Simplified methods, as the name implies, involve some simplifications or assumptions. More refined methods avoid some of these simplifications and may provide more accurate and higher resolution results, but with a commensurate higher level of input data and analysis labor. Computational fluid dynamic (CFD) models are now available for some of the hazard types addressed in this volume, but these are not necessarily more accurate. A high level of expertise is required of users of CFD models. Regardless of the model or method used, expertise is needed to properly apply the models, and results can vary significantly with the quality of input data, assumptions, applicability of models to the actual situations, and other factors.

Experimental data, accident case histories, and example problems are provided in this book to assist readers in understanding the potential consequences of flash fires, PVBs, VCEs, and BLEVEs, and to quantify results for various circumstances. These data may serve as helpful benchmarks to assist analysts in making consequence predictions.

3. CASE HISTORIES

3.1. HISTORICAL EXPERIENCE

The selection of incidents described in this chapter was based on the availability of information, the kind and amount of material involved, and the severity of damage. The incidents described in this chapter cover a range of factors:

- Materials: Histories include incidents involving hydrogen, propylene, propane, cyclohexane, ethylene oxide, and natural gas liquids.
- Event Type: Case studies include vapor cloud explosions, BLEVEs, pressure vessel bursts, and flash fires.
- Period of time: Events occurring over the period between the years 1964 and 2007 are reported.
- Quantity released: Releases ranged in quantity from 90 kg (200 lb) to 40,000 kg (85,000 lb).
- Site characteristics: Releases occurred in settings ranging from rural to very congested industrial areas.
- Availability of information: Very well-documented incidents (e.g., Flixborough, Texas City) as well as poorly documented incidents (e.g., Ufa) are described.
- Severity: Death tolls and damage vary widely in cases presented.

Documentation of flash fires is scarce. In several accident descriptions of vapor cloud explosions, flash fires appear to have occurred as well. The selection and descriptions of flash fires were based primarily on the availability of information.

3.2. FLASH FIRES

3.2.1. Donnellson, Iowa, USA: Propane Fire

During the night of August 3, 1978, a pipeline carrying liquefied propane ruptured, resulting in the release of propane. A National Transportation Safety

Board report (1979) describes a flash fire resulting from the rupture of a 20 cm (8 in) pipeline carrying liquefied propane. The section of the pipeline involved in the incident extended from a pumping station at Birmingham Junction, Iowa, to storage tanks at a terminal in Farmington, Illinois. Several minutes before midnight on August 3, 1978, the pipeline ruptured while under 1,200 psig pressure in a cornfield near Donnellson, Iowa. Propane leaked from an 838-cm (33-in.) split and then vaporized. The cloud moved through the field and across a highway following the contour of the land. The cloud eventually covered 30.4 ha (75 acres) of fields and woods, surrounding a farmhouse and its outbuildings. There was a light wind, and the temperature was about 15°C (about 59°F). At 12:02 A.M. on August 4, the propane cloud was ignited by an unknown source. The fire destroyed a farmhouse, six outbuildings, and an automobile. Two other houses and a car were damaged. Two persons died in the farmhouse. Three persons who lived across the highway from the ruptured pipeline had heard the pipeline burst and were fleeing their house when the propane ignited. All three persons received burns on over 90% of their bodies, and one later died from the burns. Fire departments extinguished smaller fires in the woods and adjacent homes.

The fire at the ruptured pipe produced flame heights of up to 120 m (400 ft). It was left burning until the valves were shut off to isolate the failed pipe section.

The investigation following the accident showed that the pipeline rupture was due to stresses induced in, and possibly by damage to, the pipeline resulting from its repositioning three months before. This work had occurred in conjunction with road work on the highway adjacent to the accident site. Th pipeline had been dented and gouged.

3.2.2.   Lynchburg, Virginia, USA: Propane Fire

On March 9, 1972, an overturned tractor-semitrailer carrying liquid propane resulted in a propane release. The National Transportation Safety Board report (1973) describes the accident involving the overturning of a tractor-semitrailer carrying liquid propane under pressure. On March 9, 1972, the truck was traveling on U.S. Route 501, a two-lane highway, at a speed of approximately 40 km/h (25 mph). The truck was changing lanes on a sharp curve while driving on a downgrade at a point 11 km (7 mi) north of Lynchburg, Virginia. Meanwhile, an automobile approached the curve from the other direction. The truck driver managed to return to his own side of the road, but in a maneuver to avoid hitting the embankment on the inside of the curve, the truck rolled onto its right side. The scene is depicted in Figure 3.1.

The manhole-cover assembly on the tank struck a rock; the resulting rupture of the tank head caused propane to escape. There were woods on one side of the road; on the other side a steeply rising embankment and trees and bushes, and then a steep drop-off to a creek.

The truck driver left the tractor, ran from the accident site in the direction the truck had come from, and warned approaching traffic. The driver of the first arriving car stopped and tried to back up his car, but another car blocked his path. The occupants of these cars got out of their vehicles. Three occupants of nearby houses at a distance of 60 m (195 ft), near the creek and about 20 m (60 ft) below the truck, fled after hearing the crash.

An estimated 4,000 U.S. gal (8,800 kg; 19,500 lb) of liquefied propane was discharged. At the moment of ignition, the visible cloud was expanding but had not reached the motorists who left their cars at a distance of about 135 m (450 ft) from the truck. The cloud reached houses about 60 m (195 ft) from the truck, but had not reached the occupants at a distance of approximately 125 m (410 ft). The cloud was ignited at the tractor-semitrailer, probably by the racing tractor engine. Other possible ignition sources were the truck battery or broken electric circuits.

The flash fire that resulted was described as a ball of flame with a diameter of at least 120 m (400 ft). No concussion was felt. The truck driver (at a distance of 80 m or 270 feet) was caught in the flames and died. The motorists and residents were outside the cloud but received serious burns.

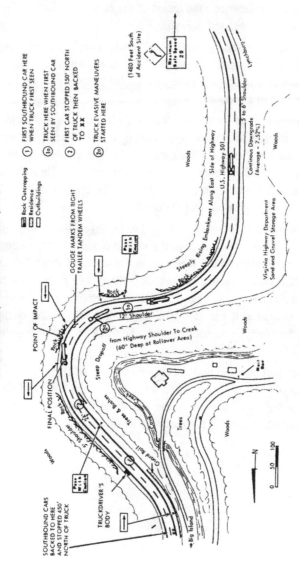

Figure 3.1. Details of Lynchburg, VA accident site.

3.2.3. Quantum Chemicals, Morris, Illinois, USA: Olefins Unit Flash Fire

On June 7, 1989, a loss of containment of a propylene/propane vapor stream resulted in a vapor cloud flash fire in an Olefins Unit in Morris, Illinois. The ensuing fire caused major plant piping, piperack, and equipment damage. There was one burn injury from the initial vapor cloud flash fire. Additional minor personnel injuries were experienced during the response to the fire.

The facility had undertaken an initiative to identify and eliminate from service threaded pipe fittings in hydrocarbon service. As part of that initiative, a vent line that recycled depropanizer distillation column overhead vapors back to the main process gas compressor had been identified to contain threaded fittings. The subject vent line contained a control valve, which was typically only opened to recycle vapors during column upsets. On the day of the incident, the vent line was removed from service for replacement of sections of the line that contained threaded fittings. The piping replacement work had not been completed at the end of the day-maintenance shift. Maintenance personnel had replaced one section of the line with a prefabricated flange-to-flange section of pipe, leaving bolts on one of the flanges finger-tight.

On the day of the event, a power outage had affected operations of the unit's amine absorber system and debutanizer distillation column. Power had been recovered on the day shift. Plant operators were working to restore normal operating conditions on the following night shift. Approximately three hours into the night shift, the control board operator made a move to open the depropanizer vent line control valve approximately 10%. Some minutes later, the board operator made an additional opening move on the depropanizer vent control valve. Very shortly thereafter, combustible gas sensor alarms from the area of the unit's Propylene Splitter distillation tower went off in the control room. Workers responded to the alarms to report a significant vapor release in the piperack. Firewater monitors were sprayed on the leak and the site emergency response was activated. Attempts were made to identify the line that was leaking and the source of the leak. Responders reported that the leak was spraying radially in a 360-degree circle from a line that could not be identified through the cold vapor cloud and firewater spray. The process unit in the area of the leak was moderately-to-highly congested. The vapor leak had been in progress for approximately 30 minutes when the vapor cloud ignited in a flash. Eyewitness accounts indicated that the ignition occurred when a firewater spray was repositioned and accidentally struck a lighting fixture, breaking the light's protective glass cover. The ignition was reported by eyewitnesses to be a flash of fire, rather than an explosion or detonation.

The resulting fire impinged upon a number of adjacent pipes in the piperack. This led to overheating, failures, and loss of containment of additional piping. The fire escalated and resulted in extensive piperack and area equipment damage (Figure 3.2). The facility was shut down for repairs for approximately three months.

Figure 3.2. Damage resulting from the Morris, Illinois flash fire.

Post-incident investigation identified that the source of the vapor release was from the vent line that had been worked on during the day shift. The line had been improperly isolated when it was released for replacement of threaded fittings. As a result of the improper isolation, the vent line control valve was the only isolation between the process and the vent line being worked on. The night shift board operator was not aware that the depropanizer vent line was not available for use, and initially made a move to open the depropanizer vent valve approximately 10%, but there was no leak at that time. It is believed that the vent control valve did not come open off of its seat at the approximate 10% signal. Later, when the board operator opened the vent valve further, it is believed the valve came open then, resulting in the vapor release through the flange with finger-tight bolts (and thus the reported radial spray).

3.3.   VAPOR CLOUD EXPLOSIONS

3.3.1.   Flixborough, UK: Vapor Cloud Explosion in Chemical Plant

On June 1, 1974, a cyclohexane vapor cloud was released after the rupture of a pipe bypassing a reactor. In total, approximately 30,000 kg (73,000 lb) of cyclohexane was released.   HSE (1975), Parker (1975), Lees (1980), Gugan (1978), and Sadée et al. (1976, 1977) have extensively described the vapor cloud explosion that occurred in the reactor section of the caprolactam plant of the Nypro Limited, Flixborough Works. The Flixborough Works is situated on the east bank of the River Trent (Figure 3.3). The nearest villages are Flixborough (800 meters or one-half mile away), Amcotts (800 meters or one-half mile away), and Scunthorpe (4.9 km or approximately three miles away).

Figure 3.3 Flixborough works prior to the explosion.

The cyclohexane oxidation plant contained a series of six reactors (Figure 3.4). The reactors were fed by a mixture of fresh cyclohexane and recycled material. The reactors were connected by a pipe system, and the liquid reactant mixture flowed from one reactor into the other by gravity. Reactors were designed to operate at a pressure of approximately 9 bar (130 psi) and a temperature of 155°C (311°F). In March, one of the reactors began to leak cyclohexane, and it was, therefore, decided to remove the reactor and install a bypass (Figure 3.5). A 0.51 m (20 in) diameter bypass pipe was installed connecting the two flanges of the reactors. Bellows originally present between the reactors were left in place. Because reactor flanges were at different heights, the pipe had a dog-leg shape (Figure 3.6).

Figure 3.4 Flixborough cyclohexane oxidation plant (six reactors on left).

Figure 3.5. Area of spill showing removed reactor.

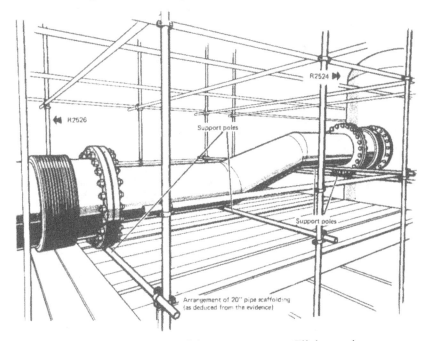

Figure 3.6. Bypass on cyclohexane reactors at Flixborough.

On May 29, the bottom isolating valve on a sight glass on one of the vessels began to leak, and a decision was made to repair it. On June 1, start-up of the process following repair began. As a result of poor design, the bellows in the bypass failed and a release of an estimated 33,000 kg (73,000 lb) of cyclohexane occurred, most of which formed a flammable cloud of vapor and mist.

After a period of 30 to 90 seconds following release, the flammable cloud was ignited. The time was then about 4:53 P.M. The explosion caused extensive damage and started numerous fires. The blast shattered control room windows and caused the collapse of its roof. It demolished the brick-constructed main office block, only 25 m (82 ft) from the explosion center. Fortunately, the office block was unoccupied at the time of the incident. None of the buildings had been constructed to protect the occupants from the effects of an explosion. Twenty-eight people died, and thirty-six were injured. Eighteen of the fatalities were in the control room at the time. None survived in the control room. The incident occurred on a Saturday. If it had occurred on a weekday, over 200 people would have been working in the main office block. The plant was totally destroyed (

Figure 3.7and Figure 3.8) and 1,821 houses and 167 shops and factories in the vicinity of the plant were damaged.

Figure 3.7 Aerial view of damage to the Flixborough works.

Figure 3.8. Damage to the Office Block and Process Areas at the Flixborough works.

Sadée et al. (1976–1977) give a detailed description of structural damage due to the explosion and derived blast pressures from the damage outside the cloud (Figure 3.9). Several authors estimated the TNT mass equivalence based upon the damage incurred. Estimates vary from 15,000 to 45,000 kg (33,000 to 99,000 lb) of TNT. These estimates were performed at a time when TNT equivalence was the predominant prediction method, which is typically not used today.

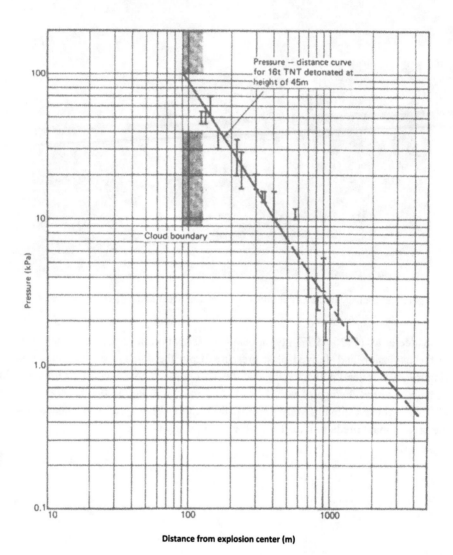

Figure 3.9. Blast-distance relationship outside the cloud area of the Flixborough explosion.
(Vertical bars were drawn based on observed damage.)

Estimates of pressures inside the cloud vary widely. Gugan (1978) calculated
that the forces required to produce damage effects observed, such as the bending
of steel, would have required local pressures of up to 5–10 bar (73-145 psi).

3.3.2.   Port Hudson, Missouri, USA: Vapor Cloud Explosion after Propane Pipeline Failure

On December 9, 1970, a liquefied propane pipeline ruptured near Port Hudson, Missouri. About 24 minutes later, the resulting vapor cloud was ignited. The pressure effects were very severe. The blast was estimated to be equivalent to a detonation of 50,000 kg (125,000 lb) of TNT.

Burgess and Zabetakis (1973) described the timeline of the Port Hudson explosion. At 10:07 P.M., an abnormality occurred at a pumping station on a liquid propane line 15 miles (24 km) downstream from Port Hudson. At 10:20 P.M., there was a sudden increase in the throughput at the nearest upstream pumping station, indicating a major break in the line. During the first 24 minutes, an estimated 23,000 kg (50,000 lb) of liquid propane escaped. The noise of escaping propane was noticed at about 10:25 P.M. A plume of white spray was observed to be rising 15 to 25 m (50 to 80 ft) above ground level.

The pipeline was situated in a valley, and a highway ran at about one-half mile (800 m) from the pipeline. Witnesses standing near a highway intersection observed a white cloud settling into the valley around a complex of buildings. Weather conditions were as follows: low wind (approximately 2.5 m/s or 8 ft/s) and near-freezing temperature (1°C or 34°F). At about 10:44 P.M., the witnesses saw the valley "lighting up." No period of flame propagation was observed. A strong pressure pulse was felt and one witness was knocked down.

In the seconds after the valley was illuminated, a fire was observed to "roll" up the sloping terrain and consume the remainder of the cloud. After the explosion and flash fire, a jet fire resulted at the point of the initial release. Buildings in the vicinity of the explosion were damaged as shown in Figure 3.10 and Figure 3.11. Damage from the blast in the vicinity was calculated to be equivalent to a blast of 50,000 -75,000 kg (125,000-165,000 lb) of TNT.

Figure 3.10. Damage to a farm 600 m (2,000 ft) from explosion center.

Figure 3.11. Damage to a home 450 m (1,500 ft) from the blast center.

The cloud was probably ignited inside a concrete-block warehouse. The ground floor of this building, partitioned into four rooms, contained six deep-freeze units. Gas could have entered the building via sliding garage doors, and ignition could have occurred at the controls of a refrigerator motor.

According to Burgess and Zabetakis (1973), the initial event at Port Hudson was a vapor cloud detonation. Their conclusion was based on the abruptness of illumination of the valley and extent of damage to buildings in the vicinity of the explosion. Very few other accidents have been reported as vapor cloud detonations. Several incidents with localized areas of high damage within the VCE combustion zone have been investigated where a small area may have detonated, but they have not been published in the open literature. Port Hudson witness accounts of a flash fire after the valley was illuminated indicates that only a portion of the cloud was involved in the detonation if, in fact, a detonation was achieved. The long ignition delay, large quantity of fuel released, topography, atmospheric conditions, and initiation within a heavy wall building combined to make Port Hudson an unusual case.

3.3.3.   Jackass Flats, Nevada, USA: Hydrogen-Air Explosion during Experiment

Reider et al. (1965) described an incident at Jackass Flats, Nevada. An experiment was conducted by Los Alamos Laboratory on January 9, 1964, to test a rocket nozzle, primarily to measure the acoustic sound levels in the test-cell area which occurred during the release of gaseous hydrogen at high flow rates. Hydrogen discharges were normally flared, but in order to isolate the effect of combustion on the acoustic fields, this particular experiment was run without the flare. Releases were vertical and totally unobstructed. High-speed motion pictures were taken during the test from two locations.

During the test, hydrogen flow rate was raised to a maximum of approximately 55 kg/s (120 lb/s). About 23 seconds into the experiment, a reduction in flow rate began. Three seconds later, the hydrogen exploded. Electrostatic discharges and mechanical sparks were proposed as probable ignition sources. The explosion was preceded by a fire observed at the nozzle shortly after flow rate reduction began. The fire developed into a fireball of modest luminosity, and an explosion followed immediately.

Walls of light buildings and heavy doors were bulged out. In one of the buildings, a blowout roof designed to open at 0.02 bar (0.3 psi) was lifted from a few of its holding clips. Outward damage can be caused by the negative phase of the generated blast wave, rebound of structural components that were deflected inward during the positive phase, or a combination thereof. Buildings are typically designed for inward loads, such as wind and snow loads, and are weaker in the outward direction; as a result, structural components may exhibit greater outward deformation than inward even though the positive phase was the dominant blast load.

High-speed motion pictures indicate that the vertical downward flame speed was approximately 30 m/s (100 ft/s); the flame was undisturbed by effluent velocity. This value is roughly ten times the burning velocity expected for laminar-flow conditions, but is reasonable because a turbulent free jet was present, thereby enhancing flame burning rate. According to Reider et al. (1965), blast pressure at 45 m (150 ft) from the center was calculated to be 0.035 bar (0.5 psi) based on explosion damage. Approximately 90 kg (200 lb) of hydrogen was involved in the explosion.

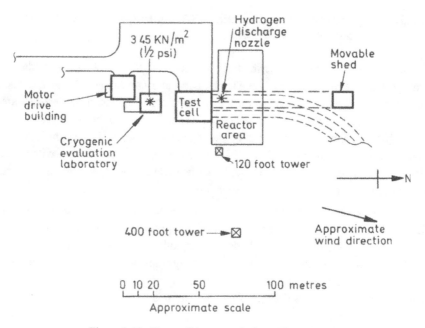

Figure 3.12. Test-cell layout at Jackass Flats, NV.

3.3.4.   Ufa, West-Siberia, USSR: Pipeline Rupture Resulting In a VCE

Lewis (1989) and Makhviladze (2002) describe an accident that occurred in Siberia on the night of June 3 and early hours of June 4, 1989. Late on June 3, 1989, engineers in charge of the 0.7 m (28 in) pipeline, which carried natural gas liquids from the gas fields in western Siberia to chemical plants in Ufa in the Urals, noticed a sudden drop in pressure at the pumping end of the pipeline. It appears that the engineers responded by increasing the flow rate in order to maintain normal pipeline pressure. The pipeline normally conveyed about 10,000 t/day (116-120 kg/s) at a working pressure of 38 bar (550 psi).

The evidence shows that a leak had formed in the pipeline between the towns of Ufa and Asma at a point 800 m (0.5 mi) away from the Trans-Siberian double railway track. The area was a wooded valley. The leak was about 900 m from the railway track, and was uphill at the head of the valley (Figure 3.13). The total area of the rupture opening was 0.77 m^2 (8.3 ft^2), twice the cross-sectional area of the pipe. Throughout the area there had been a strong smell of gas a few hours before the blast. The gas cloud was reported to have drifted for a distance of 8 km (5 mi).

Two trains coming from opposite directions approached the area where the cloud was present. Each consisted of an electrically powered locomotive and 19 coaches constructed of metal and wood. Either train could have ignited the cloud, with possible ignition sources being catenary wires which powered the locomotives and open fire heaters for boiling water in coaches.

Two explosions seem to have taken place in quick succession, and a flash fire subsequently ran down the railroad track in two directions. A considerable part of each train was derailed. Four rail cars were blown sideways from the track by the blast, and some of the wooden cars were completely burned within 10 minutes. Trees within a 2.5 km^2 (1 mi^2) around the explosion center were completely flattened (Figure 3.13, Figure 3.14), and windows up to 13 km (8 mi) were broken. By the end of June, the total death toll had reached 645.

Congestion for the VCE included trees from a dense wooded area, which was close to the railroad tracks as seen in Figure 3.14. The two trains provided additional congestion and generated turbulence, particularly where they overlapped. In addition, vapors entered coaches and locomotives, producing internal pressure that deformed the cars outward to a barrel-like shape.

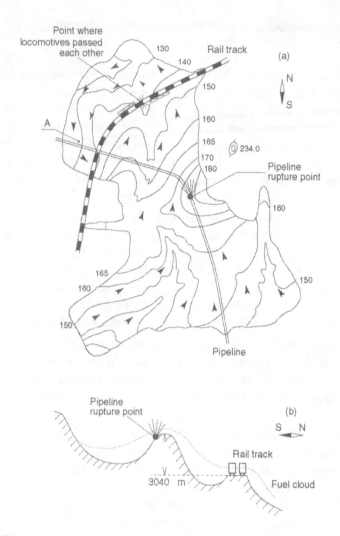

Figure 3.13 Ufa accident: (a) topographical sketch of demolished area with directions trees fell shown by arrows; (b) terrain profile (not to scale). (Makhviladze, 2002)

Figure 3.14. Aerial views of Ufa accident site: (upper) broad view of the forest and rail line (Makhviladze, 2002); (lower) closer view of the area where the trains passed. (Lewis, 1989)

3.3.5.   Phillips, Pasadena, Texas USA: Propylene HDPE Unit VCE and BLEVEs

On October 23, 1989, an explosion and a fire occurred at the Phillips 66 Company's Houston Chemical Complex located near Pasadena, Texas. This incident was caused by an accidental release of 40,000 kg (85,000 lbs) of a mixture containing ethylene, isobutane, hexene and hydrogen in a low density polyethylene unit (Figure 3.15). In this incident, 23 persons were killed and 314 people were injured. All of the fatalities were within 76 m (250 ft) of the point of the initial release, and 15 of them were within 46 m (150 ft).

Figure 3.15. Phillips Pasadena plant prior to the incident.

A 1990 OSHA report (OSHA, Apr 1990) described the accident. On Sunday, October 22, 1989, a contractor crew started the maintenance procedure on the valves of a high density polyethylene reactor. Polyethylene was produced in loop reactors, which were supported by tall steel frame structures (Figure 3.15). The maintenance procedure consisted of disassembling and clearing a leg that had become clogged with polyethylene particles. On Monday afternoon (October 23) at about 1:00 P.M., a release occurred when the valve upstream of the discharge leg was accidentally opened. Almost all the contents of the reactor, approximately 40,000 kg (85,000 lbs) of high reactivity materials, were dumped. A large vapor cloud formed in a few seconds and moved downwind through the plant. Within two minutes, this cloud was in contact with an ignition source and exploded with the force of 2,400 kg (5,300 lbs) of TNT.

Following this VCE, two other major explosions occurred. The second explosion occurring 10 to 15 minutes after the initial explosion involved BLEVEs of two 75 m³ (20,000 U.S. gal) isobutene storage tanks (Figure 3.16). The third explosion occurred 25 to 45 minutes later, which was the catastrophic failure of the ethylene plant reactor. Damage to the process unit is shown in Figure 3.17.

Figure 3.16 BLEVE at the Phillips Pasadena site.

Figure 3.17 Phillips Pasadena process area damage. (*courtesy of FM Global*)

The initial blast destroyed the control room and caused the rupture of the adjacent vessels containing flammable materials and the water lines. The proximity between the process equipment and the buildings contributed to the intensity of the blast. Twenty-two of the victims were found within 76 m (250 ft) of the release point, 15 of which were within 45 m (150 ft). Most of the fatalities were within buildings, but the actual number was not reported.

The Phillips Pasadena 1989 incident, along with the 1984 Bhopal, India, 1988 Shell Norco, 1987 Arco Channelview, and 1989 Exxon Baton Rouge incidents, triggered the development of the Process Safety Management (PSM) regulation by the U.S. Department of Labor, Occupational Safety & Health Administration.

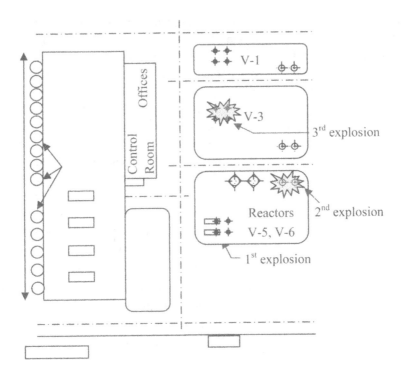

Figure 3.18. Explosion locations at Phillips Pasadena site.

3.3.6. BP, Texas City, Texas USA: Discharge from Atmospheric Vent Resulting in a VCE

On March 23, 2005, at 1:20 P.M., an explosion and fire occurred at the BP Texas City Refinery Isomerization (ISOM) plant. In this incident, 15 people were killed and 180 were injured. During the incident, a shelter-in-place order was issued that required 43,000 people in the surrounding community to remain indoors.

According to the report by BP Products North America (Mogford, 2005), and the U.S. Chemical Safety and Hazard Investigation Board (CSB, 2007), on the morning of the accident, the raffinate splitter tower in the ISOM unit was restarted after a maintenance outage. During the procedure, the night shift charged the raffinate splitter to 100% of normal operating range (equivalent to 3.1 m (10 ft 3 inches) height above tangent in the 50 m (164-ft) tall tower) and stopped flow. The day shift resumed pumping raffinate into the tower for over three hours without any liquid being removed, introducing an additional 397 m³ (105,000 U.S. gal). As a consequence, the tower was overfilled, and the liquid overflowed into the overhead pipe at the top of the tower. The pressure relief valves opened at about 1:14 P.M. for 6 minutes and discharged an estimated 175 m³ (46,000 U.S. gal) of flammable liquid to a blowdown drum with a vent stack open to the atmosphere. This blowdown drum overfilled after about 4 ½ minutes, which resulted in a geyser-like release that reached 6 m (20 ft) above the top of the stack at about 1:18 P.M. An estimated 8 m³ (2,000 U.S. gal) of the hydrocarbon liquid overflowed from the blowdown drum stack.   The flammable cloud was predominately on the west side of the unit to the south of the release point; the flammable cloud did not reach the eastern leg of the ISOM unit.

The vapor cloud was ignited at about 1:20 P.M. by an undetermined ignition source. A diesel pickup truck by the road on the north side of ISOM was observed to have its engine racing, and was a high potential ignition source.

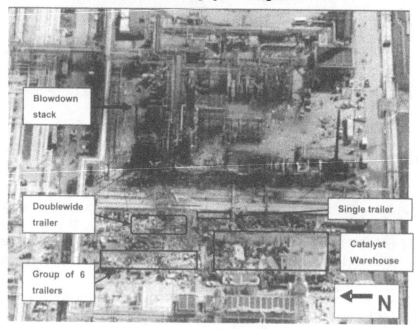

Figure 3.19. Aerial view of the ISOM unit after the explosion. (CSB, 2007)

In the explosion, 15 workers in or near trailers sited to the west of the ISOM unit were killed. Three occupants in a single-wide trailer perished, and 12 of 20 workers inside a double-wide trailer were killed; the others were seriously injured. Trailer locations are shown in Figure 3.19. These temporary office trailers were light wood construction. The cause of death for all 15 was blunt force trauma, probably resulting from being struck by structural components of the trailers. A total of 180 workers at the refinery reported injuries.

The trailers were placed about 46 m (150 ft) west of the blowdown stack in the open area next to a piperack that was about 1 m (3 ft) above grade. The piperack provided congestion between the western edge of the ISOM unit and the trailers. The flammable cloud extended west past the piperack and trailers, resulting in the trailers adjoining a congested area that was involved in the VCE.

Figure 3.20. Destroyed trailers west of the blowdown drum.
(Arrow in upper left of the figure)

3.4.   PRESSURE VESSEL BURST

3.4.1.   Kaiser Aluminum, Gramercy, Louisiana USA: Alumina Process Pressure Vessel Burst

On July 5, 1999, at approximately 5:17 A.M., an explosion occurred at the Gramercy Works Plant operated by Kaiser Aluminum and Chemical Corporation in Gramercy, Louisiana. As a result of this incident, 29 persons were injured, but no lives were lost. This plant produced alumina through a process involving the caustic leaching of bauxite at elevated temperature and pressure.

A description of the accident was provided in the Mine Safety and Health Administration report (MSHA, 1999). The explosion took place in the digestion area where the slurry passes through digesters and flash tanks and is finally moved to a "blow-off tank" (see Figure 3.21).

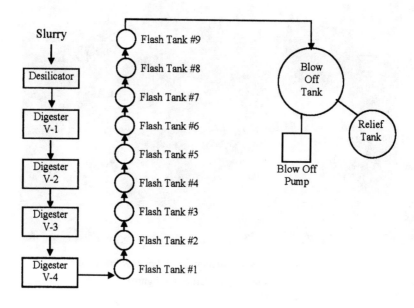

Figure 3.21. Kaiser slurry digester area flow schematic.

A power failure occurred at about 4:43 a.m., which resulted in all electrically-powered pumps stopping in the digestion area. Consequently, the movement of the slurry through the digestion process and the recirculation of liquor through a heat exchanger stopped. Pressure built up in the desilicator and digestors (because steam continued to be produced by the gas-fired boilers in the power house) and in the flash tanks (because of excessive heat).

Pressure levels exceeded the pressure relief valve settings for all the flash tanks except one (Tank #3). At the time of the explosion, the pressures in the desilicator and the digesters were between 34 bar (500 psi) and 41 bar (600 psi). For several flash tanks, the instrumentation pressure limit was reached (see Table 3.1).

Table 3.1. Pressure at time of explosion in the digestion area equipment

Equipment	Pressure at Explosion	
	Bar	Psi
Desilicator	40.3	584
Digester V1	39.0	567
Digester V2	37.5	548
Digester V3	38.5	558
Digester V4	37.0	537
Flash Tank 1	35.2*	510*
Flash Tank 2	33.1	481
Flash Tank 3	29.5	428
Flash Tank 4	21.1*	306*
Flash Tank 5	/	/
Flash Tank 6	10.5*	153*
Flash Tank 7	8.4*	122*
Flash Tank 8	4.2*	61*
Flash Tank 9	2.1*	30*

*Instrumentation limit was reached

According to MSHA, the explosion occurred as a result of a build-up of excessive pressure within a vessel or vessels in the digestion area, and the subsequent rupture of the vessel or vessels. The rupture of the vessels exposed the superheated liquid contents to atmospheric pressure, resulting in a BLEVE. Figure 3.22 (a) and (b) show the digestion area before and after the explosion.

(a) Digestion area before explosion

(b) Digestion area after explosion

Figure 3.22. Kaiser aluminum digester area before and after the explosion.
(MSHA, 1999)

The force of the explosion released more than 180,000 kg (400,000 lbs) of sodium hydroxide into the atmosphere. Residue settled on homes, buildings, and vehicles in the towns of Gramercy and Lutcher located approximately 1.6 km (3 mi) away. The explosion blew portions of the blow-off tank and four flash tanks hundreds of feet from the digestion area.

The dome of Flash Tank No. 6, which weighed approximately 3,500 kg (7,600 lbs), was propelled about 900 m (3,000 ft) from its original location. The explosion also propelled sections of steel pipe, structural supports, and valves hundreds of feet in several different directions. The concussion from the blast wave and flying debris damaged equipment and structures throughout the plant (including three switch houses; the maintenance trailer; sections of the office buildings; the power house; several nearby substations; the maintenance warehouse; and the digestion control room) and also caused glass breakage and minor damage to structures in Gramercy and Lutcher.

3.4.2.   Union Carbide Seadrift, Texas USA: Ethylene Oxide Distillation Column Pressure Vessel Burst

On March 12, 1991, at approximately 1:18 A.M., an explosion occurred at the No. 1 Ethylene Oxide Redistillation Still (ORS) at Union Carbide Corporation's Seadrift, Texas plant. The blast and the ensuing fire killed one person and caused extensive damage to the plant.

A description of this accident was given in a presentation by G.A. Viera and P.H. Wadia (1993). Several days before the accident, the Seadrift Ethylene Oxide Unit was shut down for maintenance and repair. The start-up of the unit began late in the afternoon on March 11. A problem with the No. 1 ORS occurred involving an increase in pressure, which activated the shutdown system. Eventually, the problem was solved and the No. 1 ORS was re-started at about midnight. About one hour later (at about 1:18 A.M., on March 12) it exploded.

The No. 1 ORS distillation column was 40 m (120 ft) high and 2.5 m (8.5 ft) in diameter, with a design maximum allowable working pressure (MAWP) of 6 bars (90 psig). The explosion resulted from the autodecomposition of the ethylene oxide; this reaction does not require oxygen and releases a large amount of energy by producing water, carbon monoxide, carbon dioxide and methane. A unique set of coincidences allowed this reaction to occur, as the ethylene oxide vapor must be heated at about 500°C (well above the normal operating temperature for the No. 1 ORS). Since this temperature was reached at a localized region in the reboiler, a flame front was initiated and progressed into the column base section, accelerated upwards, causing the explosion. The pressure buildup in the No. 1 ORS reached four times the design MAWP, causing a ductile failure. The column shell fragmented over the upper 2/3 of its height, with fragment size being large near the bottom and small near the top. The bottom section of the column is shown in Figure 3.23.

Figure 3.23. Remaining No. 1 ORS base section and skirt with the attached vertical thermosyphon reboiler.

3.4.3.   Dana Corporation, Paris, Tennessee USA: Boiler Pressure Vessel Burst

On June 18, 2007 at about 1:50 P.M., a boiler pressure vessel burst at the Dana Corporation plant located in Paris, Tennessee, causing extensive damage to the facility and surrounding area, and seriously injuring one employee.

According to the Department of Labor report (DOL, 2007), the plant maintenance man was making his rounds of the plant boiler room at about 1:50 pm. When entering the room, he noticed that one of the boilers (Boiler #1) was in low water condition. He proceeded to the feed tank and pump area. Feed water was then introduced back to the boilers. At the time of water introduction, the water level was not indicated properly for Boiler #2 and the water level was very low. Inoperative controls and safety devices allowed Boiler #2 to continue to fire even with the low water level. The dry-fire state of Boiler #2 caused a thermal shock from contact between the feed water and the overheated surfaces of the boiler. A rapid gas expansion then occurred, contributing to the explosion of Boiler #2.

The boiler was propelled through the industrial roll-up door, knocking down a portion of the wall and coming to rest over 30 m (100 ft) from its original location at the center of the manufacturing room floor (see Figure 3.24 and Figure 3.25).

Figure 3.24. Final location of the boiler after explosion.

Figure 3.25. Hole created by boiler through roll-up door wall (west wall).

The explosion hurled the rear door of the boiler through the opposite cement block wall, creating a 30-foot hole in the outer wall of the plant (Figure 3.26). The rear door and debris damaged many vehicles in the parking lot and a pedestrian walkway before coming to rest in a ditch 30 m (100 ft) from the boiler room (Figure 3.27). The internal south wall of the boiler room also collapsed during the explosion (Figure 3.28), causing additional damage to property and equipment.

Figure 3.26. Damaged exterior wall viewed from inside boiler room (east wall).

Figure 3.27. View of east wall from outside plant (note rear boiler door in ditch).

Figure 3.28. Interior wall of boiler room (south wall).

3.5.   BLEVE

3.5.1.   Procter and Gamble, Worms, Germany: Liquid CO_2 Storage Vessel Explosion

On November 21, 1988, an explosion resulted from the catastrophic failure of a liquid CO_2 storage vessel at the citrus process plant owned by Procter & Gamble GmbH in Worms, Germany. This plant produced household cleaning products. Three employees were killed and ten were injured in the incident.   Two neighboring units were destroyed, and fragments of the vessel weighing more than 100 kg were thrown over 500 m.

The incident was described in a Process Safety Progress article (Clayton, et al. 1994). The storage vessel involved in the accident was leased by Procter & Gamble from the CO_2 supplier, which was not a member of the European Industry CO_2 Association.   The horizontal tank involved in the incident had a 30 m^3 (66,000 lbs) capacity, with a 2.6 m (8.5 ft) diameter and an overall length of 6.5 m (21 ft). This vessel burst resulted from an overpressure, most likely caused by a combination of the internal heater failure, the dysfunction of the relief valve (ice formation blocking the valve) and the absence of an alarm indicating the overpressure condition. An assessment of the failure pressure suggested that it would have been between 35 and 51 bars, corresponding to 1.75 to 2.5 times the original design pressure of the vessel.

The explosion created two small fragments; the head of the tank and a small part of the side. A large piece that represented about 80% of the vessel rocketed into the Rhein River about 300 m (1,000 ft) away from its original position.

3.5.2.   San Juan Ixhuatepec, Mexico City, Mexico: Series of BLEVEs at LPG Storage Facility

On November 19, 1984, an initial leak and flash fire of liquefied petroleum gas (LPG) resulted in the destruction of a large storage facility and a portion of the community surrounding the storage facility. Approximately 500 people were killed and approximately 7,000 were injured. The storage facility and the community near the facility were almost completely destroyed.

Pietersen (1988) describes the San Juan Ixhuatepec disaster. The storage site consisted of four spheres of LPG with a volume of 1600 m^3 (422,000 U.S. gal) and two spheres with a volume of 2400 m^3 (634,000 U.S. gal). An additional 48 horizontal cylindrical tanks of various dimensions were present (Figure 3.29). At the time of the disaster, the total site inventory may have been approximately 11,000–12,000 m^3 (2,900,000 – 3,170,000 U.S. gal) of LPG.

Early in the morning of November 19, 1984, large quantities of LPG leaked from a pipeline or tank. The LPG vapors dispersed over the 1 m (3 ft) high dike wall into the surroundings. The vapor cloud had reached a visible height of about 2 m (6 ft) when it was ignited at a flare pit.

At 5:45 A.M., a flash fire resulted. The vapor cloud is assumed to have penetrated houses, which were subsequently destroyed by internal explosions. A violent explosion, probably involving the BLEVE of several storage tanks, occurred one minute after the flash fire. It resulted in a fireball and the propulsion of one or two cylindrical tanks. Heat and fragments resulted in additional BLEVEs. The explosion and fireball completely destroyed the four smaller spheres. The larger spheres remained intact, although their legs were buckled. Only 4 of the 48 cylindrical tanks were left in their original position. Twelve of the ruptured cylindrical tanks traveled distances of more than 100 m (330 ft), and one reached a distance of 1,200 m (3,900 ft). Several buildings on the site collapsed and were destroyed completely. Residents living as far away as approximately 300 m (1,000 ft) from the center of the storage site (Figure 3.30) were killed or injured.

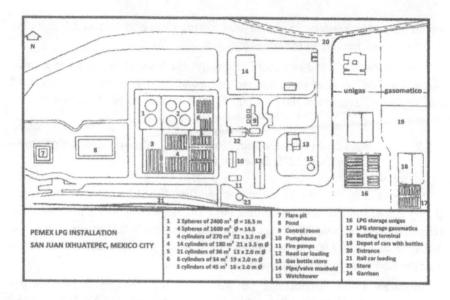

| PEMEX LPG INSTALLATION
SAN JUAN IXHUATEPEC, MEXICO CITY | 1 2 Spheres of 2400 m³ ∅ = 16.5 m
2 4 Spheres of 1600 m³ ∅ = 14.5
3 4 cylinders of 270 m² 32 x 3.5 m ∅
4 14 cylinders of 180 m² 21 x 3.5 m ∅
5 21 cylinders of 36 m² 13 x 2.0 m ∅
6 6 cylinders of 54 m² 19 x 2.0 m ∅
   3 cylinders of 45 m³ 16 x 2.0 m ∅ | 7   Flare pit
8   Pond
9   Control room
10 Pumphouse
11 Fire pumps
12 Road car loading
13 Gas bottle store
14 Pipe/valve manhold
15 Watchtower | 16 LPG storage unigas
17 LPG storage gasomatica
18 Bottling terminal
19 Depot of cars with bottles
20 Entrance
21 Rail car loading
23 Store
24 Garrison |

Figure 3.29. Installation layout at San Juan Ixhuatepec, Mexico.

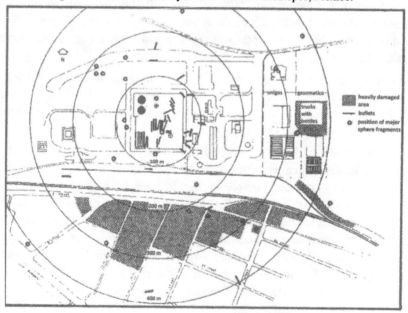

Figure 3.30. Area of damage at San Juan Ixhuatepec, Mexico.

Pietersen compared damage results to effect and damage models that were available at the time. The main findings were:

- Overpressure effects due to the vessel failure appear to be explained by gas expansion, not by flash vaporization.
- Very rapidly expanding ground-level fireballs occurred whenever vessels failed.

Spherical tanks fragmented into ten to twenty pieces, whereas cylindrical vessels fragmented into two pieces. Because cylinders at the storage site had been stored parallel to each other, their fragments were launched in specific directions (

- Figure 3.31).

Pettitt, Harms and Woodward evaluated the BLEVE using updated models in 1994 (Pettit, 1994). They found that the largest explosion, BLEVE of a 1600 m³ (420,000 U.S. gal) LPG sphere, gave an estimated fireball radius of 183 m (600 ft) with 410,00 kg (900,000 lb) flammable mass, which closely matched witness accounts of a 185 m radius fireball. The estimated fireball duration was 46 seconds.

San Juan Ixhuatepec is at an altitude of about 2,250 m (7,400 ft), which affects both blast overpressure and fragment throw. Blast overpressure is a multiple of ambient pressure; the lower ambient pressure at high altitude reduces the blast overpressure compared to lower altitudes. Lower air density at higher altitude results in less air drag for fragments, and therefore longer fragment throw distances.

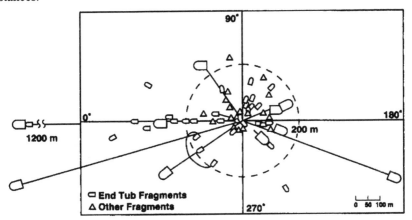

Figure 3.31. Directional preference of projected cylinder fragments
of cylindrical shape.

3.5.3.   San Carlos de la Rapita, Spain: Propylene Tank Truck Failure

Stinton (1983) and Lees (1980) describe this accident. On July 11, 1978, at 12:05 P.M., the loading of a tank truck with propylene was completed. According to weight records obtained at the refinery exit after loading, it had been overloaded; head space was later calculated to be inadequate. The truck scale recorded a weight for the load of 23,470 kg (52,000 lb) — well over the maximum allowable weight of 19,099 kg (42,000 lb). The tank truck was not equipped with a pressure relief valve.

The tank truck was en route to Valencia, but traveled on a back road instead of the highway in order to avoid tolls. It was a hot summer day. As it passed through the village of San Carlos de la Rapita, observers noticed that the tank truck sped up appreciably and was traveling at an excessive speed.

The tank truck left the road near a campsite and crashed at 4:29 P.M. (Figure 3.32). Propylene seems to have been released. The resulting vapor cloud was ignited, possibly by camp cooking fires. One or two explosions then occurred (some witnesses heard two explosions).

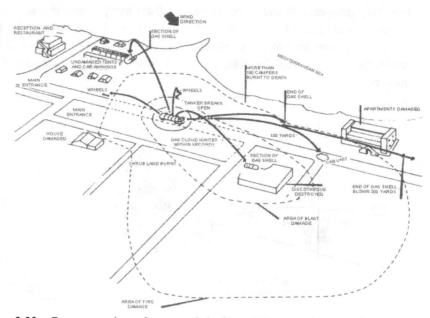

Figure 3.32.   Reconstruction of scene of the San Carlos de la Rapita campsite disaster.

About three minutes after the initial explosion or fire, the tank failed and produced fragments and a fireball. Blast effects were more severe in the tank truck longitudinal directions as depicted in Figure 3.32. About 75 m (250 ft) from the explosion center, a single-story building was completely demolished. This failure resulted in the death of four people. In the opposite direction, a motorcycle was still upright after the incident at a distance of only 20 m (65 ft) from the blast origin. About 500 people were at the campsite at the time of the incident. Deaths, primarily from engulfment in the fireball, totaled 211.

3.5.4.   Crescent City, Illinois, USA: LPG Rail Car Derailment

At 6:30 A.M. on June 21, 1970, fifteen railroad cars, including nine cars carrying LPG, derailed in the town of Crescent City, Illinois. The derailment caused one of the tanks to be punctured, which then released LPG. The ensuing fire, fed by operating safety valves on other cars, resulted in ruptures of tank cars, followed by projectiles and fireballs. No fatalities occurred, but 66 people were injured. There was extensive property damage.

A National Transportation Safety Board report (1972), Eisenberg et al. (1975), and Lees (1980) each describe the accident. At 6:30 A.M. on June 21, 1970, 15 rail cars, including nine cars carrying LPG, derailed in the town of Crescent City, Illinois. The force of the derailment propelled the 27th car in the train over the derailed cars in front of it (Figure 3.33). Its coupler then struck the tank of the 26th car and punctured it. The released LPG was ignited by some unidentified source, possibly by sparks produced by the derailing cars. The resulting fireball reached a height of several hundred feet and extended into the part of the town surrounding the tracks. Several buildings were set on fire.

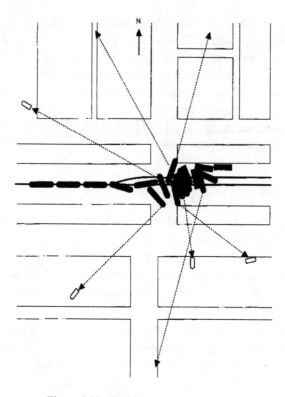

Figure 3.33. Derailment configuration.

The safety valves of other cars opened, thereby releasing more LPG. At 7:33 A.M., the 27th car ruptured with explosive force. Four fragments were hurled in different directions (Figure 3.34). The east end of the car dug a crater in the track structure, and was then hurled about 180 m (600 ft) eastward. The west end of the car was hurled in a southwesterly direction for a total distance of about 90 m (300 ft). This section struck and collapsed the roof of a gasoline service station. Two other sizable portions of the tank were hurled in a southwesterly direction and came to rest at points 180 m (600 ft) and 230 m (750 ft) from the tank.

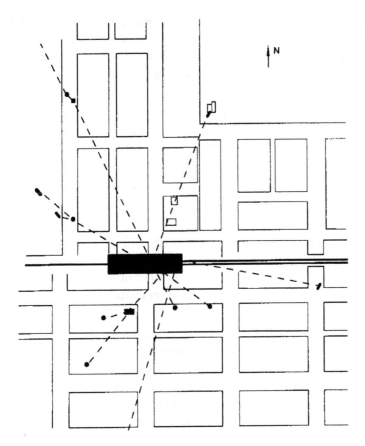

Figure 3.34. Trajectories of tank car fragments.

At about 9:40 A.M., the 28th car in the train ruptured. The south end of this car was hurled about 60 m (200 ft) southward across the street, where it entered a brick apartment building. The north end of the car was hurled through the air in a northwesterly direction over the roofs of several houses, landed in an open field, and rolled until it had traveled over 480 m (1,600 ft).

At 9:45 A.M., the 30th car in the train ruptured. The north end of the car, which included about one-half of the tank, was propelled along the ground in a northeasterly direction for about 180 m (600 ft). It destroyed two buildings and came to rest in a third.

At about 10:55 A.M., the 32nd and 33rd car ruptured almost simultaneously. One of them split longitudinally but did not separate into projectiles. The second one was hurled in the direction of the 34th car and punctured its head, resulting in further LPG releases. The other end of the car also struck the 34th car, ricocheted, and then struck the protective housing of the 35th car. The housing and valves of the 35th car broke off, permitting more LPG to be released. Fires continued for a total of 56 hours.

In all, sixteen business establishments were destroyed and seven others were damaged. Twenty-five residences were destroyed, and a number of others were damaged. Sixty-six people were injured. Due to prompt evacuation, no deaths occurred.

3.5.5. Kingman, Arizona USA: LPG Railroad Tank Car BLEVE

On July 5, 1973, at approximately 2:10 A.M., the BLEVE of a railroad tank car containing LPG occurred at the Doxol Gas Distribution Plant in Kingman, Arizona. In this incident, 13 people were killed (12 of whom were firefighters responding to the scene) and 95 people (spectators clustered along the highway) were injured.

This accident was described in an article in the Fire Journal (Sherry, 1974). The Kingman plant consisted of a railroad siding, a tank car unloading rack, two horizontal storage tanks, three truck and cylinder-loading racks, a storage yard and a small office. A highway was located about 21 m (70 ft) north of the plant office. On July 5, a transfer of LPG from a tank car to the storage tank took place at about 1:30 P.M.. The capacity of the tank car was 130 m^3 (33,900 U.S. gal), its test pressure 23.4 bar (340 psi), and its design burst pressure 34.5 bars (500 psi).

Two employees at the site were in charge of the LPG transfer process. One of them thought he detected a leak, and attempted to solve it by tightening the connections by hitting them with a large aluminum alloy wrench. This procedure was done several times for each connection, until a fire resulted at the westerly liquid connection. Both employees fell down and were burned; one of them died later of severe burns. Firefighters arrived on the site at about 2:00 P.M., while the fire gained intensity. The firefighters intended to cool the tank with two one-inch booster lines and a deluge gun. About 19 minutes after the start of the fire (about 2:10 P.M.), the tank ruptured. A ground level fireball (Figure 3.35 (a)) expanded about 45-60 m (150-200 ft) on all sides, followed immediately by fireball liftoff (Figure 3.35 (b)) with a diameter of approximately 250-300 m (800-1,000 ft). The tank split into two portions: the westerly half of the tank rocketed about 370 m (1,200 ft) away from its original position, propelled by the expanding LPG. The easterly portion did not rocket, but split open and flattened out. There were 13 firefighters within 45 m (150 ft) of the tank car when the explosion occurred: 12 of them died of extensive burns. Most of the 95 injured people were spectators located along the highway at about 300 m (1,000 ft) from the tank.

(a) Extension of fireball just after tank
rupture.

(b) Fireball liftoff.

Figure 3.35. Kingman explosion fireball. (Sherry, 1974)

The office building and most of the vehicles in the immediate vicinity of the fireball were destroyed. Several small LPG cylinders in the storage yard had burned at their relief valves, and one of them ruptured. Three businesses located at 180 m (600 ft), 240 m (800 ft) and 270 m (900 ft) east of the explosion were ignited by flaming debris and heat radiating from the fireball.

4. BASIC CONCEPTS

Accident scenarios leading to vapor cloud explosions, flash fires, and BLEVEs were described in the previous chapter. Blast effects are a characteristic feature of VCEs, PVBs and BLEVEs. Fireballs and flash fires cause damage primarily from heat effects caused by thermal radiation. This chapter describes the basic concepts underlying these phenomena.

Section 4.1 treats atmospheric dispersion in just enough detail to permit understanding of its implications for VCEs and flash fires. Section 4.2 describes typical ignition sources and the ignition characteristics of typical fuel-air mixtures. Section 4.3 introduces basic concepts of thermal radiation modeling. Section 4.4 covers the evolution from slow, laminar, premixed combustion to an intense, explosive, blast-generating process, and introduces the concepts of deflagration and detonation. Section 4.5 covers the physical concepts of blast and blast loading and describes how blast parameters can be established and scaled.

4.1. ATMOSPHERIC VAPOR CLOUD DISPERSION

Vapor cloud factors such as flammable cloud size, homogeneity, and location are largely determined by the release conditions and turbulent dispersion following release. Vapor cloud dispersion is affected by numerous factors including wind speed, wind direction relative to the discharge direction, discharge orientation, weather stability, local turbulence, vapor density (initially and after mixing), heating/cooling with air, and jet mixing at the source. Vapor cloud ignition, burning, and explosion depends on such material properties as the flammable limits, the flash point, the autoignition temperature, and laminar burning velocity, as are listed in Table 4.1.

Vapor clouds develop a concentration profile in the horizontal and vertical directions. The flammable mass in a vapor cloud is found by integrating the mass between flammable limits. Flammability limits of flammable gases and vapors in air have been published, for example by Nabert and Schön (1963), Coward and Jones (1952), Zabetakis (1965), and Kuchta (1985).

Liquid does not burn; rather, the evaporated vapor above the liquid burns but only if the vapor concentration is between the lower flammable limit (LFL) and upper flammable limit (UFL). An indication of the volatility of fuels is the flash point, which is the minimum temperature at which a liquid gives off vapor in sufficient concentration to form an ignitable mixture with air near the surface of the liquid. The flash point is, at least theoretically, the temperature at which the vapor pressure in units of atmospheres equals the LFL. Vapors from low vapor pressure liquids such as kerosene at ambient temperature are not above LFL and are not ignitable; however, they can form flammable vapor-air mixtures if the liquid is at an elevated temperature.

Combustion properties change with oxygen concentration. Oxygen enrichment, which refers to oxygen concentrations greater than 21% volume, expands the flammability limits, reduces the ignition energy, and increases burning velocity.

In addition, a fuel at a temperature below its flash point released under pressure may form a flammable aerosol (or mist) mixture. That is, the smaller aerosol droplets can ignite, and there is a separate LFL for such aerosols. Unfortunately, aerosol flammable limits are not well known, although Burgoyne (1963) showed that the LFL on a weight basis of hydrocarbon aerosol mixtures is in the same range as those of gas-air or vapor-air mixtures; namely, about 50 g/m^3.

Burning velocity is the speed at which a flame front propagates relative to the unburned gas. Burning velocity differs from flame speed. The laminar burning velocity is the speed at which a laminar (planar) flame front propagates relative to the unburned gas mixture ahead of it. The fundamental burning velocity is similar, but generally not identical to, the observed laminar burning velocity. This is because fundamental burning velocity is a characteristic parameter referring to standardized unburned gas conditions (normally 760 mm-Hg and 25°C), and which has been corrected for nonidealities in the measurement.

Laminar burning velocities are measured primarily to characterize the reactivity of a fuel. As seen in Table 4.1, laminar burning velocities are less than 1 m/s for most fuels, but are up to 3 m/s for the most reactive fuels. By contrast, turbulent flame speeds develop in flash fires (at about 10 m/s), in deflagration explosions (at about 100-300 m/s), and in detonation explosions (up to 2,000 m/s).

Table 4.1. Explosion properties of flammable gases and vapors in air
at standard pressure [a]

Gas or Vapor	Flammability Limits (vol. %)	Flash Point (°C)	Autoignition Temperature (°C)	Laminar Burning Velocity (m/s)
Methane	5.0–15.0	—	595	0.448
Ethane	3.0–15.5	—	515	0.476
Propane	2.1–9.5	—	470	0.464
Ethylene	2.7–34	—	425	0.735
Propylene	2.0–11.7	—	455	0.512
Hydrogen	4.0–75.6	—	560	3.25
Acetone	2.5–13.0	–19	540	0.444
Diethyl ether	1.7–36	–20	170	0.486
Acetylene	1.5–100	—	305	1.55
Ethanol	3.5–15	12	425	—
Toluene	1.2–7.0	—	535	—
Cyclohexane	1.2–8.3	–18	260	—
Hexane	1.2–7.4	–15	240	—
Xylene	1.0–7.6	30	465	—

[a] Nabert and Schön (1963), Coward and Jones (1952), Zabetakis (1965), and Gibbs and Calcote (1959).

Turbulence affects the vapor cloud properties before and after ignition. Before ignition the degree of homogeneity of a fuel-air mixture largely determines whether the fuel-air mixture is able to perpetuate and accelerate the combustion process. After ignition, flame turbulence is a primary determinant of possible blast effects produced by a vapor cloud explosion. Thus, it is useful to briefly discuss turbulent dispersion.

Turbulence may be described as a random motion superimposed on the mean flow. Many aspects of turbulent dispersion are reasonably well-described by a simple model in which turbulence is viewed as a spectrum of eddies in an extended range of length and time scales (Lumley and Panofsky, 1964). In shear layers, large-scale eddies extract mechanical energy from the mean flow. This energy is continuously transferred to smaller and smaller eddies. Such energy transfer continues until energy is dissipated into heat by viscous effects in the smallest eddies of the spectrum.

The extent of the flammable vapor cloud is dependent on many factors including release conditions, material properties, topography, process unit geometry, and atmospheric conditions. Several models for calculating release and dispersion effects have been developed. Hanna and Drivas (1987) provide guidance on model selection for various accident scenarios. CFD models may also be applied to model dispersion effects. More recently, Hanna (2006) summarized the characteristics of currently developed dispersion models.

4.2.   IGNITION

Ignition of vapor clouds can occur by various means such as open flame, electric spark, or by contact with a hot surface. In each case, there is a minimum contact time and a minimum energy transfer needed to form a self-perpetuating burning volume, or ignition kernel. Minimum ignition energy (MIE) is the lowest energy of an electrical spark discharge to ignite flammable fuel-air mixture under defined test conditions. Table 4.2 provides MIE values for a number of fuels. Also presented in Table 4.1 is Autoignition Temperature (AIT), the minimum temperature at which a fuel-air mixture will spontaneously ignite. In the measurement protocol to determine AIT, an exposure duration is inherently applied.

Depending on source properties, ignition can lead to deflagration or detonation. Deflagration is by far the more likely mode of flame propagation to occur immediately upon ignition. Deflagration ignition energies are on the order of 10^{-4} J, whereas direct initiation of detonation requires energy of approximately 10^6 J. Table 4.2 gives initiation energies for deflagration and detonation for some hydrocarbon-air mixtures (initiation energies for immediate detonation were converted from explosive charge weights shown in the table). Considering the high energy required for direct initiation of a detonation, it is a very unlikely occurrence.

Table 4.2. Initiation energies for deflagration and detonation
for some fuel-air mixtures[a]

Gas Mixture	Ignition Energy for a Deflagration (mJ)	Charge Weight for Immediate Detonation (g Tetryl)	Initiation Energy for Immediate Detonation (mJ)
Acetylene-Air	0.01	0.04	1.8×10^5
Propane-Air	0.25	90	4.1×10^8
Methane-Air	0.21	22,000	9.9×10^{10}
Hydrogen-Air	0.016	1.15	5.2×10^6

[a] Data from Britton (1982), Lewis and von Elbe (1987), Knystautas et al. (1983), Guirao (1982), Bull (1978)

In practice, vapor cloud ignition can be the result of a sparking electric apparatus or hot surfaces present in a chemical plant, such as extruders, hot steam lines or friction between moving parts of machines. Another common source of ignition is open fire and flame, for example, in furnaces and heaters. Motorized equipment and vehicles also present potential ignition sources. Mechanical sparks, for example, from the friction between moving parts of machines and falling objects, are also frequent sources of ignition. Many metal-to-metal combinations result in mechanical sparks that are capable of igniting gas or vapor-air mixtures (Ritter 1984).

4.3. THERMAL RADIATION

Released heat from combustion is transmitted to the surroundings by conduction, convection and thermal radiation. For large fires, thermal radiation is the main hazard, causing secondary fires and burn hazards to people in the area.

Thermal radiation is electromagnetic radiation covering wavelengths from 2 to 16 im (infrared). It is the net result of radiation emitted by radiating substances such as H_2O, CO_2, and soot (often dominant in fireballs and pool fires), absorption by these substances, and scatter. This section presents general methods to describe the radiation effects at a certain distance from the source of thermal radiation. Two different methods are used to describe the radiation from a fire: the point-source model and the surface-emitter, or solid-flame model.

4.3.1. Point-Source Model

In the point-source model, it is assumed that a fraction (f) of the heat of combustion is emitted as radiation in all directions from a central point in the flame. The radiation per unit area and per unit time received by a target (q) at a distance (x) from the point source is, therefore, given by:

$$q = \frac{f\dot{m}H_c\tau_a}{4\pi x^2}$$

(Eq. 4.1)

where:

$\dot{m}$	=	rate of combustion	(kg/s)
H_c	=	heat of combustion per unit of mass	(J/kg)
τ_a	=	atmospheric attenuation of thermal radiation (transmissivity)	(–)
q	=	radiation per unit area per unit time (flux)	(W/m²)
x	=	distance from the point source	(m)

It is assumed that the target surface faces toward the radiation source so that it receives the maximum incident flux. The rate of combustion depends on the release. For a pool fire of a fuel with a boiling point (T_b) above the ambient temperature (T_a), the combustion rate can be estimated by the empirical relation:

$$\dot{m} = \frac{0.0010H_cA}{H_v + C_v(T_b - T_a)}$$

(Eq. 4.2)

where:

$\dot{m}$	=	combustion rate	(kg/s)
H_c	=	heat of combustion	(J/kg)
A	=	pool area	(m²)
H_v	=	heat of vaporization	(J/kg)
C_v	=	specific heat of fuel	(J/kg/K)
T_b	=	boiling temperature	(K)
T_a	=	ambient temperature	(K)
0.0010	=	a constant	(kg/s/m²)

The fraction of combustion energy dissipated as thermal radiation (f) is the unknown parameter in the point-source model. This fraction depends on the fuel and on dimensions of the flame. Measurements give values for this fraction ranging from 0.1 to 0.4 (Mudan 1984; Duiser 1989). Raj and Atallah (1974) measured the fraction of radiation from 2- to 6-m pool fires of LNG and found values between 0.2 and 0.25. The data from Burgess and Hertzberg (1974) for methane range from 0.15 to 0.34, and for butane, from 0.20 to 0.27. The highest value they found, 0.4, was for gasoline. Roberts (1982) analyzed the data from fireball experiments of Hasegawa and Sato (1977) and found values of 0.15 to 0.45.

The point-source model can be inaccurate for target positions close to emitting surfaces. It is of value primarily for estimating radiation exposure in the far field where the fire geometry appears approximately like a point source.

4.3.2.   Solid-Flame Model

The solid-flame model can be used to overcome the inaccuracy of the point-source model. This model assumes that the fire can be represented by a solid body of a simple geometrical shape, and that all thermal radiation is emitted from its surface. Rather than calculating the total thermal energy of combustion and then seeking to find the fraction of this energy that is radiated, a surface emissive power, E, is correlated from radiometer measurements of fires. This has the advantage of being based on direct measurements of radiation and it can also describe the structure of a fire in various degrees of detail.

This model applies in the near field, and has the advantage that the geometries of the fire and target, as well as their relative positions, are taken into account. That is, only a portion of the radiation from each point on the fire reaches the target (with or without shadowing obstructions). This geometric relationship is accounted for by a "view factor" which integrates the area of the target "seen" by each point on the fire surface. The view factor is the fraction of the total emitted radiation falling on the receiving target. The view factor depends on the shapes of the fire and receiving target, and on the distance between them, by a purely geometric analysis independent of fire temperature or radiation properties. Tables of view factors are formulated for idealized geometries such as a cylindrical or cone-shaped fire and a tilted plane or spherical target.

The incident radiation per unit area and per unit time (q) is given by:

$$q = FE\tau_a \qquad \text{(Eq. 4.3)}$$

where:

q	=	radiation received by receptor	(W/m^2)
F	=	view factor	(–)
E	=	surface emissive power	(W/m^2)
τ_a	=	atmospheric attenuation factor (transmissivity)	(–)

Transmissivity is the fraction of radiated energy that is absorbed or scattered in the atmosphere before reaching the target. This fraction increases with distance to the target and with the factors that affect absorption and scattering, mainly relative humidity.

Emissive Power

Emissive power is the radiative flux leaving the surface of the fire (radiation per unit area and per unit time). If correlated with narrow-angle pyrometers, E varies locally and provides detail to flame structure. If correlated with wide-angle pyrometers, it provides an average over the entire flame. Theoretical values of emissive power can be calculated by use of Stefan-Boltzmann's law, which gives the radiation of a black body in relation to its temperature. Because the fire is not a perfect black body, the emissive power is a fraction (ε) of the black body radiation:

$$E = \varepsilon\sigma T^4 \qquad \text{(Eq. 4.4)}$$

where:

E	=	the emissive power	(W/m^2)
T	=	temperature of the fire	(K)
ε	=	emissivity	(–)
σ	=	Stefan-Boltzmann constant = 5.67×10^{-8}	(W/m^2/K^4)

Again, the temperature can vary over the fire structure, $T(x,z)$, or can be taken as an average over the fire.

The use of Stefan-Boltzmann's law to calculate radiation requires the knowledge of the fire's temperature and emissivity. Turbulent mixing causes fire temperature to vary. Therefore, in practice it is best to rely solely on measured radiation values.

The surface-emissive powers of fireballs depend strongly on fuel quantity and pressure just prior to release. Fay and Lewis (1977) found small surface-emissive powers for 0.1 kg (0.22 pound) of fuel (20 to 60 kW/m^2; 6300 to 19,000 Btu/hr/ft^2). Hardee et al. (1978) measured 120 kW/m^2 (38,000 Btu/hr/ft^2). Moorhouse and Pritchard (1982) suggest an average surface-emissive power of 150 kW/m^2 (47,500 Btu/hr/ft^2), and a maximum value of 300 kW/m^2 (95,000 Btu/hr/ft^2), for industrial-sized fireballs of pure vapor. Experiments by British Gas with BLEVEs involving fuel masses of 1000 to 2000 kg of butane or propane revealed surface-emissive powers between 320 and 350 kW/m^2 (100,000–110,000 Btu/hr/ft^2; Johnson et al. 1990).

Emissivity

The fraction of black-body radiation actually emitted by flames is called emissivity. Emissivity is determined first by adsorption of radiation by combustion products (including soot) in flames and second by the radiation wavelength. These factors make emissivity modeling complicated. By assuming that a fire radiates as a gray body, in other words, that extinction coefficients of the radiation adsorption are independent of the wavelength, a fire's emissivity can be written as:

$$\varepsilon = 1 - \exp(-kx_f) \qquad \text{(Eq. 4.5)}$$

where:

ε	=	emissivity	(–)
x_f	=	beam length of radiation in flames	(m)
k	=	extinction coefficient	(m^{-1})

For a fireball, x_f can be replaced by the fireball diameter (Moorhouse and Pritchard 1982). Hardee et al. (1978) reported, for optically thin LNG fires, a value of $k = 0.18$ m^{-1}. The emissivity of larger fires approaches unity.

Transmissivity

Atmospheric attenuation is the consequence of absorption and scattering of radiation by the air between emitter and receiver. Atmospheric absorption is primarily due to water vapor and, to a lesser extent, to carbon dioxide. Absorption also depends on radiation wavelength, and consequently, on fire temperature since the proportion of radiation at the absorbed wavelengths varies with fire temperature. Duiser (1989) approximates transmissivity as:

$$\tau_a = 1 - \alpha_w - \alpha_c \qquad \text{(Eq. 4.6)}$$

where:

τ_a	=	transmissivity	(–)
α_w	=	radiation absorption factor for water vapor	(–)
α_c	=	radiation absorption factor for carbon dioxide	(–)

Both factors depend on the respective partial vapor pressures of water and carbon dioxide and upon the distance to the radiation source. The partial vapor pressure of carbon dioxide in the atmosphere is fairly constant (30 Pa), but the partial vapor pressure of water varies with atmospheric relative humidity. Duiser (1989) published graphs plotting absorption factors (α) against the product of partial vapor pressure and distance to flame (Px) for flame temperatures ranging from 800 to 1800 K.

Moorhouse and Pritchard (1982) presented the following relationship to approximate transmissivity of infrared radiation from hydrocarbon flames through the atmosphere:

$$\tau_a = 0.998^x \qquad \text{(Eq. 4.7)}$$

where:

τ_a	=	transmissivity	(–)
x	=	the distance to the source	(m)

This equation is valid for distances up to 300 m.

Simpson (1984) provides plots and correlations that account for both absorption and scattering, and are consequently considered definitive for the most accurate work.

Raj (1982) presents graphs for transmissivity depending only on the relative humidity of air. His graphs can be approximated by:

$$\tau_a = \log(14.1 RH^{-0.108} x^{-0.13})$$ (Eq. 4.8)

where:

τ_a	=	transmissivity	(–)
x	=	distance	(m)
RH	=	relative humidity	(%)

This equation should not be used for relative humidities of less than 20%. The transmissivity calculated by Raj's method agrees, for distances up to 500 m, with the values calculated according to the procedure suggested by Duiser (1989).

Lihou and Maund (1982) define attenuation constants for hydrocarbon flames through the atmosphere, which can vary from 4×10^{-4} m^{-1} (for a clear day) to 10^{-3} m^{-1} (for a hazy day). The mean value suggested by the authors is 7×10^{-4} m^{-1}, which gives a transmissivity of:

$$\tau_a = \exp(-0.0007x)$$ (Eq. 4.9)

where:

τ_a	=	transmissivity	(–)
x	=	distance	(m)

This equation gives higher transmissivity values than those calculated with methods described earlier. Presumably, Lihou and Maund's transmissivity is to be used for conditions of low relative humidity, in which dust particles (haze) are the main cause of attenuation. A conservative approach is to assume $\tau_a = 1$.

View Factor

Let F_{12} be the fraction of radiation impinging directly on a receiving surface. If the emitting surface equals A_1, the incident radiation on the target's receiving area A_2 follows from:

$$A_1 E F_{12} = A_2 q_2$$ (Eq. 4.10)

where:

E	=	emissive power of emitting surface	(W/m^2)
q_2	=	incident radiation receiving surface	(W/m^2)

Application of the reciprocity relation $(A_1F_{12} = A_2F_{21})$ allows the fraction of radiation received by the target (apart from atmospheric attenuation and emissivity) to be expressed as:

$$q_2 = F_{21}E \qquad\qquad \text{(Eq. 4.11)}$$

where:

F_{21}	=	view factor or geometric configuration factor	(–)
E	=	emissive power of emitting surface	(W/m²)
q_2	=	incident radiation-receiving surface	(W/m²)

The view factor depends on the shape of the emitting surface (plane, cylindrical, spherical, or hemispherical), the distance between emitting and receiving surfaces, and the orientation of these surfaces with respect to each other. In general, the view factor from a differential plane (dA_2) to a flame front (area A_1) on a distance L is determined by:

$$F_{dA_2-A_1} = \int_{A_1} \frac{\cos\Theta_1 \cos\Theta_2}{\pi L^2} dA_1 \qquad\qquad \text{(Eq. 4.12)}$$

The geometry pertinent to this equation is represented in Figure 4.1.

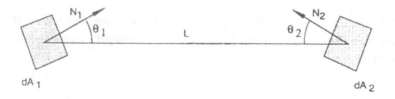

Figure 4.1. Configuration for radiative exchange between two differential elements.

where:

L	=	length of line connecting elements dA_1 and dA_2	(m)
Θ_1	=	angle between L and the normal to dA_1	(deg)
Θ_2	=	angle between L and the normal to dA_2	(deg)
A_1	=	surface area flame front	(m^2)
dA_2	=	differential plane	(m^2)

A fireball is represented as a solid sphere with a center height H and a diameter D. Let the radius of the sphere be R ($R = D/2$). (See Figure 4.2.) Distance x is measured from a point on the ground directly beneath the center of the fireball to the receptor at ground level. When this distance is greater than the radius of the fireball, the view factor can be calculated.

For a vertical surface:

$$F_v = \frac{x(D/2)^2}{(x^2+H^2)^{3/2}}$$
(Eq. 4.13)

For a horizontal surface:

$$F_h = \frac{H(D/2)^2}{(x^2+H^2)^{3/2}}$$
(Eq. 4.14)

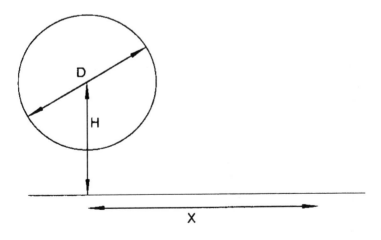

Figure 4.2. View factor of a fireball.

For a vertical surface beneath the fireball $(x < D/2)$, the view factor is given by:

$$F_v = \frac{1}{2} - \frac{1}{\pi} \sin^{-1}\left[\frac{(x_r^2 + H_r^2 - 1)^{1/2}}{H_r}\right] + \frac{x_r}{\pi(x_r^2 + H_r^2)^{3/2}} \cos^{-1}\left[-\frac{x_r(x_r^2 + H_r^2 - 1)^{1/2}}{H_r}\right]$$
$$- \frac{(1 - x_r)^{1/2}}{\pi(x_r^2 + H_r^2)^{1/2}}$$

(Eq. 4.15)

where:

x_r	=	reduced length x/R	(–)
H_r	=	reduced length H/R	(–)

For a flash fire, the flame can be represented as a plane surface. Appendix A contains equations and tables of view factors for a variety of configurations, including spherical, cylindrical, and planar geometries.

4.4.   EXPLOSIONS — VCE

4.4.1.   Deflagration

The mechanism of flame propagation into an unburned fuel-air mixture in a deflagration is determined largely by conduction and molecular diffusion of heat and species. Figure 4.3 shows the change in temperature across a laminar flame, whose thickness is on the order of one millimeter.

Heat is produced by chemical reaction in a reaction zone. The heat is transported ahead of the reaction zone into a preheating zone in which the mixture is heated and thereby preconditioned for reaction. Since molecular diffusion is a relatively slow process, laminar flame propagation is slow. Table 4.1 gives an overview of laminar burning velocities of some of the most common hydrocarbons and hydrogen.

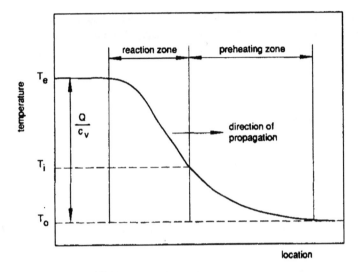

Figure 4.3. Temperature distribution across a laminar flame.

Transformation from laminar to turbulent combustion is most strongly influenced by turbulence, and secondarily by combustion instabilities. A laminar-flame front propagating into a turbulent mixture is strongly affected by the turbulence. Low-intensity turbulence will only wrinkle the flame front and enlarge its surface area. With increasing turbulence intensity, the flame front loses its more-or-less smooth, laminar character and assumes a more intense and efficient combustion form. In an intensely turbulent mixture, combustion takes place in an extended zone in which the combustion products and unreacted mixture are efficiently mixed. High combustion rates can result because, within the combustion zone, the reacting interface between combustion products and reactants can become very large.

The interaction between turbulence and combustion plays a key role in the development of a VCE. Laminar combustion following ignition generates expansion and produces a flow field. If the boundary conditions of the expansion flow-field are such that turbulence is generated, the flame front, which is convected by expansion flow, will interact with the turbulence. The combustion rate is now increased by this turbulence.

As more fuel is converted into combustion products per unit of volume and time, expansion flow becomes stronger. Higher flow velocities go hand in hand with more intense turbulence. This process feeds on itself; that is, a positive feedback coupling comes into action. In the turbulent stage of flame propagation, a gas explosion may be described as a process of combustion-driven expansion flow with the turbulent expansion-flow structure acting as an uncontrolled positive feedback (Figure 4.4).

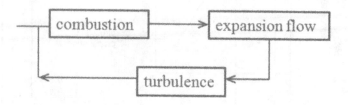

Figure 4.4. Positive feedback, the basic mechanism of a gas explosion.

If such a process continues to accelerate, the combustion mode may suddenly change drastically from deflagration to detonation. The reactive mixture just in front of the turbulent combustion zone is preconditioned for reaction by a combination of compression and of heating by turbulent mixing with combustion products. If turbulent mixing becomes too intense, the combustion reaction may quench locally. A very local, nonreacted but highly reactive mixture of reactants and hot products is the result. The intensity of heating by compression can raise temperatures of portions of the mixture to levels above the autoignition temperature. These highly reactive "hot spots" eventually react very rapidly, resulting in localized, constant-volume sub-explosions (Urtiew and Oppenheim 1966; Lee and Moen 1980). If the surrounding mixture is sufficiently close to autoignition as a result of blast compression from one of the sub-explosions, a detonation wave results.

4.4.2.   Detonation

The two basic modes of combustion—deflagration and detonation—differ fundamentally in their propagation mechanisms. In deflagrative combustion, the reaction front is propagated by molecular-diffusive transport of heat and turbulent mixing of reactants and combustion products. In detonative combustion, on the other hand, the reaction front is propagated by a strong shock wave that compresses the mixture beyond its autoignition temperature. At the same time, the shock is maintained by the heat released from the combustion reaction behind it.

To understand the behavior of detonation, some basic features of a detonation must be understood; these are briefly summarized in the next few paragraphs. Various properties of detonation are reflected by different models (Ficket and Davis 1979). Fairly accurate values of overall properties of a detonation, including, for example, wave speed and pressure, may be computed from the Chapman-Jouguet (CJ) model (Nettleton 1987). In this model, a detonation wave is simplified as a reactive shock in which instantaneous shock compression and the combustion front coincide. Zero induction time and an instantaneous reaction are inherent in this model (Figure 4.5). For stoichiometric hydrocarbon-air mixtures, the detonation wave speed is in the range of 1700–2100 m/s and corresponding detonation wave overpressures are in the range of 18–22 bars.

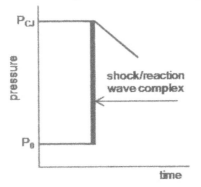

Figure 4.5. The CJ-model.

A slightly more realistic concept is the Zel'dovich-Von Neumann-Döring (ZND) model. In this model, the fuel-air mixture does not react on shock compression beyond autoignition conditions before a certain induction period has elapsed (Figure 4.6). The pressure behind the nonreactive shock (P_{vN}) is much higher than the CJ detonation pressure. The CJ pressure is not attained until the reaction is complete. The duration of the induction period at the nonreactive, postshock state is on the order of microseconds. As a consequence, nonreactive, postshock pressure—the "Von Neumann spike"— is difficult to detect experimentally, and decays immediately to the CJ pressure.

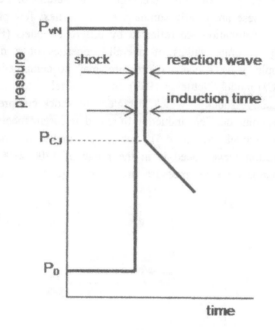

Figure 4.6. The ZND-model.

The one-dimensional representation described above is too simple to describe the behavior of a detonation in response to boundary conditions. Denisov et al. (1962) showed that the ZND-model of a detonation wave is unstable. Figure 4.7 shows how a plane configuration of a shock and a reaction wave breaks up into a cellular structure. Detonation is not a steady process, but a highly fluctuating one. Its multidimensional cyclic character is determined by a process of continuous decay and reinitiation. The collision of transverse waves plays a key role in the structure of a detonation wave. The nature of this process has been described in detail many times, for example, Denisov et al. (1962); Strehlow (1970); Vasilev and Nikolaev (1978). In this cyclic process, a characteristic length scale or cell size can be distinguished, at least on the average (Figure 4.7 and Figure 4.8). The characteristic cell size is specific to a fuel-oxidizer mixture. Some guide values taken from Bull et al. (1982), Knystautas et al. (1982), and Moen et al. (1984) are given in Table 4.3 for stoichiometric fuel-air mixtures.

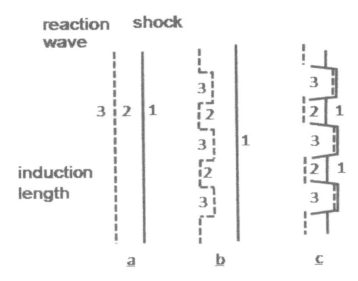

Figure 4.7. Instability of ZND-concept of a detonation wave.

(a) planar structure, (b) reaction wave breaking up into cells, (c) fully developed cellular structure; (1) undisturbed fuel-air mixture, (2) post-shock nonreactive mixture, (3) post-reaction products.

Figure 4.8. Cellular structure of a detonation.

Cell size depends strongly on the fuel and mixture composition; more reactive mixtures result in smaller cell sizes. Table 4.3 shows that a stoichiometric mixture of methane and air has an exceptionally large cell size compared to other hydrocarbon-air mixtures, indicating that methane-air is less reactive and more difficult to detonate than other hydrocarbon-air mixtures.

Table 4.3. Characteristic detonation cell sizes for some
stoichiometric fuel-air mixtures[a]

Fuel	Cell Size (cm)
methane	33
ethane	5
propane	5.4
n-butane	6.5
ethylene	2
acetylene	0.5
hydrogen	1.5

[a] Data from Guirao, Knystautas and Lee (1982) and Knystautas, Guirao, Lee and Sulmistras (1984).

4.5.   BLAST EFFECTS

4.5.1.   Manifestation

A characteristic feature of explosions is airblast. Gas explosions are characterized by rapid combustion in which high-temperature combustion products expand and affect their surroundings. In this fashion, the heat of combustion of a fuel-air mixture (chemical energy) is partially converted into expansion (mechanical energy). Mechanical energy is transmitted by the explosion process into the surrounding atmosphere in the form of a blast wave. This process of energy conversion is very similar to that occurring in internal combustion engines. Such an energy conversion process can be characterized by its thermodynamic efficiency. At atmospheric conditions, the theoretical maximum thermodynamic efficiency for conversion of chemical energy into mechanical energy (blast) in gas explosions is approximately 40%. Thus, less than half of the total heat of combustion produced in explosive combustion can be transmitted as blast-wave energy.

In the surrounding atmosphere, a blast wave is experienced as a transient change in gas-dynamic-state parameters: pressure, density, and particle velocity.

Generally, these parameters increase rapidly, then decrease less rapidly to sub-
ambient values (i.e., develop a negative phase). Subsequently, parameters slowly
return to atmospheric values (Figure 4.9). The shape of a blast wave is highly
dependent on the nature of the explosion process.

The initial portion of the blast wave in which pressure is above ambient
pressure is referred to as the positive phase. The latter portion in which pressure
is below ambient is the negative phase. The term "overpressure" refers to the
magnitude of blast pressure above ambient. Durations of positive and negative
phases are defined by the times at which pressure is at or crosses through ambient.
Impulse is the integration of the pressure-time history for either the positive or
negative phase.

Figure 4.9. Blast wave shapes.

If the combustion process within a gas explosion is relatively slow, then
expansion is slow, and the blast consists of a low-amplitude pressure wave that is
characterized by a gradual increase in gas-dynamic-state variables (Figure 4.9a).
If, on the other hand, combustion is rapid, the blast may achieve a sudden increase
in the gas-dynamic-state variables: a shock (Figure 4.9b). The shape of a blast
wave changes during propagation because the propagation mechanism is
nonlinear. Initial pressure waves tend to steepen and in some circumstances may
result in shock waves in the far field. Wave durations tend to increase with
propagation distance.

4.5.2.   Blast Loading

An object struck by a blast wave experiences a loading. This loading has two
aspects. First, the incident wave induces a transient pressure distribution over the
object which is highly dependent on the shape of the object. The complexity of
this process can be illustrated by the phenomena represented in Figure 4.10 (Baker
1973). As the incident wave encounters the front wall, the portion striking the

wall is reflected and builds up a local, reflected overpressure. For weak waves, the reflected overpressure is slightly greater than twice the incident (side-on) overpressure. As the incident (side-on) overpressure increases, the reflected pressure multiplier increases.

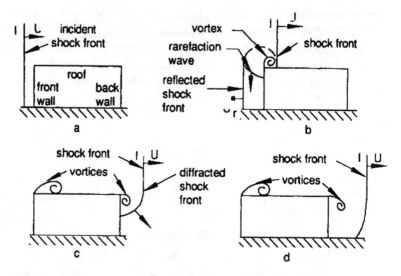

Figure 4.10. Interaction of a blast wave with a rigid structure (Baker, 1973).

In Figure 4.10b, the reflected wave moves to the left. Above the structure, the incident wave continues on relatively undisturbed. As the reflected wave moves back from the front wall, a rarefaction front moves down the front face of the structure (Figure 4.10b). A rarefaction is an expansion wave that reduces the pressure. In the same way, rarefaction waves move across the front face from the two sides. The top face of the structure experiences no more than the side-on wave overpressure. As the incident shock passes beyond the rear face of the structure, it diffracts around this face, as shown in   Figure 4.10c. At the instant shown in Figure 4.10c, the reflected overpressure at the front face has been completely attenuated to the side-on pressure level by the rarefactions. Subsequently, the incident shock has passed beyond the structure, the diffraction process is over, and the structure is immersed in the particle-velocity flow-field behind the leading shock front. At this stage, the structure experiences the blast wave as a gust of wind that exerts a drag force.

The blast loading of objects with substantial lateral dimensions is largely governed by the overpressure aspect of a blast wave. On the other hand, slender objects—lampposts, for example—are hardly affected by the overpressure aspect of blast loading.

The second component of blast loading is a drag force induced by particle velocity in the blast wave. Drag force magnitude is determined by the object's frontal area and the dynamic pressure of flow after the leading shock. The blast loading of slender objects is largely governed by the dynamic pressure (drag) aspect of a blast wave.

Detailed estimates of the full loading of an object by a blast wave are possible by use of multidimensional gas-dynamic numerical models, also known as computational fluid dynamic (CFD) models. However, if the problem is sufficiently simplified, analytic methods may do as well. For such methods, it is sufficient to describe the blast wave somewhere in the field in terms of the side-on peak overpressure and the positive-phase impulse or duration. Blast models used for vapor cloud explosion blast modeling (Section 6.3) give the distribution of these blast parameters in the explosion's vicinity.

4.5.3.   Ground Reflection

Explosions above the ground produce a reflected blast wave from the ground surface similar to the process described above for reflection from a building surface. A rigid non-moving ground plane will reflect a shock wave with little loss in strength of the wave. As a result, an object near a blast source can experience two blast waves in close succession: the incident wave and the reflected wave.

Shock waves coalesce when the angle between the incident and reflected waves is below a critical angle, at which point a third shock wave is formed called a Mach stem (Baker, 1973). Since the Mach stem is the resulted of two shock waves coalescing, its strength is greater than either of the original two shock waves.

The Mach stem forms at an angle that depends on incident wave strength. For an elevated explosion, the angle to the point on the ground where the Mach stem forms is between about 40 and 80 degrees from vertical (TM5-1300, 1990). For example, the Mach stem forms at an angle of incidence of about 54 deg for a 34 kPa (5 psi) incident shock, and 65 deg for 7 kPa (1 psi) incident shock.

VCEs are generally ground level or "surface burst" events. PVBs and BLEVEs may be elevated, but the height is limited by process structures. Buildings within or very near to a process unit in which there is PVB or BLEVE may experience blast loads from an elevated burst, which produce reflected blast loads on the roof, and ground reflections that follow the incident wave. Buildings located at distances beyond the formation of the Mach stem will be loaded by a single wave, the Mach stem. A surface burst will produce only a single wave.

Baker (1973) illustrated that the ground reflection from an elevated burst can be simulated using an imaginary explosion that is below the ground surface by the same distance as the actual explosion above ground. The imaginary explosion has the same characteristics as the explosion above ground. This leads to a common simplification to deal with ground reflections, which is to double the explosion energy to account for ground reflection when using blast curves for explosion in free-air.

Most buildings in chemical processing plants are beyond the angle at which a Mach stem forms from an elevated burst. The simplification of treating the explosion as a surface burst with double energy is applicable to these cases.

Blast curves have been developed for surface bursts. They are usually designated "hemispherical surface burst" blast curves. Ground reflection is inherently included in these blast curves since they were produced with a reflection plane incorporated in the model. Explosion energy should not be doubled when using surface burst blast curves.

4.5.4.   Blast Scaling

The upper half of Figure 4.11 represents how a spherical explosive charge of diameter d produces a blast wave of side-on peak overpressure P and positive-phase duration t^+ at a distance R from the charge center. Experimental observations show that an explosive charge of diameter Kd produces a blast wave of identical side-on peak overpressure p and positive-phase duration Kt^+ at a distance KR from the charge center. This situation is represented in the lower half of Figure 4.11. Consequently, charge size can be used as a scaling parameter for blast.

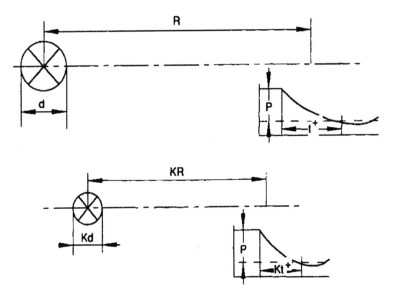

Figure 4.11. Blast-wave scaling. (Baker, 1973)

Charge size, however, is not a customary unit for expressing the power of an explosive charge; charge weight is more appropriate. Therefore, the cube root of the charge weight, which is proportional to spherical charge diameter, is used as a scaling parameter. If the distance to the charge, as well as the duration of the wave, are scaled with the cube root of the charge weight, the distribution of blast parameters in a field can be graphically represented, independent of charge weight. This technique, which is common practice for high-explosive blast data, is called the Hopkinson scaling law (Hopkinson, 1915).

It is more complete, however to scale a problem by full non-dimensionalization. To achieve this, all governing parameters, such as the participating energy E, the ambient pressure P_0, and the ambient speed of sound a_0 (ambient temperature), should be taken into account in dimensional analysis. The result is Sachs' scaling law (Sachs 1944), which states that the problem is fully described by the following dimensionless groups of parameters:

Scaled side-on overpressure:

$$\overline{P} = \frac{\Delta P_s}{p_0}$$

(Eq. 4.16)

Scaled specific impulse:

$$\overline{i} = \frac{ia_0}{p_0^{2/3} E^{1/3}}$$

(Eq. 4.17)

Scaled duration:

$$\overline{t} = \frac{ta_0^{1/3} p_0^{1/3}}{E^{1/3}}$$

(Eq. 4.18)

Scaled standoff distance:

$$\overline{R} = \frac{Rp_0^{1/3}}{E^{1/3}}$$

(Eq. 4.19)

where:

ΔP_s	=	side-on peak overpressure	(Pa)
i_s	=	*side-on specific impulse*	*(Pa/s)*
t^+	=	blast wave duration	(s)
R	=	distance from blast center	(m)
E	=	amount of participating energy	(J)
p_0	=	ambient pressure	(Pa)
a_0	=	ambient speed of sound	(m/s)

5. FLASH FIRES

Ignition of an accidental release of flammable vapors or volatile liquids can generate different events:

- Flash fire
- Pool fire
- Jet fire
- Fireball
- Explosion

All have in common a dispersed concentration of vapors within flammable limits. A discharge can include aerosols developed from mechanical breakup of a discharged liquid, or from the flash vaporization of a superheated liquid.

A flash fire is the combustion of a flammable gas/air mixture that produces relatively short term thermal hazards with negligible overpressure (blast wave). A flash fire burns through a pre-mixed vapor cloud largely horizontally as a vertical "wall of flame." If a flash fire burns back to the source, it can become a jet fire or a pool fire. Large and destructive flash fires occurred at Port Newark, New Jersey in 1951, and at Mexico City, Mexico in 1984. These incidents are described in Lees (2005, Section 16.38 and Appendix 4).

Flash fires, in common with fireballs and explosions, expand the cloud volume by combustion about a factor of eight. That is, x, y, and z linear dimensions increase by a factor proportional to their dimensions. A vapor cloud is usually low, wide, and of varying length depending on the duration of the release. This expansion can push flammable vapors ahead, enlarging the visible flame, as discussed later under flash fire models.

A jet fire is a pressurized steady-state or slowly-decaying fire. It can have more severe effects than a transient flash fire, especially if it impinges on an object. A pool fire has a vertical or tilted flame that has a rather random pattern. Jet fires and pool fires are not covered in this book.

A fireball generally has an initial vertical velocity and rises by buoyancy while burning. A BLEVE may result in a fireball originating from a bursting pressure vessel containing a pressurized ignitable liquid.

Vapor cloud explosions are reviewed in Chapter 6. Here, the conditions for ignition resulting in an explosion include:
- flammable vapors within congested obstacles that generate turbulence, or
- jet release, or
- high-energy ignition.

These conditions are usually absent with a flash fire.

Figure 5.1 is an idealized illustration that the direction of a flash fire depends upon the point of ignition. The velocity of a flame front is slower when burning against the wind and faster with the wind. In fact, a flash fire may not advance into the wind.

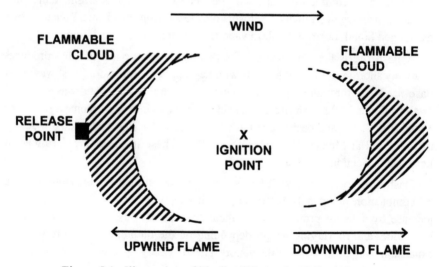

Figure 5.1. Illustration of idealized flame fronts for a flash fire.

Reproduced by permission of Cracknell and Carsley (1997)

The idealized diagram in Figure 5.1 does not account for concentration profiles through a plume and flame speed variability with concentration. For example, a test performed the Health and Safety Laboratory (HSL, 2001) using a LPG vapor cloud with a discharge rate of 2.6 kg/s, wind speed of 2.0 m/s, ignition source 25 m downwind of the source showed that the flame front did not burn into the upwind plume. The flame burned faster on the edges of the plume and finally engulfed the remaining downwind plume vapors.

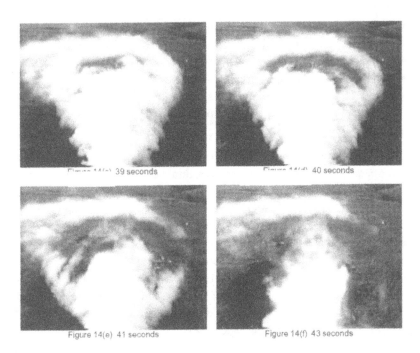

Figure 5.2. Flame front progression in LPG vapor cloud (2.0 m/s wind, 2.6 kg/s discharge
for 51 s, ignition 25 m from source, [HSL, 2001).

Reproduced by permission of the HSL.

The literature provides few experiments and models for the properties and
effects of flash fires, probably because transient radiation hazards are less harmful
than jet fires, fireballs, or explosions. The major hazard of a flash fire is from the
transient heat from thermal radiation as the fire passes a given point. Flash fire
exposure normally lasts no more than a few tens of seconds. (See Table 5.1 for
experiment durations.) Therefore, the total radiation intercepted by an object near
a flash fire is substantially lower than that from a pool or jet fire the same distance
away, although because of duration, a flash fire may produce more total radiation
than a fire ball.

The properties of a flash fire required for analysis of fire radiation effects
include:

- Flammable vapor cloud area, length, and width at time of ignition and
  transiently during combustion
- Flammable vapor cloud height

- Flame speed of flash fire (added or subtracted from the wind speed)
- Surface emissive power (SEP) of the flame, a function of flame temperature
- Transmissivity of the atmosphere (atmospheric attenuation)
- View factor from flame front to a target (a transient and a function of the burning path through the vapor cloud)

These properties are discussed later in the chapter.

The cloud border to the Lower Flammability Limit (LFL) defines a hazard zone for personnel safety; i.e. any person within the cloud LFL contour is likely to be in danger. Additionally, fuel-air clouds are often non-homogeneous in concentration and pockets of higher than average concentration can ignite outside the LFL border as estimated by a dispersion model using a prescribed averaging time. To account for such pockets and for uncertainties in modeling including plume averaging, a value of 0.5 LFL is typically considered as the border for an ignitable cloud. Lees (2005) attributes Feldbauer, et al. (1972) for first substituting 0.5 LFL for the LFL to account for the cloud area susceptible to ignition. Personnel outside these boundaries could also experience burns by radiation from a flash fire as the fire expands during combustion.

Dispersion models are needed to define hazardous zones and flammable mass for risk analysis. Discussion of dispersion modeling, even if specific for flash fires, is beyond the scope of the present guideline. Dispersion modeling is treated in other CCPS guidelines (Guidelines for Use of Cloud Dispersion Models, Wiley-AIChE, 22 Ed., 1996) and reviewed by Lees (2005) and by P.J. Rew, Deaves D.M., Hockey S.M., and Lines I.G., (1996). Flash fires are also treated in HSE Contract Research Report No. 94/1996. Calculation of the flammable mass in a vapor cloud and flame characteristics are treated in the CCPS Concept book by Woodward (1998).

CFD models solve the time averaged partial differential Navier Stokes equations, and based upon a grid, allow treatment of obstacles and terrain and impingement. Where such features are important, they may be the preferred approach. KAMELION (Velde, 1998) is one well-known CFD code for jet fires used extensively in the offshore oil industry. CFD models are touted as being more fundamental than the other approaches. However, additional CFD submodels need to be developed to solve key mechanisms (including soot formation). For this reason, TNO notes that the CFD approach is not necessarily any more fundamental than the simpler approaches.

5.1.   OVERVIEW OF EXPERIMENTAL RESEARCH

Full-scale experiments on flame propagation in fuel-air clouds are laborious
and expensive, so experimental data on the dynamics of flash fires and attendant
thermal radiation are scarce. This section discusses the following tests:

- China Lake cryogenic liquids tests
- Maplin Sands LNG and LPG tests
- Musselbanks propane tests
- HSE, HSL LPG tests

5.1.1.   China Lake and Frenchmen Flats cryogenic liquid tests

Schneider (1980), Urtiew (1982), Hogan (1982) and Goldwire et al. (1983)
reported on liquefied natural gas mixture (LNG), liquefied methane, and liquefied
nitrogen spill experiments at China Lake, California. The facility could hold up to
40 m^3 of liquefied gas released on a water test basin. In total, ten experiments
were performed; five primarily for the study of dispersion and burning clouds,
and five for investigating the occurrence of explosions exhibiting rapid phase
transitions.

All burn tests were performed in an unconfined environment with LNG except
for one with liquid methane. The burning process was measured by ionization
gauges (for three-dimensional measurement of local flame speed and direction),
calorimeters (to measure local heat release), thermometers (to measure local flame
temperatures), radiometers (to measure radiation intensity), and infrared (IR)
imaging from a helicopter overhead. These instruments were all located
downwind of the spill pond.

Rodean et al (1984) reported the Coyote test series at Frenchmen Flats,
Nevada with LNG. Transient flame speeds from jet plumes were measured up to
30 and 50 m/s. Transient flame speeds decay in the far field and were found to
vary primarily with fuel composition, as follows (Cracknell and Carsley, (1997):

- For LNG (liquefied natural gas) and natural gas: 6 m/s
- For propane and LPG (liquefied propane gas):   12 m/s

Heat-flux data from radiometers revealed maximum heat fluxes of
160-300 kW/m^2.

5.1.2.   Maplin Sands Tests

The Maplin Sands tests [Puttock et al. (1984), Blackmore et al. (1982), Hirst and Eyre (1983), data summarized in Ermak et al (1988)] spilled 20 m³ LNG and refrigerated liquid propane on the surface of the sea in the Thames estuary with both instantaneous and continuous releases. The major objective was to obtain data to assess the accuracy of existing models of cloud dispersion. The combustion experiments were designed to complement this objective by measuring flash fire characteristics.

Cloud dispersion and combustion were observed by instrumentation deployed on 71 floating pontoons. On the masts of 20–30 selected pontoons were mounted 27 wide-angle radiometers (to measure average incident radiation) and 24 hydrophones (to measure flame-generated overpressures). Another two special pontoons provided platforms for meteorological instruments. The instruments provided vertical profiles of temperature and wind speed up to 10 m above sea level, together with measurements of wind direction, relative humidity, solar radiation, water temperature, and wave height.

Combustion behavior differed in some respects between continuous and instantaneous spills, and also between LNG and refrigerated liquid propane. For continuous spills, a short period of premixed burning occurred immediately after ignition. This was characterized by a weakly luminous flame, followed by combustion of the fuel-rich portions of the plume, which burned with a rather low, bright yellow flame. As soon as the fire burned back to the liquid pool at the spill point, flame height increased markedly and assumed the tilted, cylindrical shape that is characteristic of a pool fire.

Following instantaneous spills, clouds had time to spread and move with the wind away from the spill point before ignition. In these tests, combustion was mostly of the premixed type. Since the pool completely evaporated before the fire burned back, pool fires did not occur. The highest measured flame speeds occurred during the premixed stage of combustion. In propane tests, average flame speeds of up to about 12 m/s were observed. Higher transient flame speeds (up to 28 m/s in one instance) were detected, but there was no sustained acceleration.

Similar behavior was observed for LNG clouds, but average flame speeds were lower than those for LPG. The maximum speed observed in any of the tests was 10 m/s. Following premixed combustion, the flame burned through the fuel-rich portion of the cloud, where the rate of flame propagation was very low. In one of the continuous LNG tests, a wind speed of 4.5 m/s (measured 10 m above the surface) was sufficient to hold the flame stationary at a point some 65 m from the spill point for almost 1 minute. The wind speed is measured at 10 m. The plume height and flame height are not given, but are likely less than 10 m high, so a lower wind speed and correspondingly lower flame speed likely apply.

Radiation effects for LNG and refrigerated liquid propane cloud fires exhibited similar surface emissive power values, averaging about 173 kW/m^2.

5.1.3.   Musselbanks Propane Tests

Zeeuwen et al. (1983) observed the atmospheric dispersion and combustion of large spills of propane (1000–4000 kg) in open and level terrain on the Musselbanks, located on the south bank of the Westerschelde estuary in The Netherlands. Thermal radiation effects were not measured. Tests were performed in open terrain.

Under unconfined conditions, flame-front velocities were highly directional and dependent on wind speed. Flame behavior was very similar to that observed in the Maplin Sands tests for propane. Average flame-front velocities of up to 10/ms were measured. In one case, however, a transient maximum flame speed of 32 m/s was observed.

Flame height appeared to be highly dependent on mixture composition: the leaner the mixture, the lower the flame height. In mixtures with compositions within flammability limits, flame heights were about 1–2 m. In mixtures with compositions exceeding the upper flammability limit, average flame heights of 2–5 m were observed. Flame heights of up to 15 m were observed, but only near the point of release. Video shots showed that the combustion products do not rise vertically after generation. Rather, they flow horizontally toward existing plumes, join them, and then rise.

Figure 5.3 shows a moment of flame propagation in an unconfined propane cloud. On the left side, a flame is propagating through a premixed portion of the cloud; its flame is weakly luminous. In the middle of the photograph, fuel-rich portions of the cloud are burning with higher flames in a more-or-less cylindrical, somewhat tilted, flame shape.

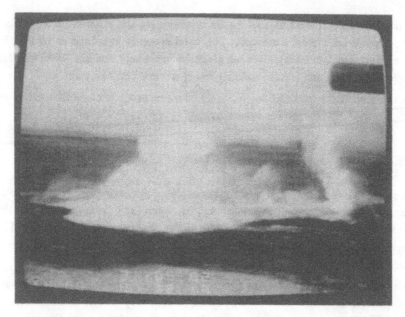

Figure 5.3. Moment of ignition in a propane-air cloud. (Zeeuwen et al., 1983)

5.1.4.   HSE LPG Tests of Flash Fires and Jet Fires

For a Joint Industry Project with partners in Canada, France, Japan, and Spain, the Health and Safety Laboratory (HSL) of the Health and Safety Executive (HSE) in England performed experiments with the ignition of vapor clouds of LPG. These formed from horizontal jet releases at discharge rates from 2.4 kg/s to 4.9 kg/s (Butler and Royle, 2001). An objective was validation of a new vapor cloud fire model. Ignition points were varied. Most tests were in unobstructed terrain, but a few were run with a fence positioned normal to the flow.

The tests obtained flame speeds and a few heat flux measurements. In some cases, isolated pockets ignited but the flame did not propagate. A 1 m high fence produced significant decreases in the vapor concentration downwind of the fence. Table 5.1 lists key experimental conditions and flame speeds. The reported flame speed is the measured downwind flame velocity minus the wind speed. In these data, there is a rough trend to lower flame speeds at higher wind speeds.

Figure 5.4 plots radiant energy measured on the side of the passing flame front for Test 14. This is both a flash fire and a jet fire. The wind speed plus or minus the flame speed was 10 or 4 m/s, the source duration was 82 seconds, and the fire duration was about 40 seconds, or only half of the discharge duration. These facts are consistent with late ignition for a flame burning into the wind direction. The radiant energy dose reaching the sensor point is obtained by integrating the curve in Figure 5.4. Isolated ignited pockets reached temperatures of 200-250 °C. The jet fires were generally 20 m long, and lifted off from buoyancy after about 10 m. The temperature was above 1000 °C at the 20 m distance in the jet fire.

Table 5.1. Experimental Conditions and Flame Speeds for HSL LPG Tests

Test	Wind Spd, m/s	Disch. Rate, kg/s	Durat. s	Ignition Pt, m	Flame Spd, m/s
17	2.0	3.0	160	23	11.3
18	2.0	2.4	143	30	7.1
21	2.0	2.8	60	25	7.9
23	2.0	2.6	51	25	11.4
4	2.5	3.0	35	15	9.0
15	2.5	3.0	41	15.3	8.7
20	3.0	3.2	148	39	9.8
8	3.0	5.0	131	30	4.3
9	3.0	2.7	78	5	-
10	5.0	3.4	141	15	7.2
16	5.0	2.6	116	17.5	6.2
6	5.5	4.9	66	20	4.5
7	6.0	3.2	59	20	6.5
14	7.0	3.8	82	14.7	3.2

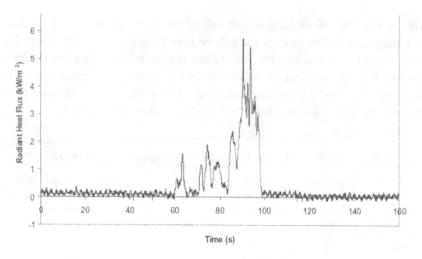

Figure 5.4. Radiant heat flux from HSL LPG flash fire test 14.

5.2.   FLASH-FIRE RADIATION MODELS

Despite the development of complex numerical tools and the increased use of Computational Fluid Dynamics, the model of Raj and Emmons (1975) is still commonly used. This considers the flash fire as a two-dimensional, turbulent flame propagating at a constant speed, based on the following experimental observations:

- the cloud is consumed by a turbulent flame front which propagates at a velocity roughly proportional to ambient wind speed;
- when a cloud burns, there is always a leading flame front propagating with uniform velocity into the unburned cloud. The leading flame front is followed by a burning zone;
- when fuel concentrations are rich, burning is characterized by the presence of a tall, diffusion-limited, flame plume. At points where the cloud has already mixed sufficiently with air, the vertical depth of the visible burning zone is about equal to the initial, visible depth of the cloud.

The model is a straightforward extension of a pool-fire model developed by Steward (1964).   The model assumes that the combustion process is fully convection-controlled, and therefore, determined by entrainment of air into the buoyant fire plume which sets the turbulent flame speed. Figure 5.5 illustrates the model, consisting of a two-dimensional, turbulent-flame front propagating at a constant velocity S into a stagnant mixture of depth d. The flame base of width

W is dependent on the combustion process in the buoyant plume above the flame base. The fire plume is fed by an unburned mixture that flows in with velocity u_0 (larger than the wind speed if the source is a jet, less than or equal to the wind speed if the source is an evaporating pool). The burning zone thickness is indicated by the sloped line in Figure 5.5. If the unburned fuel concentration is lean (between the LFL and the stoichiometric) the flame does not require further diffusive input of air, so the flame front is roughly as high as the plume. Otherwise, since a rich fuel mixture requires diffusive input of air, the burning front is higher than the original plume because the adiabatic expansion of hot combustion products pushes unburned fuel upward. This case is drawn in Figure 5.5 with an elevated burning zone.

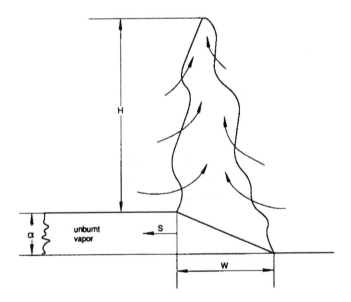

Figure 5.5. Schematic representation of unconfined flash fire.

The application of conservation of mass, momentum, and energy over the plume results in a relation between visible-flame height and the upward velocity of gases u_1 at the flame base. The theoretical solution to this simplified problem is corrected on the basis of empirical data on flame heights of diffusion flames (Steward, 1964). In free-burning clouds, the upward flame-base velocity u_1 is unknown, but experimental observations indicate a nearly proportional relation between the visible flame height H (m) and flame base width W (m), namely, H/W = 2. With this empirical observation, it is possible to relate visible flame height to burning velocity S *(m/s)* by a mass balance for the triangular area bounded by the flame front and flame base (Figure 5.5).   This results in the following approximate, semi-empirical expression for the visible flame height H that applies if w > 0:

$$H = 20d\left[\frac{S^2}{gd}\left(\frac{\rho_0}{\rho_a}\right)^2\frac{wr^2}{(1-w)^3}\right]^{1/3} \qquad \text{(Eq. 5.1)}$$

where d is the cloud depth (m), g *is* the gravitational acceleration (m/s2), r is stoichiometric mixture air-fuel mass ratio, w is defined later, and ρ_0 and ρ_a are respectively the air and the fuel-air mixture density (kg/m^3), whose ratio can be calculated from molecular weights, M_{air} and M_{fuel} (kg/kmole):

$$\frac{\rho_0}{\rho_a} = \frac{(1-\phi)M_{air} + \phi M_{fuel}}{M_{air}} \qquad \text{(Eq. 5.2)}$$

where ϕ is the fuel-air mixture mole fraction composition.

The stoichiometric air-fuel mass ratio r (dimensionless) can be calculated from the stoichiometric mixture composition, ϕ_{st}, and air and fuel molecular weights:

$$r = \frac{(1-\phi_{st})M_{air}}{\phi_{st}M_{fuel}} \qquad \text{(Eq. 5.3)}$$

The term w (dimensionless) represents the inverse of the volumetric expansion due to combustion in the plume, defined as:

$$w = \frac{\phi-\phi_{st}}{\alpha(1-\phi_{st})} \quad \textit{for } \phi > \phi_{st}$$
$$w = 0 \qquad\qquad \textit{for } \phi \le \phi_{st} \qquad \text{(Eq. 5.4)}$$

where α is the constant pressure expansion ratio for stoichiometric combustion (typically 8 for hydrocarbons). The expression for w given here differs from that recommended by Raj and Emmons (1975) and is highly dependent on the cloud's composition. If the cloud consists of pure hydrocarbon, w represents the inverse of the volumetric expansion resulting from constant-pressure stoichiometric combustion. If the mixture in the cloud is lean, the combustion occurs in the plume without appreciably elevating the plume. That is, the flame height is about equal to the cloud depth, and $w = 0$.

The model gives no solution for the dynamics of a flash fire, and requires a value for the burning speed S. From a few experimental observations, Raj and Emmons (1975) found that burning speed was roughly proportional to ambient wind speed at the elevation of the cloud vertical center of mass u_w (m/s):

$$S = 2.3u_w$$

(Eq. 5.5)

At any time t (s), knowing the visible flame height and the radiative power per unit area intercepted by some plane in the environment, $q(W/m^2)$ can be computed from the well known equation:

$$q = EF\tau_a$$

(Eq. 5.6)

where E is the emissive power in W/m^2, F is the dimensionless geometric view factor for a vertical-plane emitter, and τ_a is atmospheric attenuation (dimensionless transmissivity).

The value for emissive power can be from observed values or computed approximately from ideal black body radiation with flame temperature and emissivity. Emissivity is primarily determined by the presence of non-luminous soot within the flame. Published values for emissive power were observed for pool fires such as in the China Lake tests, in the Maplin Sands experiments, and in the Montoir LNG fire test (Nedelka, 1990). The reported wide-gauge radiometer values (average over area) are given in Table 5.2. Small pool fires are usually bright and are applicable to flash fires. Large pool fires can be smoky, and the smoke decreases the emissive power.

Table 5.2. Wide-Gauge Radiometer Measurements of Surface Emissive Power
for Flash and Pool Fires

Test	Pool Size, m or Spill, m^3	E, kW/m^2	u_F, m/s	At u_w m/s
China Lake	15	220±47 (Wide Angle)		
Maplin Sands		178-248	10	
LNG	20	203± 35		
LPG		(ave)	12-20	
Montoir	35	165 ± 10 (Wide Angle)		
Musselbanks LPG	2-7 m^3		10	
HSE LPG	jets		3.2-11.4	3-7

The atmospheric attenuation factor τ_a takes into account the influence of absorption and scattering by water, carbon dioxide, dust, and aerosol particles. Transmissivity values are provided by Raj et al (1979) using regions of high transmission between absorption bands. Simpson (1984) also tabulates and plots a set of transmissivity curves, allowing not only for absorption but also for scattering. As a simple expedient, transmissivity can be calculated by the following equation:

$$\tau_a = \log(14.1 \, RH^{-0.108} x^{-0.13})$$

(Eq. 5.7)

where RH is the relative humidity in percent, and x is the distance in meters between flame and object. One can assume, as a conservative position, a clear, dry atmosphere for which $\tau_a = 1$. However, Simpson (ibid) points out that the correction for transmissivity will give, for typical weather conditions, a reduction in hazard range of 10-40%.

The geometric view factor F can be determined from the relative positions and orientations of the receiving and the transmitting surfaces. Appendix A reports tabulated and graphical representation of the view factors for cylindrical and plane vertical transmitters and for various orientations of receiving surfaces.

The total radiation intercepted by an object equals the summation of contributions by all successive flame positions during flame propagation. For flash fires, radiation heat flux is strongly time dependent, because both the flame surface area and the distance between the flame and intercepting surfaces vary during the course of a flash fire. The analysis is often difficult with simplified approaches as there are many uncertainties (e.g., fire area and shape, flame speed, composition profiles, location of ignition site), which may greatly influence the final result. A conservative approach is recommended for practical applications, such as:

- assuming that during flash fire propagation, the location of the cloud is stationary, and its composition is fixed and homogeneous, or
- assuming that the flame surface area dependence on time is approximated by a plane cross-section moving at burning speed through the stationary cloud.

Simplifying assumptions about flame shape may result in erroneous radiation levels. Some examples are shown in Figure 5.6(a) and (b), where the flame shape is assumed to be flat, whereas in Figure 5.6(c) the flame could be cylindrical. In these figures, D and R are the cloud diameter and radius, L is the length to the ignition point, t is the time, W is the flame width, and s is the flame velocity.

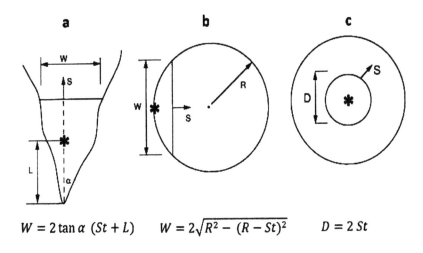

$$W = 2\tan\alpha\,(St + L) \qquad W = 2\sqrt{R^2 - (R - St)^2} \qquad D = 2\,St$$

Figure 5.6. Flame shape assumptions. (*= ignition source)

5.3.　　　　SAMPLE CALCULATIONS

A massive amount of propane is instantaneously released in an open field. The cloud assumes a flat, circular shape as it spreads.　When the average fuel concentration in the cloud is about 10% by volume, the cloud's dimensions are approximately 1 m deep and 100 m in diameter. At that time, the cloud reaches an ignition source at its edge.　Because turbulence-inducing effects are absent in this situation, a blast is not to be expected.　Therefore, thermal radiation and direct flame contact are the only hazardous effects to be analyzed. Compute the incident heat flux as a function of time through a vertical surface at x=100 m distance from the center of the cloud.

Data

Propane:　　　M_{fuel} =44 kg/kg-mol;

Air:　　　　　M_{air}= 29 kg/kg-mol

Average plume concentration: φ φ= 0.10

Stoichiometric concentration ϕ_{st} = 0.040 mole fraction

Cloud depth: d= 1 m

Wind speed: u_w = 2 m/s

Relative Humidity: RH = 50

Calculation

Calculate the flame speed S on the basis of the wind speed (Eq. 5.5):

$$S = 2.3 \times u_w = 2.3 \times 2 = 4.6 \text{ m/s}$$

Calculate the square of mixture-air density ratio Eq. (5.2):

$$\left[\frac{\rho_o}{\rho_a}\right]^2 = \left[\frac{0.9 \times 29 + 0.1 \times 44}{29}\right]^2 = 1.11$$

Calculate the air-fuel mass ratio r from the stoichiometric mixture composition ϕ_{st} and the densities of air and fuel (Eq. 5.3):

$$r = \frac{(1-0.04)29}{0.04 \times 44} = 15.8$$

Calculate w from the actual mixture composition ϕ, the stoichiometric mixture composition ϕ_{st}, and the expansion ratio for stoichiometric combustion α(Eq. 5.4):

$$w = \frac{0.1 - 0.04}{8(1 - 0.04)} = 0.0078$$

Calculate the flame height using the cloud depth d, gravity constant g, S $(\rho/\rho_a)^2$, w, and r as follows:

$$H = 20 \times 1 \times \left[\frac{4.6^2}{9.81 \times 1} \times 1.11 \times \frac{0.0078 \times 15.8^2}{(1 - 0.0078)^3} \right]^{1/3} = 33 \text{ m}$$

To calculate the heat flux, the flash-fire dynamics (shape and position of the flame dependent on time) should first be specified. For simplification, assume the cloud to be stationary during the full period of flash-fire propagation. As a conservative starting point, assume that the transmitting (flame) and receiving surfaces are vertical and parallel during the full period of flame propagation, as indicated in Figure 5.6(b).

The thermal radiation received by an object in the environment may now be computed if it is assumed that the flame appears as a flat plane, 33 m high, which propagates at a constant speed of 4.6 m s^{-1} during the full period of flame propagation (100/4.6 = 21.7 s). During this period, flame width varies from 0 to 100 m and back, according to Figure 5.6(b):

$$W = 2[R^2 - (R - St)^2]^{0.5} = 2[50^2 \ (50 - 4.6t)^2]^{0.5}$$

Radiation heat flux is strongly time dependent because both flame surface area and distance from the flame to the intercepting surface vary during the course of a flash fire. For example, after 5 seconds, the view factor of flame propagation can be calculated as follows:

Flame width:

$$W = 2[50^2 \times (50 \times 4.6t)^2]^{0.5} = 84\text{m}$$

Distance between the object and the flame:

$$x = 150 - (5 \times 4.6) = 127 \text{ m}$$

If it is assumed that the receptor's location is such that parts I and II of the flame (Figure 5.7) are equal, then r_I equals r_{II}. Values for x_r and h_r are calculated for the portions of the flame on each side from the center of the receptor onto the flame surface. That is, to calculate x_r and h_r in this case, divide the total flame width in half to determine portions on either side of the normal on the flame surface:

$$x_r = x/r = x/0.5W = 127/(0.5 \times 84) = 3.02$$
$$h_r = h/r = h/0.5W = 43/(0.5 \times 84) = 1.02$$

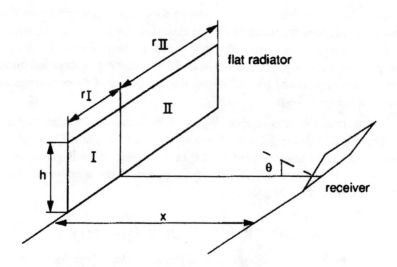

Figure 5.7. Definition of view factors for a vertical, flat radiator.

At this point, the problem focuses on the determination of the geometric view factor. Calculate the view factor using the equation given in Appendix A for a vertical plane surface emitter, or else read the view factor from Table A-2 of Appendix A for the appropriate x_r and h_r.

This results in F = 0.062 for each portion of the flame surface, and implies a total view factor of:

$$F = 2 \times 0.062 = 0.12$$

Transmissivity by Eq. (5.7) is:

$$\tau_a = \log(14.1 * 50^{-0.108} 100^{-0.13}) = 0.71$$

Radiant energy at receptor by Eq. (5.6) is:

$$q = \tau_a FE = 0.706 \times 0.12 \times 173 = 14 \, kW \, m^{-2}$$

The results for selected points in time and distances are summarized in Table 5.3 and plotted in Figure 5.8.

Table 5.3. Results of calculations

t(s)	W(m)	X(m)	h_r	X_r	F	τ_a	q (kW/m²)
0	0	150					0
5	84	127	1.02	3.02	0.12	0.69	14
10	100	104	0.86	2.08	0.20	0.70	25
15	92	81	1.07	1.76	0.30	0.72	37
20	54	58	1.59	2.94	0.30	0.74	38
21	36	53	2.39	2.94	0.26	0.74	33
21.7	0	50				0	

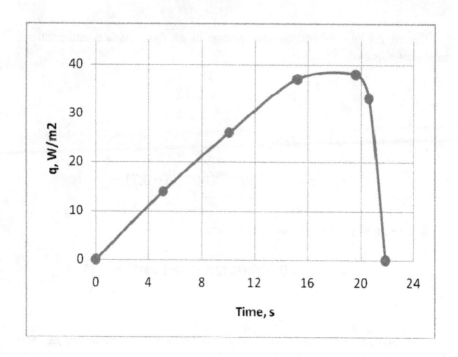

Figure 5.8. Graphical presentation for sample problem of the radiation heat flux as a function of time.

The total amount of radiation heat from a surface can be found by integration of the radiation heat flux over the time of flame propagation, that is, the area under the curve in Figure 5.8.

There are numerous uncertainties in the calculation of radiation dosage. The flame will probably not burn as a flat front, but more in the shape of a horseshoe. In addition, it could consist of several plumes with heights in excess of those assumed in the model and probably with time-varying flame radiation. Finally, wind direction and speed changes will have a considerable influence on flame shape and cloud position.

6. VAPOR CLOUD EXPLOSIONS

6.1. INTRODUCTION

6.1.1. Organization of Chapter

This chapter is organized as follows. First, a brief introduction of vapor cloud explosion (VCE) phenomena and terminology is given.

Second, an overview of experimental research on vapor cloud deflagrations and detonations is presented. The deflagration section starts with a brief description on several basic concepts, such as laminar burning velocity and flame speed. Then, a brief introduction is given to the mechanisms of flame acceleration, which is essential for a VCE to occur. Three important factors affecting flame acceleration, confinement, congestion, and fuel reactivity are discussed in order to better understand the large amount of experimental results on VCE deflagrations.

For detonations, the presentation is relatively straightforward because detonation properties are uniquely determined once the cloud composition is specified. Consequently, blast effects from VCE detonations are well defined. Experimental measurements and empirical equations on blast effects from VCE detonations are presented.

Prediction methods are discussed and arranged into three categories: TNT equivalency methods, blast curve methods, and numerical simulations. Sample problems are provided to illustrate the methodology.

6.1.2. VCE Phenomena

A vapor cloud explosion is the result of a release of flammable material in the atmosphere, dispersion of the flammable material in air, and, after some delay, ignition of the flammable portion of the vapor cloud. Experimental, analytical, and computational fluid dynamic studies on vapor cloud explosions have been pursued by two groups of investigators with totally different objectives. One group is focused on the development of weapon systems, called FAE, for Fuel Air Explosions. The other group is concerned with avoiding severe accidents, termed VCE for Vapor Cloud Explosions. In an FAE weapon system, the fuels are usually sensitive to shock initiation, while fuels in accidental VCEs can be any gas or flash-evaporating liquid fuels.

The formation of an explosive vapor cloud in an FAE weapon system is due to the intentional dispersion of ignitable gas or mist in the atmosphere by a designed explosive charge. In an accident scenario, the formation of a flammable vapor cloud is the most likely effect of a release of an ignitable gas or liquid with the discharge occurring over a period of time rather than instantaneously. Turbulence in jetting release promotes mixing of vapors with air, as do mechanical sources such as fans. Failure of a pressure vessel, pipe or equipment under pressure can lead to rapid dispersal (sometimes called explosive dispersal) of ignitable gas or mist. FAE dispersion has been engineered to get a uniform cloud at near stoichiometric conditions whereas VCE releases are not uniform, and FAE weapon systems use high explosive initiation charges compared to "soft" ignition sources in a VCE.

The accidental release of gaseous or liquid fuel may not necessarily lead to an explosion. If the released material ignites promptly, a fire is the outcome. Delayed ignition of a vapor cloud in an uncongested area leads to a flash fire. Some releases do not ignite, averting any fire or explosion consequences (Davenport, 1977).

Certain conditions must be met in order for a VCE to occur. First, there must be a release of flammable material into a confined/congested area. Second, ignition must be delayed long enough to allow the formation of the ignitable mixture, with the fuel-air concentration lying within the flammable limits. Third, there must be an ignition source of sufficient energy to ignite the fuel-air mixture.

Once the above conditions are met and a VCE is initiated, the following effects to the surroundings may include:
- A wide spectrum of air blast effects, ranging from minimal to catastrophic.
- A fireball.
- Throw of lightweight materials such as insulation and thin metal sheathing within the explosion zone and immediate surrounding area.
- Dispersal of very light materials carried upward in the fireball or secondary fire updraft and carried downwind.
- Secondary fire at the initial release sources, and often other release sources caused by displacement of equipment.

In this chapter, particular attention will be given to blast effects, since fragmentation effects are often negligible and radiation effects are discussed in Chapter 5.

6.1.3. Definition of VCE

6.1.3.1. Vapor Cloud Explosion (VCE)

A VCE is one type of fuel-air explosion. Historically, this phenomenon was referred to as "unconfined vapor cloud explosion (UVCE)," to emphasize that the incidents are outdoor events. But the term "unconfined" is a misnomer, since a truly unconfined scenario will not result in detectable damage to the surroundings. It is more accurate to call this type of explosion simply a "vapor cloud explosion (VCE)." Internal vapor explosion is another class of fuel-air explosion that refers to an explosion inside of an enclosure such as a building (room) or vessel. The presence of the enclosure and turbulence created by failure of any portion of the enclosure affects the combustion process. Prediction of internal vapor explosions is beyond the scope of this book.

Like other types of explosions, VCEs can also be categorized into two modes, deflagration and detonation, according to propagation mechanisms.

6.1.3.2. Vapor Cloud Deflagration

In a vapor cloud deflagration, the flame propagates through the unburned fuel-air mixture at a burning velocity less than the speed of sound. The overpressure generated in a VCE deflagration varies with combustion rate: minimal overpressure is produced at low flame speed, and high overpressures are produced at higher flame speeds. Consequently, the damage to the surroundings caused by VCE deflagration ranges from minimal to more severe. VCE detonations are typically more severe than deflagrations due to the high overpressure generated by a supersonic wave. The situation for VCE deflagrations is complex because the flame speed and pressure buildup in the deflagration are not unique for a given cloud composition, but vary in a wide spectrum depending on many factors. Moreover, the composition of fuel and combustion products at the flame front, within the cloud, which supports the deflagration, changes continuously. The vast majority of accidental VCEs are vapor cloud deflagrations.

6.1.3.3. Vapor Cloud Detonation

In a vapor cloud detonation, the combustion wave propagates at supersonic velocity through the unburned fuel-air mixture. A detonation is the most violent form of vapor cloud explosions and can cause the most severe damage. While the detonation mode is the expected result in FAE weapon systems, it is very unlikely to occur in accidental vapor cloud explosions. As with other types of explosions,

VCE detonations can be achieved through either direct initiation or the transition from a deflagration. One method of direct initiation used in research testing has been a "bang box," which is a strong enclosure inside of which an internal VCE is initiated, and allowed to propagate through an opening or breached wall to provide high energy initiation of the external vapor cloud. A bang box must be capable of withstanding high explosion pressures, and such strong enclosures are typically not found in chemical processing plants. There are restrictive conditions that must be met if a detonation is to propagate as previously described in Section 4.4.2.

6.1.3.4. Similarities and Differences between Detonations and Deflagrations

Detonation and deflagration are both explosions propagating in reactive media, and both are supported by the energy released from chemical reactions in the media. However, the propagation mechanism, range of burning rates, peak overpressure, and duration differ substantially from one another. The most important difference between vapor cloud detonations and deflagrations is the propagating mechanism. For VCE deflagration, chemical reaction propagation is due to the diffusion of heat and chemical species from the reaction zone to the unburned material. Thus, the propagation velocity is limited by the molecular diffusivities. For VCE detonation, chemical reaction propagation is due to adiabatic shock compression of the unburned material, where the flame burns extremely rapidly in highly compressed and preheated gases. In a stoichiometric vapor cloud, which is near optimal conditions, the detonation wave may propagate at the speed of about Mach 5 for most hydrocarbon fuels.

The peak overpressure of a detonation exceeds that of a weak to moderate flame speed deflagration. Assuming that the explosion energy is constant, the duration of a detonation will be shorter than a deflagration. Fast deflagrations, approaching the speed of sound, have lower pressures than detonations in the combustion zone, but converge to detonation overpressures a short distance outside of the combustion zone.

A deflagration can be initiated by a weak energy source with only a few millijoules of energy for ideal mixtures. Direct initiation of a vapor cloud detonation requires a strong ignition source with the energy being several orders of magnitude greater than a deflagration. However, detonation can be achieved not only from direct initiation, but also from the transition from a deflagration.

It should also be noted that while detonation is a unique state with detonation properties being determined by the fuel-air mixture composition, there is a continuous spectrum of burning rates for deflagration, ranging from low flame speed to near sonic velocities.

6.1.4.   Confinement and Congestion

The physical layout of a process area within a flammable cloud has a direct bearing on the outcome of a VCE. The layout is described by two parameters: confinement and congestion. Confinement refers to solid surfaces that prevent movement of unburnt gases and a flame front in one or more dimensions. For example, a solid deck in a process structure prevents upward expansion when combustion takes place beneath the deck, thereby eliminating one dimension for expansion. Combustion inside a pipe can only expand in one dimension (axially), thereby eliminating two dimensions of expansion. The degree of confinement is based on the dimensions available for expansion, referred to as three-dimensional (3-D), two-dimensional (2-D) and one-dimensional (1-D). The degree of confinement is lessened with a partial surface or perforated surface. Examples of confinement are provided below.

- Three-dimensional (3-D) flame expansion is characterized by an "unconfined" volume such as an open field or process area with no horizontal spatial covering (Figure 6.1). Process areas with no solid surfaces, a tank farm, and a series of vessels on a pad with no solid roof or horizontal flat obstacles are examples of geometries that allow 3-D flame expansion, which allow almost completely free expansion of the flame front in three directions.

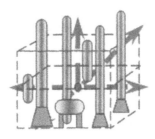

Figure 6.1. Three dimensional (3-D) flame expansion geometry *(courtesy of FM Global)*

- Two-dimensional (2-D) flame expansion is characterized by surfaces that prevent the flame front from expanding in one dimension such as shown in Figure 6.2. Other examples are elevated storage tanks, compressor houses or multi-story solid floored buildings similar to open-sided parking garages, and elevated fin-fan coolers where the flame can expand in two horizontal dimensions but not in the vertical dimension. Grating used as decking is not a confining plane, but is an obstacle that should be evaluated as congestion.

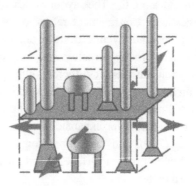

Figure 6.2.   Two dimensional (2-D) flame expansion geometry
(courtesy of FM Global)

- One-dimensional (1-D) flame expansion is characterized by solid surfaces that prevent flame expansion in two dimensions such as the interior of pipes, tunnels, enclosed corridors, or sewage systems where the flame front can expand only out the ends of structure.

Figure 6.3. One dimensional (1-D) flame expansion geometry

Congestion refers to obstacles in the path of the flame that generate turbulence. Turbulence produces flame folding and wrinkling, increasing flame surface area and combustion rate. Repeated, closely spaced obstacles may create conditions favorable in creating a VCE. Congestion is defined by measures indicative of obstacle density and number of obstacle layers along the flame path, including:

- Obstacle diameter D – obstacle diameter;
- Pitch P – spacing between successive obstacles or obstacle rows, often expressed as multiples of obstacle diameter (e.g., P=6D)
- Area blockage ratio ABR – ratio of area blocked by a plane of obstacles to the total area;
- Volume blockage ratio VBR – ratio of volume blocked by all obstacles in a volume to the total volume.

The word "blockage" is used to describe the part of the flow field area occupied by obstacles. Obstacles in process units are not arranged in systematic planes, nor are they uniform diameter. As a result, average values are typically used for obstacle diameter, pitch and area blockage ratio. Volume blockage ratio needs no average, but does not characterize the obstacle environment in terms of the size and number of obstacles.

The selection of 2-D confinement depends not only on the presence of confining surfaces but also the height of the flammable cloud. A cloud that is flammable only near the ground surface when confining planes are well elevated above ground level does not constitute a 2-D confinement situation for a VCE since the confining surface does not affect flame expansion. Baker (1997) recommended that confining surfaces be more than 3.2 times the flammable cloud height to treat the VCE as 3-D confinement. When confining surfaces are less than 3.2 times flammable cloud height, the VCE should be analyzed as 2-D confinement.

6.2.   VAPOR CLOUD DEFLAGRATION THEORY AND RESEARCH

6.2.1.   Laminar Burning Velocity and Flame Speed

Burning velocity is the speed of the flame front with respect to unburned gas. Laminar burning velocity, a fundamental property of a material, is the propagation velocity of a laminar flame front when measured in a standardized test apparatus. Laminar burning velocity depends only on initial conditions of a given mixture. Some models correlate VCE fuel reactivity to laminar burning velocity. Fuel reactivity is the propensity of a flame of a given fuel-air mixture to accelerate due to the effects of confinement and congestion in a VCE. Although in general the higher the laminar burning velocity, the higher the reactivity of the mixture, some fuels with similar laminar burning velocities have significantly different reactivity. Laminar burning velocity varies with fuel equivalence ratio and typically reaches its maximum at slightly rich mixture for hydrocarbon fuels where the maximum flame temperature occurs. Some species such as acrolein and allyl chloride reach maximum burning velocity slightly lean.

Flame speed is the propagation velocity of the flame front with respect to a fixed observer and differs from burning velocity, which is relative to the unburned gas.. Flame speed is the sum of the burning velocity and the gas flow velocity and is usually one to two orders of magnitude larger than laminar burning velocity because expansion of the combustion products accelerates unburned gas ahead of the flame. Figure 6.4 shows the laminar burning velocity, S_u, gas flow velocity, S_g, and flame velocity, S_s, for methane-air mixture with various equivalence ratios. These velocities reach a maximum with slightly rich mixtures (equivalence ratio about 1.1) with laminar burning velocity peaking at nearly 50 cm/s and flame speed peaking at about 260 cm/s in an uncongested, quiescent mixture.

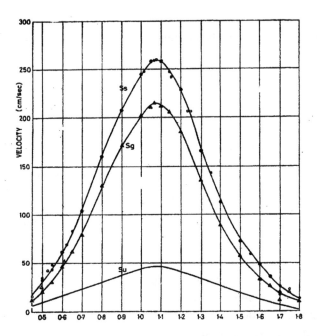

Figure 6.4. Flame speed S_s, gas flow velocity S_g, and laminar burning velocity, S_u, for various methane-air mixture equivalence ratios at 1 atm and 298° K (Andrews, 1997).

The expansion ratio E is defined as the ratio of densities of the products of combustion behind the flame front relative to the unburned fuel-air mixture ahead of the flame front (Phylaktou, 1994). Expansion ratio is generally a specific value for each fuel-air mixture at a defined temperature and pressure. The expansion ratio E is:

$$E = \frac{\rho_b}{\rho_u} \qquad \text{(Eq. 6.1)}$$

where:

ρ_b = the density of burned gas

ρ_u = the density of unburned gas.

For initial and final pressure being equal, expansion ratio can be expressed as:

$$E = \frac{T_b}{T_u} \frac{N_b}{N_u} \qquad \text{(Eq. 6.2)}$$

where:

T_b = the absolute temperature to which burnt gases are raised during combustion

T_u = the initial absolute temperature of the fuel-air mixture

N_b = the number of moles of product gases after combustion

N_u = the number of moles of reactant gases before combustion

If the chemical composition of the fuel and the oxidizer are known, the expansion ratio can be computed for stoichiometric combustion under ideal conditions. The expansion ratio is highest in fuel-air mixtures slightly above the stoichiometric concentration. This expansion ratio is directly proportional to the volume and temperature of the products of combustion. For most commonly encountered gaseous fuels, the expansion factors are between 7 and 8.

Laminar flame speed is the product of the burning velocity and the expansion ratio given by:

$$S_b = S_u \frac{\rho_u}{\rho_b} \qquad \text{(Eq. 6.3)}$$

where:

S_b = laminar flame speed

S_u = laminar burning velocity

ρ_u = the density of unburned gas

ρ_b = the density of burned gas

Computer codes are available for the calculation of the laminar burning velocity, and experimentally measured data are readily available. For example, a collection of data on laminar burning velocity for commonly used hydrocarbons can be found in NFPA, "Standard on Explosion Protection by Deflagration Venting," National Fire Protection Association, Quincy, MA, 2007.

Although laminar burning velocity is a fundamental property of a flame of specific composition, flame propagation is influenced by many factors, such as the flow field ahead of the flame and flame instabilities induced by aerodynamic and diffusional-thermal influences. In essence, a deflagration is different from a detonation in that the unburned mixture ahead of a deflagration wave has been disturbed by the expansion of the combustion products.

6.2.2. Mechanisms of Flame Acceleration

Laminar flames are inherently unstable and are strongly influenced by many factors, the most important of which are congestion (i.e., process equipment, piping etc.) and confinement (i.e., decks, building walls, etc). Consequently, this chapter will focus on confinement and congestion as flame acceleration mechanisms. Other flame acceleration mechanisms include flame instabilities (Istratov, 1969), and flame enhancement due to flame-shock interactions, called Markstein-Taylor instability (Markstein, 1964). Fuel reactivity, defined as the propensity for flame acceleration in a VCE, is the third dominant factor that will be discussed (Baker, 1994).

The 3-D geometry has the lowest degree of confinement. It has the highest rate of divergence in the flow field and is most common in outdoor areas with no confining surfaces. In this geometry, the overall flame surface area increases with the square of the distance from the ignition point. Flow velocities are relatively low and the influence of the disturbances induced by obstacles on flame speed is small, as only a small portion of the original flame surface area is affected. This situation is most common in outdoor areas with no confining surfaces other than the ground plane.

The 2-D geometry has a higher degree of confinement and a lower rate of divergence than 3-D, can be found in the presence of solid decks, where decks and the ground act as two confining planes, and is often referred to as "partially confined." In this geometry, the overall flame surface area is proportional to the distance from the ignition point, and the flame induced flow field can decay only in two directions. Obstacles will be more effective in inducing disturbances and result in higher flame acceleration.

The 1-D geometry allows flame expansion in only one dimension as in pipes, tubes and channels, and the projected flame surface area is constant. There is little flow decay because there is no flow field divergence, and flame deformations have a very strong effect on flame acceleration, producing the strongest positive feedback mechanism on flame acceleration.

The effect of flame speed on overpressures generated in vapor clouds for the above three geometries were predicted analytically by Kuhl et al. (1973), and are presented in Figure 6.5. Flame propagation was measured by Stock et al. (1989) in two apparatus: (1) a channel-like geometry (0.25m high, 0.5m wide and 2.0m long) and (2) a sector of a circle (top angle 30^0, 0.25m high and 2.0m in length), both with similar obstacle arrangement and fuel type (Figure 6.6). Figure 6.5 and Figure 6.6 clearly show that flame speeds and overpressures are significantly larger with a higher degree of confinement.

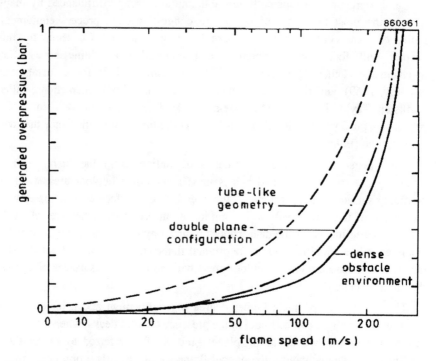

Figure 6.5. Overpressure as a function of flame speed for three geometries. (Tube-like geometry is 1-D; double plane is 2-D, and dense obstacle environment is 3-D confinement). (Kuhl et al. 1973)

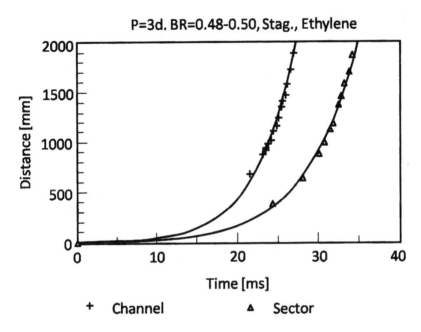

Figure 6.6. Flame propagation in 1-D (channel) and 2-D (sector) geometries.
(Stock et al. 1989)

The effect of congestion can be illustrated by two flame acceleration mechanisms related to obstacles: turbulence and flow velocity gradients. The presence of obstacles in the flow field ahead of the flame will generate flow velocity gradients. When the flame reaches the obstacle, the flame front will be stretched and folded, resulting in an increase in flame surface area as shown in Figure 6.7. Thus, the amount of gas consumed by the flame per unit time increases, causing an increase of the heat release rate. Consequently, the flame will propagate at higher velocity. Flow velocity gradients will also occur near the rigid boundary of a confinement surface.

Figure 6.7. Flow visualization image sequence of flame propagation over rectangular, square and cylindrical obstacles with stoichiometric fuel-air mixtures. Left frame, t=32ms after ignition; time between frames is 1.66 ms. (Hargrave, 2002)

When there is flow around an obstacle, a vortex is generated behind it. Initially, discrete vortices will be formed. The flow becomes turbulent as combustion progresses and the flame accelerates. A shear layer becomes the separation between the turbulent region behind the obstacle and the main flow between obstacles. A flame entering the turbulent region increases in surface area and reaction rate.

At some distance from an obstacle, the distortions of the flow field will decay and hence the flame front velocity will decay. In other words, if only one obstacle is present, flame acceleration is temporary and local, and at some distance from the obstacle, flame folds burn themselves out and turbulent intensity decreases.

On the other hand, if obstacles are repeated at separations small enough to prevent deceleration, the flame may maintain speed or continue accelerating depending on other geometry and fuel reactivity considerations. In the case of continued acceleration, the flow velocity gradients after the second row of obstacles will be larger and the turbulent intensity in the wake of the second row of obstacles will be stronger due to the higher velocity of incoming flow. This will cause a further increase in flame velocity and the flow velocity ahead of the flame. This feedback mechanism will remain effective as long as the obstacles are repeated at appropriate distances.

From the above discussion it is clear that the feedback mechanism for flame acceleration due to interaction with repeated obstacles is very effective. Flame acceleration depends largely on the congestion configuration, such as the size and shape of the obstacle, and distance between two successive obstacles. The geometry in which the obstacles are placed also plays an important role, i.e., the degree of confinement determines the effectiveness of obstacles on flame acceleration. For instance, obstacles inside 1-D geometries are most effective, and the same obstacles in 3-D geometries, least effective. Congestion and confinement are always coupled with each other.

6.2.3.   Effect of Fuel Reactivity

Van Wingerden (1989) reported the effects of fuel reactivity on flame acceleration. The fuels tested were stoichiometric methane, propane, and ethylene with air mixtures. The maximum flame speeds occurred for ethylene in all test conditions. In general, the higher the fuel reactivity, the higher the flame speeds and overpressures. Van Wingerden demonstrated in Figure 6.8, where flame speeds were non-dimensionalized by dividing by the laminar flame speeds of each fuel, that the relative increase of flame speeds with distance is similar for all three fuels.

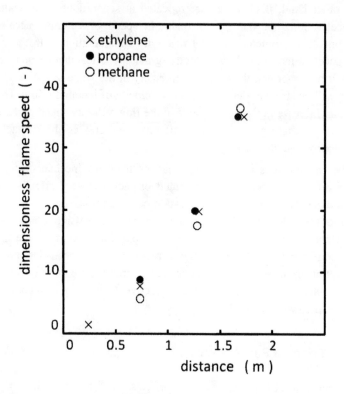

Figure 6.8. Flame speeds versus distance, non-dimensionalized with respective laminar flame speeds (fixed test conditions P = 6D, ABR = 0.5, H = 2D).

It should be noted that Figure 6.8 is for low flame speeds. When the flame speeds are high and flame propagation is dominated by turbulent combustion, the scaling with respect to laminar flame speeds is not valid. Deviations from the above findings were reported (van Wingerden and Zeeuwen, 1983) at higher flame speeds. Experiments by Hjertager et al. (1984) showed that a 2.5 m diameter × 10 m long tube with internal obstructions resulted in a clear difference between methane and propane at all blockage ratios (0.1, 0.3 and 0.5) despite their similar laminar burning velocities.   Overpressure for propane reached 7.1 bar, whereas methane reached 4.2 bar at the same area blockage ratio of 0.5.

Mercx (1992) presented flame propagation for methane, propane and ethylene mixtures in TNO large scale tests in Figure 6.9. Maximum flame speeds occurred with ethylene as expected. The sharp drop in flame velocity for ethylene is apparently due to the premature venting. However, there is no explanation why flame speeds were higher for methane than propane. For pitch P of three obstacle diameters (3d) configuration, flame speeds for propane were higher than methane at all distances.

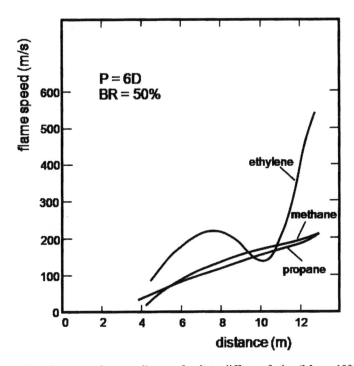

Figure 6.9. Flame speed versus distance for three different fuels. (Mercx 1992)

6.2.4. Effect of Confinement

6.2.4.1. Blast Effects Produced in 1-D Configurations

A 1-D configuration has the highest degree of confinement, which provides the lowest degree of divergence for flame expansion. When the flame consumes the unburned gas, combustion products expand to multiples of the initial volume. For example, expansion ratios for stoichiometric mixtures of alkanes are about 8 times the initial volume. Since there is no flow field divergence in a 1-D geometry, the unburned gas will be strongly pushed ahead of the flame, and a turbulent flow field will be generated. When the flame propagates into a turbulent flow field, the burning rate will increase dramatically. This enhanced burning rate will further increase the flow velocity and turbulence of the unburned gas ahead of the flame. This positive feedback loop for flame acceleration is most effective in a 1-D configuration. Consequently, very high flame speeds, even transition to detonation, can be achieved.

Tubes or Channels with Solid Walls

Several small and large scale experiments have been conducted by various researchers in a tube-like apparatus. Results are summarized in Table 6.1. For tubes open only at the end opposite to the ignition location and with very large length/diameter ratios, the positive feedback mechanism resulted in DDT for many fuels, even without obstructions. In an obstacle-free channel 30.5 m long × 2.44 m × 1.83 m, hydrogen-air mixtures detonated, both with a completely closed top and with a top opening of 13% (Sherman et al. 1985).

Table 6.1. Test Results of VCE Deflagration in Tubes

Reference	Configuration	Fuel	Max. Flame Speed (m/s)	Max. Overpressure (bar)
Chapman and Wheeler (1926, 1927)	2.4 m long pipe, D = 50 mm with orifice plates	CH_4	420	3.9
Dörge et al. (1981)	2.5 m long pipe, D = 40 mm with orifice plates	CH_4	770	12.0
Chan et al. (1980)	0.45 m, 63 mm ID pipe and 1.22 m, 152 mm ID pipe both with orifices	CH_4	550	10.9
Moen et al. (1982), Hjertager et al. (1984, 1988)	10 m, 2.5 m ID pipe with orifices	CH_4 C_3H_8	500 650	4.0 13.9
Lee et al. (1984)	11 m, 50 mm ID pipe with orifices or spirals	H_2	DDT	DDT

Effect of Venting in 1-D Geometries

Flame acceleration can to some extent be reduced by venting the confined/congested volume. In experiments in which channels were open on top, or varying perforations on one side of a channel allowed venting, far lower flame speeds were measured.

Urtiew (1981) performed experiments in a chamber 30 cm high × 15 cm wide × 90 cm long. Obstacles of several different heights were introduced into the test chamber. Due to the top venting, maximum flame speeds were only on the order of 20 m/s for propane-air mixtures.

Chan et al. (1983) studied flame propagation in an obstructed channel in which the degree of confinement could be varied by adjustment of the perforations in the top. Its dimensions were 1.22 m long and 127 × 203 mm in cross section. Results showed that reducing top confinement greatly reduced flame acceleration. When the channel's top confinement was reduced to 10% of the top surface area, the maximum flame speed produced for methane-air mixtures dropped from 120 m/s to 30 m/s.

Elsworth et al. (1983) reported experiments performed in an open-topped channel 52 m long × 5 m high. Width was variable from 1 to 3 m. Experiments were performed with propane, both premixed as vapor and after a realistic spill of liquid within the channel. In some of the premixed combustion tests, baffles 1–2 m high were inserted into the bottom of the channel. Ignition of the propane-air mixtures revealed typical flame speeds of 4 m/s for the spill tests, and maximum flame speeds of 12.3 m/s in the premixed combustion tests.

It is apparent that the reduction of the degree of confinement resulted in a significant decrease of the effectiveness of the positive feedback mechanism for flame acceleration.

Effect of Congestion in 1-D Geometries

Obstacles used in 1-D experiments were generally orifice plates and helical plates. In general, higher blockage ratios result in higher turbulence intensity, and therefore higher flame speeds and overpressures. There is usually an optimum value of the pitch regarding flame acceleration. When P is too large, i.e., obstacles are placed too far apart, the flame will burn out the deformations of the flame front resulting from the previous obstacle, then the flame slows down before the next obstacle is reached. On the other hand, if P is too small, i.e., the obstacles are placed too close to each other, the gas pockets in between the obstacles are relatively unaffected by the outer flow; as a result, the influence of obstacles on flame acceleration is less effective. Other obstacle parameters, such as the shape or arrangement of the obstacles, are of secondary importance regarding flame acceleration.

The dramatic influence of repeated obstacles in a 1-D configuration on flame acceleration, and hence pressure development, was well demonstrated by Moen et al. (1982). Without any obstacles, the maximum overpressure developed in a 10 m long and 2.5 m diameter open-end tube in stoichiometric methane/air mixture was of the order of 0.1 bar, and the air blast overpressure at 10 m from the open end of the vessel was 0.03 bar. However, for the most violent case of six equally spaced orifice plates with $ABR = 0.3$ (orifice diameter $d = 2.1$ m), the peak overpressure was 8.86 bar, and the external air blast overpressure at 10 m from the open end of the vessel was 0.46 bar. Overpressures between these two extreme cases (i.e., between 0.1 bar and 8.86 bar) were measured for obstacles with different blockage ratios and different pitches.

6.2.4.2.   Blast Effects Produced in 2-D Configurations

Small-scale 2-D Experiments

Van Wingerden and Zeeuwen (1983) conducted experiments to study the influence of confinement, obstruction and gas reactivity. The experimental set-up is shown in Figure 6.10. Obstacles were vertical cylinders 1 cm in diameter and 9 cm in height. The pitch, distance between two adjacent obstacle rows, was $P = 3$ cm, and the area blockage ratio was $ABR = 0.31$. Ignition was at the center of the plate.

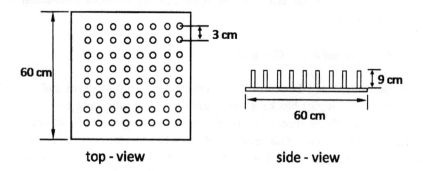

Figure 6.10. Experimental set-up for TNO small scale tests.

Confinement was introduced by mounting a parallel top plate. Four configurations were tested:

Configuration a:	single plate without sticks, 60 × 60 cm
Configuration b:	two parallel plates without sticks, 9 cm apart
Configuration c:	single plate with sticks, vertical sticks 1 cm in diameter, 9 cm height, P = 3 cm, ABR = 0.31
Configuration d:	two parallel plates with sticks, vertical sticks 1 cm in diameter, 9 cm height, P = 3 cm, ABR = 0.31

Flame speeds recorded by a high-speed camera are summarized in Table 6.2. For low flame speeds, the values were averaged over several frames to increase the accuracy. The accuracy is ± 1 m/s.

Table 6.2. Maximum flame speeds for various fuels and configurations (Van Wingerden and Zeeuwen, 1983)

Mixture	Maximum flame speed, m/s (Radius, cm)			
% fuel-air	a (3-D)	b (2-D)	c (3-D + Obstacles)	d (2-D + Obstacles)
10% methane	3.67	3.67	6.94 (23.2)	26 (19.2)
4% propane	5	5	13 (26)	38 (21.2)
8% ethylene	9.5	9.5	20 (23.6)	37.2 (17.6)

Experimental results were for four stoichiometric fuel-air mixtures:
- 10% methane/air with laminar flame velocity 2.1 m/s,
- 4% propane/air with laminar flame velocity 2.6 m/s,
- 8% ethylene/air with laminar flame velocity 5.0 m/s, and
- 10% acetylene/air with laminar flame velocity 13.5 m/s.

For configurations a and b (3-D and 2-D without obstacles), flame velocities were low; only slight flame acceleration was observed in acetylene mixture. For configuration c (3-D with obstacles), slight flame acceleration was observed in methane and propane mixtures, and significant acceleration was measured for ethylene and acetylene mixtures. For configuration d (2-D with obstacles), flame velocity increased exponentially, even for methane and propane.

Hjertager (1984) reported overpressures of 1.8 bar and 0.8 bar for propane and methane-air mixtures, respectively, between parallel 0.5-m radius disks with repeated obstacles. The results are presented in Table 6.3 along with the van Wingerden and Zeeuwen (1983) data.

Table 6.3. Small scale test results on VCE deflagration in 2-D configuration

Reference	Configuration	Fuel	Max. Flame Speed (m/s)	ÄPmax (bar)
van Wingerden and Zeeuwen (1983)	Two plates 0.6 × 0.6 m with forest of cylindrical obstacles	CH_4	26	—
		C_3H_8	38	—
		C_2H_4	37	—
		C_2H_2	30	
Hjertager(1984)	Radial disk 0.5 m radius with pipes and flat-type obstacles	C_3H_8	225	1.8
		CH_4	160	0.8

Van Wingerden (1989) reported experimental results in small and large scales. The experimental set-up with vertical cylinders as obstacles in small scale is shown in Figure 6.10. The experimental set-up with horizontal cylinders as obstacles is shown in Figure 6.11. The cylinders were placed parallel to each other on the ground surface or at some distance h above it. Pitch P was variable, and the number of obstacles depended on the pitch. Confinement was introduced by mounting a top plate over the obstacles and the height between two parallel plates was variable. Ignition was at the center of the array. In small scale tests, obstacle diameters D of 0.01, 0.02 and 0.03 m were used, and in large scale tests, the diameter of the cylinder was 0.1 m.

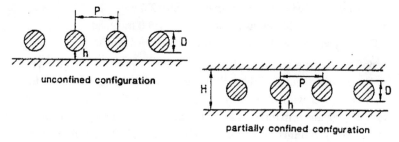

Figure 6.11. Experimental set-up for TNO tests with horizontal obstacles. (van Wingerden, 1989)

The effect of confinement is clearly shown in Figure 6.12, where blast overpressure $\bar{P}$ was plotted against energy scaled distance $\bar{R}$ for various heights between two parallel plates. The mixture was 7-8 % ethylene in air. Blast overpressures were higher for the smaller height, which corresponded to the higher degree of confinement.

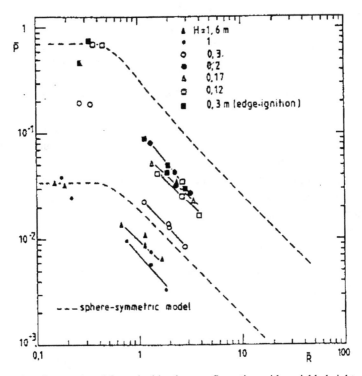

Figure 6.12. Blast produced from double plate configuration with variable heights.

Effect of Venting in 2-D Confinement Geometry

The effect of the degree of confinement was also presented by van Wingerden (1989). The experimental set-up is shown in Figure 6.13. The 2-D configuration was constructed with two parallel plates with dimensions 4m × 2m, with a wall of symmetry. The obstacle array consisted of concentric circles of vertical cylinders 8 cm in diameter. Ignition was at the center. The degree of confinement was varied using perforated top plates, with porosities varying between 0 % and 100 %. During these tests, the obstacle height was $H = 2D$, pitch $P = 3D$, $ABR = 0.5$, and the fuel was ethylene-air.

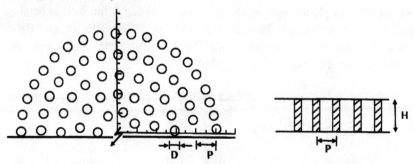

Figure 6.13. Experimental set-up for 2-D configuration (van Wingerden, 1989).

Experimental results (van Wingerden, 1989) of flame speed versus distance are shown in Figure 6.14. It can be seen that a slight reduction of the top confinement resulted in a considerable influence on terminal flame speeds. With the top plate open only 5%, flame speed was reduced to less than half of the values measured with no openings in the top plate at the same distance from ignition. With the top plate open 25%, flame speeds were similar to those without a top plate.

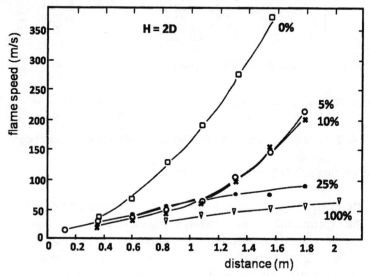

Figure 6.14. Flame speeds versus distance for various porosities.
(van Wingerden 1989)

2-D Experiments with Variable Congestion

Moen et al. (1980b) published experimental results for flame propagation between two plates, 2.5 × 2.5 m, in 2-D configurations shown in Figure 6.15. The gas was usually ignited in the center. Tube spirals (diameter $H = 4$ cm) were introduced between the plates (plate separation D) as obstacles. The pitch P was held constant at $P = 3.8$ cm.

In methane/air mixtures, flame speeds up to 400 m/s, accompanied by an overpressure of 0.64 bar, were produced. The effect of obstruction is clearly shown in Figure 6.16, where higher flame speeds were measured for higher degrees of obstruction.

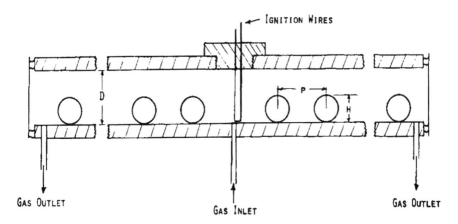

Figure 6.15. Experimental setup to study flame propagation in a cylindrical geometry.
(Moen, 1980b)

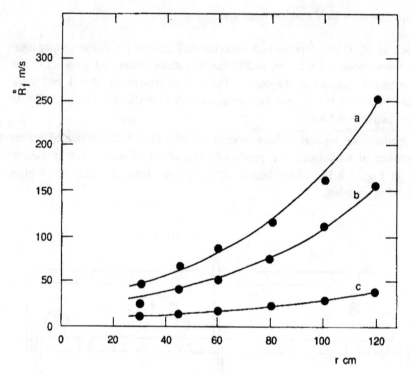

Figure 6.16. Flame speed-distance relationship of methane-air flames in a double plate
geometry (2.5 × 2.5 m), by Moen et al. (1980b).
(a) $H/D = 0.34$; (b) $H/D = 0.25$; (c) $H/D = 0.13$

Van Wingerden (1989a) reported a comprehensive study into the effects of forest-like obstacle arrays (0.08-m diameter obstacles) on the propagation of flames in a rectangular, double-plate apparatus of 2 × 4 m. Flames propagated over distances of up to 4 m from the point of ignition in some configurations. Ethylene-air mixtures generated flame speeds of up to 685 m/s and pressures of up to 10 bar inside an obstructed area. Obstacle parameters were varied over a wide range; flame speeds increased with blockage ratio and pitch.

Some of these tests were repeated by van Wingerden on a larger scale (scaling factor 6.25) with ethylene, propane, and methane as fuels (van Wingerden, 1989b). Figure 6.17 depicts the test setup. Findings from the small-scale tests were generally confirmed. However, flame speeds and overpressures were higher than those found in the equivalent small-scale tests, and ethylene tests resulted in detonations.

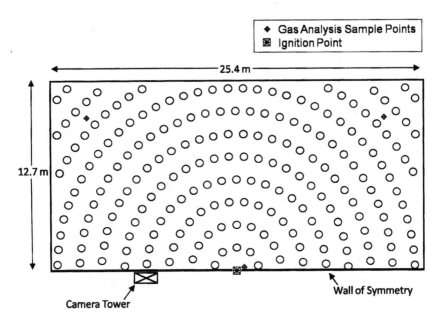

Figure 6.17. Large-scale test setup for investigation of flame propagation in a cylindrical geometry. Dimensions: 25 m long; 12.5 m wide; and 1 m high. Obstacle diameter 0.5 m.

Results from large scale experiments for 2-D configurations are summarized in Table 6.4.

Table 6.4. Large scale test results on VCE deflagration in 2-D configuration

Reference	Configuration	Fuel	Max. V_f (m/s)	$\ddot{A}P_{max}$ (bar)
Moen et al. (1980a,b)	Two plates diam. 2.5 m with spiral tube obstacles	CH_4 H_2S	400 50	0.64 —
van Wingerden (1989a)	Two plates vertical cylinders in concentric circles (2 × 4 m)	C_2H_4	685	10.0

2-D Experiments in Topside Structures

Large scale experiments were conducted in a test structure designed to simulate the topside structure of an offshore platform (Walker, 2002). The test structure represented one bay of a topside structure, measuring 28 m long, 8 to 12 m wide and 8 m high, with a solid floor and roof. One objective of the tests was to increase the scale (size) of tests, and generate data for use in validation of CFD programs. The structure also had side wall panels on the two long sides, which were used in various configurations in the tests, including fully open on one side, partially open on one side, and fully enclosed on both sides. One side was fully covered in all tests. The structure also had a solid floor, and in the majority of tests a solid roof as well. Congestion was configured like typical of offshore platforms. The combination of congestion configuration and side wall enclosure resulted in geometries that are not representative of most on-shore chemical processing plants. The tests may be relevant for some enclosed processes that are highly congested.

6.2.4.3.   Blast Effects Produced in 3-D Configurations

A 3-D geometry has the lowest confinement which has the highest degree of flow field divergence. The overall flame surface area is proportional to the square of distance from the ignition point. Unburned fuel-air mixtures and combustion products are allowed to expand freely in three directions. The positive feedback mechanism on flame acceleration is least effective in 3-D configurations.

3-D Configuration without Obstacles

Experiments were conducted in small laboratory scale, with soap bubbles ranging from 4 to 40 cm in diameter (Deshaies, 1981; Okasaki et al., 1981). With no internal obstacles, flame speeds were very close to laminar flame speeds. For example, maximum flame speeds for ethylene of 4.2 m/s and 5.5 m/s were found in cylindrical and hemispherical bubbles, respectively. Obstacles introduced in unconfined cylindrical bubbles resulted only in local flame acceleration. Pressures measured at some distance from the cylindrical bubble were, in general, two to three times the pressure measured in the absence of obstacles.

Large-scale balloon experiments with flammable gases in air were carried out by several groups of investigators (Lind, 1977; Broussard, 1985; Schneider, 1981; Harris, 1989). No obstacles were placed in the balloons. In all of these tests, flame acceleration was minimal or absent and showed no significant overpressures. The highest flame speed of 84m/s was measured in a 20-m diameter balloon with a hydrogen-air mixture. For other mixtures, flame speeds were no more than 40 m/s. It was reported that the flame acceleration was entirely due to intrinsic flame instability. To investigate whether the flame would accelerate when allowed to propagate over greater distances, tests were carried out in an open-sided test apparatus 45 m long (Harris and Wickens 1989). Flame acceleration was found to be no greater than in the balloon experiments.

It has been demonstrated that low flame speeds and insignificant overpressures result from 3-D configurations without obstacles, even at very large scale. Results are summarized in Table 6.5. Such low flame speed/low overpressure events would not be classified as VCEs; rather, they would be classified as flash fires.

Table 6.5.　Results of experiments under unconfined conditions without obstacles

Reference	Configuration	Fuel	V_{fmax} (m/s)	$\ddot{A}P_{max}$ (bar)
Deshaies and Layer (1981)	Hemispherical soap bubbles (D = 4–40 cm)	CH_4	3.0	—
		C_3H_8	4.0	—
		C_2H_4	5.5	—
Okasaki, et al. (1981)	Cylindrical soap bubbles (D = 44 cm)	C_2H_4	4.2	—
Lind and Whitson (1977)	Hemispherical balloons (D = 10–20 m)	C_4H_6	5.5	—
		CH_4	8.9	—
		C_3H_8	12.6	—
		C_2H_4	17.3	—
		C_2H_4O	22.5	—
		C_2H_2	35.4	—
Brossard et al. (1985)	Spherical balloons (D = 2.8 m)	C_2H_4	24	0.0125
		C_2H_2	38	—
Schneider and Pförtner(1981)	Hemispherical balloon (D = 20 m)	H_2	84	0.06
Harris and Wickens (1989)	Spherical balloons (D = 6.1 m)	Natural gas	7	—
		LPG	8	—
		C_6H_{12}	8	
		C_2H_4	15	—
Harris and Wickens (1989)	45 m long open-sided tent	Natural Gas	8	—
		LPG	10	—
		C_6H_{12}	10	—
		C_2H_4	19	—
		C_2H_4	19	—

3-D Configuration with Low Congestion

The introduction of obstacles within "unconfined" vapor clouds produced flame acceleration. On a small scale, an array of vertical obstacles mounted on a single plate (60 × 60 cm) resulted in flame accelerations within the array (van Wingerden and Zeeuwen 1983). Maximum flame speeds of 52 m/s for acetylene-air were measured over 30 cm of flame propagation, versus 21 m/s in the absence of obstacles.

Harris and Wickens (1989) report large-scale tests in an open-sided 45 m long apparatus incorporating grids and obstructions. Maximum flame speeds were approximately ten times those found in the absence of obstacles.

The influence of hemispherical wire mesh screens (obstacles) on the behavior of hemispherical flames was studied by Dörge et al. (1976) on a laboratory scale. The dimensions of the wire mesh screens were varied. Maximum flame speeds for methane, propane, and acetylene are given in Table 6.6.

Harrison and Eyre (1986, 1987) studied flame propagation and pressure development in a segment of a cylindrical cloud both with and without obstacles, and with jet ignition (Figure 6.18). The sector was 30 m long and 10 m high, and its top angle was 30°. The obstacles, when introduced, consisted of horizontal pipes of 0.315 m in diameter, arranged in grids. These experiments (Table 6.6) demonstrated the following points:

- Low-energy ignition of unobstructed propane-air and natural gas-air clouds does not produce damaging overpressures.

- Combustion of a natural gas-air cloud in a highly congested obstacle array leads to flame speeds in excess of 100 m/s (pressure in excess of 200 mbar).

- High-energy ignition of an unobstructed cloud by a jet flame emerging from a partially confined explosion produces a high combustion rate in the jet-flow region.

- Interaction of a jet flame and an obstacle array can result in an increase of flame speed and production of pressures in excess of 700 mbar.

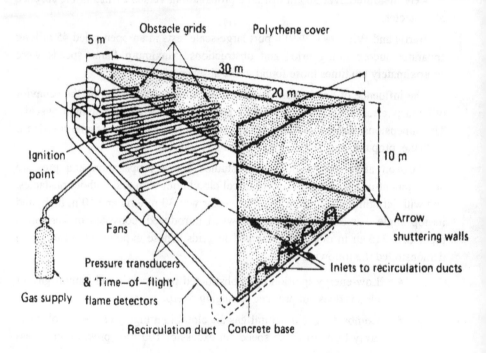

Figure 6.18. Experimental apparatus for investigation of effects of pipe racks on flame propagation. (Harrison and Eyre, 1986 and 1987)

Table 6.6. Experimental results on VCE deflagration under
unconfined conditions with obstacles (low congestion)

Reference	Configuration	Fuel	V_{fmax} (m/s)	$\ddot{A}P_{max}$ (bar)
van Wingerden and Zeeuwen (1983)	60 × 60 cm plate with 1 cm vertical obstacles on top	CH_4	7	—
		C_3H_8	13	—
		C_2H_4	20	—
		C_2H_2	52	—
Harris and Wickens (1989)	45-m-long open-sided tent with obstructions	Natural gas	50	0.03–0.07
		C_4H_{10}	65	0.03–0.07
		C_6H_{12}	70	0.03–0.07
Dörge et al. (1976)	Spherical grids in a 0.6 m cube	C_2H_4	>200	0.8
		C_2H_2	150	—
		C_2H_4	30	—
		C_3H_8	16	—
Harrison and Eyre (1986, 1987)	Sector with pipework	Natural gas	119	0.208
		C_3H_8	—	0.052
Harrison and Eyre (1986, 1987)	Sector with pipework and jet flame ignition	Natural gas	170	0.710

The results in Table 6.5 and Table 6.6 demonstrated that weak ignition of an unconfined cloud in an unobstructed environment will generally not result in a damaging explosion, even for high reactivity fuels such as acetylene and ethylene. The introduction of low-density obstacles results in some flame acceleration, especially for high reactivity fuels. The more reactive the fuel, the more effect obstacles have on flame acceleration (Harris and Wickens 1989). However, for 3-D geometry with low congestion, the flame acceleration is not significant. The highest flame speed observed was 84 m/s, and it was accompanied by an overpressure of 60 mbar for hydrogen-air in a 10-m radius balloon (Schneider and

Pförtner 1981). For all other fuels, flame speeds were below 40 m/s and corresponding overpressures were below 35 mbar.

It should be noted that many 3-D experiments conducted before approximately 1990 used low congestion and small scale. As a result, flame acceleration and pressure buildup were not significant.

3-D Configuration with High Congestion

A number of large-scale experiments demonstrated that very high flame speeds and overpressures were achieved in 3-D configurations with high obstacle density. Mercx (1995) demonstrated that flame speeds of several hundred meters per second and overpressures of the order of 1 bar were produced in 3-D configuration with a moderate to very high degree of obstruction. DDT occurred with high reactivity fuels in the MERGE program. Some of the MERGE test results on flame speeds and overpressures measured in 3-D configurations are presented in Table 6.7.

Thomas (2003) and Pierorazio (2004) performed experiments at larger scale in elongated geometries of dimensions shown in Table 6.7. Tests by Thomas involved lean (5.9%) to rich (9.3%) ethylene-air mixtures. Within this range of fuel-air ratios, rapid flame acceleration near the end of the congested volume and DDT occurred. Leaner and richer mixtures did not undergo DDT. Pierorazio used the same test setup as Thomas and tested stoichiometric methane and propane in air mixtures. Maximum flame speeds were 115 and 170 m/s, respectively.

The MERGE experiments presented in Table 6.7 used extremely dense obstacle arrays with pipes running in three orthogonal directions. Some of the blockage ratios were considerably in excess of practical plant geometries.

Table 6.7. Flame speed and overpressure from 3-D configurations

Reference	Fuel [a] %	Obstacle parameters			V_f (max) m/s	Δp (max) bar
		P [b]	ABR %	VBR %		
Mercx 1995, MERGE 3D, 4×4×2 m, tube 3 direction d = 4.1cm	CH$_4$	4.65 d	35	10	250	0.66
	C$_3$H$_8$	4.65 d	35	10	397	1.48
	C$_2$H$_4$	4.65 d	35	10	-----	10.8
Mercx 1995, MERGE 3D, 4×4×2 m, tube 3 direction d = 4.1cm	CH$_4$	3.25 d	45	20	-----	1.60
	C$_3$H$_8$	3.25 d	45	20	-----	3.88
	C$_2$H$_4$	3.25 d	45	20	-----	13.3
Thomas 2003, ERC 14.6×3.7×1.8 m, vertical tube d = 5.08 cm	C$_2$H$_4$ (7.3)	4.5d	23	4.2	1768	>6.9
	C$_2$H$_4$ (7.3)	4.5d	23	4.2	1768	>6.9
	C$_2$H$_4$ (5.9)	4.5d	23	4.2	>340	>6.9
	C$_2$H$_4$ (9.3)	4.5d	23	4.2	>340	>6.9
Pierorazio 2004, ERC 14.6×3.7×1.8 m, vertical tube d = 5.08 cm	CH$_4$	4.3d	23	4.3	78	-----
	CH$_4$	3.1d	23	5.7	115	-----
	C$_3$H$_8$	7.6d	13	1.5	37.5	-----
	C$_3$H$_8$	4.3d	23	4.3	150	-----
	C$_3$H$_8$	3.1d	23	5.7	170	-----

[a]Fuel concentration is stoichiometric unless otherwise noted.

[b]Pitch P is shown as multiples of the obstacle diameter d.

3-D Configuration with Variable Congestion

Experiments reported by Harris and Wickens (1989) involved three levels of confinement/congestion. They modified the experimental apparatus shown in Figure 6.18 — a 45 m long, open-sided apparatus. The first 9 m of the apparatus was modified by the fitting of solid walls to its top and sides in order to produce a confined region. Thus, it was possible to investigate whether a flame already propagating at high speed could be further accelerated in unconfined parts of the apparatus, where obstacles of pipework were installed. The initial flame speed in the unconfined parts of the apparatus could be modified by introduction of obstacles in the confined part.

Experiments were performed with cyclohexane, propane, and natural gas. In a cyclohexane experiment, the flame emerged from the confined region at a speed of approximately 150 m/s, and progressively accelerated through the unconfined region containing obstacles until transition into a detonation occurred. Detonation continued to occur in the unconfined region. A similar result was found for propane, in which flames emerged from the confined area at speeds of 300 m/s.

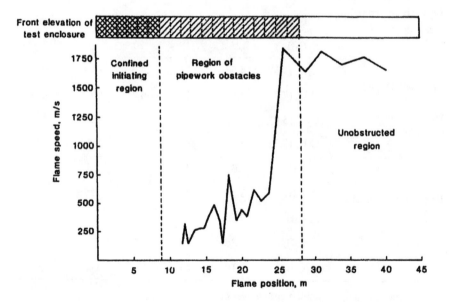

Figure 6.19. Flame speed-distance graph showing transition to detonation in a cyclohexane-air experiment. (Harris and Wickens 1989)

Experiments performed with natural gas yielded somewhat different results. Flames emerged from the confined portion of the apparatus at speeds below 500 m/s, and then decelerated rapidly in the unconfined portion with obstacles. On the other hand, flames emerging from the confined portion at speeds above 600 m/s continued to propagate at speeds of 500–600 m/s in the obstructed, unconfined portion of the cloud. There were no signs of transition to detonation. Once outside the obstructed region, the flame decelerated rapidly to speeds of less than 10 m/s.

6.2.5. Effect of Congestion

The two most important congestion parameters are blockage ratio and pitch. The word "blockage" is used to describe the flow field occupied by obstacles. The term "area blockage ratio" is defined as the ratio of area blocked by obstacles to the cross-sectional area. The term "volume blockage ratio" is defined as the ratio of the volume occupied by obstacles to the total volume of the congested zone.

6.2.5.1. Effect of Obstacle Pitch

Van Wingerden (1989) also demonstrated the effects of blockage ratio and pitch on flame acceleration using the TNO medium scale test platform, as shown previously in Figure 6.13. Experimental results of flame speed versus distance are presented in Figure 6.20 for four different pitches. It is shown that at the same distance of flame propagation the maximum flame speed occurred at the smallest pitch, $P = 1.5$ D. This may be explained by the smallest pitch allowing the flame to pass more obstacles.

In Figure 6.21, flame speeds were plotted against a non-dimensional distance, distance divided by pitch, R/P. Hence, at a given value of R/P, the flame has passed a fixed number of obstacles and the $p = 6D$ can be seen as the optimum pitch, which corresponds to the greatest flame speed. The ultimate flame speed is higher for smaller pitch, because the flame has passed more obstacles before reaching the edge of the test apparatus.

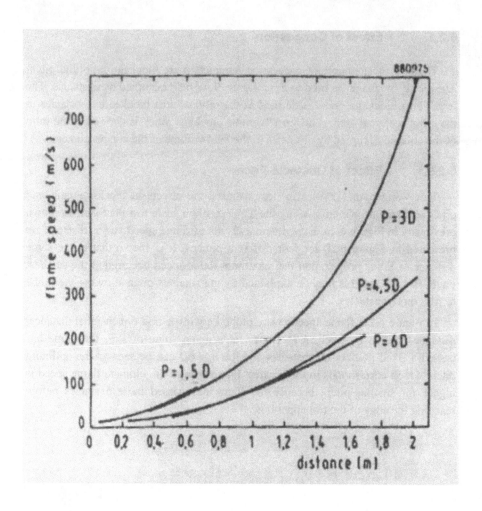

Figure 6.20. Effect of obstacle pitch on flame speed (dimensional distance).
(van Wingerden 1989)

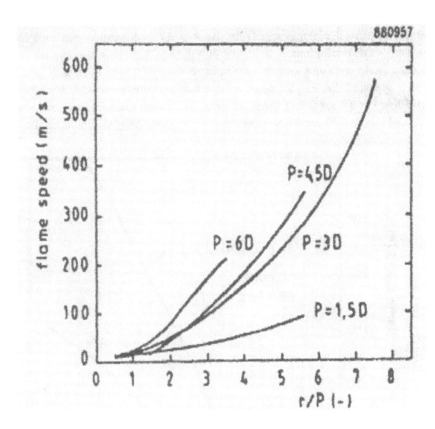

Figure 6.21. Effect of obstacle pitch on flame speed (non-dimensional distance).
(van Wingerden 1989)

Similar pitch effects were reported by Mercx (1992) from TNO large scale experiments, as shown in Figure 6.22 and Figure 6.23. In Figure 6.22, flame development was presented over propagation distance r, for $P = 3D$ and $P = 6D$. In Figure 6.23, flame speeds were plotted against non-dimensional distance r/P, where r is flame propagation distance and P is pitch, the distance between two adjacent rows of obstacles. Hence, at a fixed value of r/P, the flame has passed a fixed number of obstacles.

Figure 6.22 shows that higher flame speeds were obtained for $P = 3D$ than $P = 6D$ configuration, which is similar to the medium scale experiments. The reason is that more obstacles were passed at the same distance for smaller pitch. In addition, higher volume blockage ratio results from smaller pitch for the same area blockage ratio. On the other hand, comparing flame speeds after passing the same number of obstacles, flame speeds for the $P = 6D$ configuration exceeded those for the $P = 3D$ configuration, as shown in Figure 6.23. This is also similar to the medium scale tests and may be attributed to the effectiveness of each obstacle row. It appeared that $P = 6D$ was a more effective distance between two obstacle rows than $P = 3D$ regarding flame acceleration through each obstacle row. Figure 6.22 and Figure 6.23 are for propane; tests for methane showed similar results.

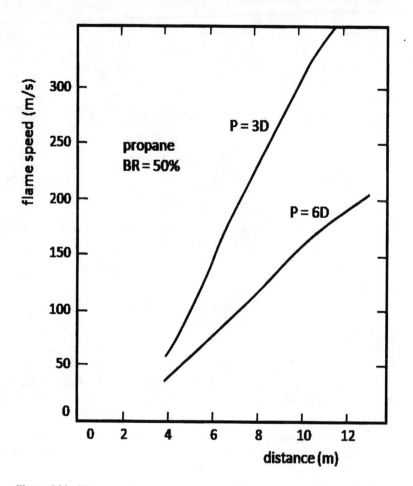

Figure 6.22. Flame speed versus distance for different pitches (Mercx, 1992).

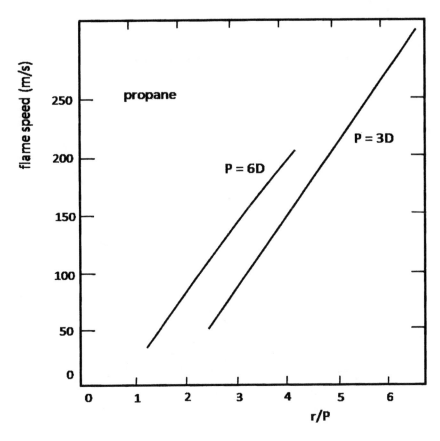

Figure 6.23. Flame speed versus dimensionless distance R/P.
(Mercx, 1992)

6.2.5.2.   Effect of Obstacle Blockage Ratio

The effect of blockage ratio, *ABR* of the obstacle array is presented in Figure 6.24, given by van Wingerden (1989). Ethylene-air mixtures were confined between 4m × 4m parallel plates, and 8cm vertical cylinders were placed in concentric circles around the ignition source. It is clearly shown that higher flame speeds were produced by larger *ABR*; the maximum flame speeds were found for the largest blockage ratio, *ABR* = 0.7.

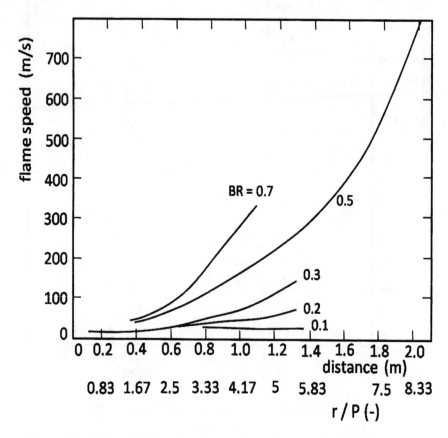

Figure 6.24. Effect of ABR on flame speed. (van Wingerden and Hjertager, 1991)

The effect of blockage ratio was also reported by Mercx (1992) from TNO large scale experiments. The dimensions of the two parallel plates were 25.4 m × 12.7 m with a solid vertical plate as a plane of symmetry. Similar to medium scale test results, higher flame speeds and overpressures were obtained for higher values of area blockage ratio. Higher volume blockage ratios also resulted from higher area blockage ratios. As depicted in Table 6.8 in test No. 4 and 5, the same mixture of 6.85 % ethylene/air was used, and pitch $P = 3D$ for both tests; the only difference was blockage ratio. The overpressures found for $ABR = 30\%$ are approximately twice the values for $ABR = 20\%$.

Table 6.8. Effect of blockage ratio (Mercx, 1992)

Test Conditions			Maximum Overpressure (bar at Various Transducer Locations)				
Test No	ABR %	VBR %	T1	T2	T5	T6	T12
4	30	8.7	1.04	0.43	0.90	1.15	20.53
5	20	5.8	0.59	0.28	0.48	0.54	11.44

6.2.5.3.   Effect of Obstacle Shape and Obstacle Arrangement

Hjertager (1984) reported the influence of obstacle shape on flame propagation and pressure development in a lab-scale 2-D configuration. Two types of obstacles were tested, sharp-edged and cylindrical obstacles.

Figure 6.25 gives the results of peak overpressure as a function of blockage ratio. Sharp-edged obstacles produced pressures approximately two times higher than round edged obstacles because higher intensity turbulence is generated at sharp-edged obstacles. The results are consistent with the turbulence produced by sharp-edged obstacles shown in Figure 6.7.

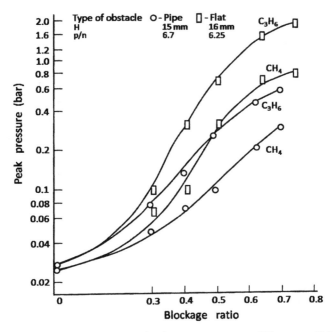

Figure 6.25. Effect of obstacle shape on pressure. (Hjertager 1984)

Chan et al. (1983) showed in their channel experiments that when blockage ratio and pitch were kept constant, different flame speeds could be obtained just due to a different arrangement of the obstacles in the channel. Apparently, the obstacle position in a channel (against a wall or in the center of the channel) can cause differences in flame speeds. This might be due to the different number of shear layers associated with each obstacle.

6.2.6.   Effects of Other Factors

In addition to the three important factors: confinement, congestion, and fuel reactivity, there are several other factors that also influence flame propagation, and hence pressure development, to some extent. Following are discussions based on available data.

6.2.6.1.   Effect of Unburned Mixture Displacement

The Research to Improve Guidance on Separation Distance for the Multi-energy Method (RIGOS) report (van den Berg and Mos, 2005) suggests that the actual explosion energy is less than the nominal energy, (full combustion energy originally present within the obstructed region assuming stoichiometry) since unburned fuel-air mixture is pushed out of the obstructed region as the combusting gases expand. Some data is given based on MERGE experiments, van den Berg indicates that the amount of energy required to model the blast when overpressure is high is approximately 100% of the nominal energy, and if the overpressure is low, no more than 10 to 20% of the nominal energy is present in the congested area. The "Guidance for Application of the Multi-Energy method" (GAMES) report (Mercx, 1998) states that taking 100% of the energy is conservative for low overpressures and that the "efficiency" is less than 20% for source overpressures less than 0.5 bar. For clouds smaller than the obstructed region, GAMES recommends taking an efficiency factor of 100%.

EMERGE tests in which DDT occurred were found to need energy for the Multi-energy Method predictions greater than the nominal energy to match observed overpressures (Eggen, 1998). DDT was removed by Eggen from the energy efficiency analysis, and the above rules should not be applied for DDT.

6.2.6.2.   Effect of Scale

Experiments and accidents have shown that the combustion rate of VCEs increase with the size of the congested zone. Scaling techniques have been investigated to adjust VCE blast predictions for size effects. These techniques typically take into account the overall size of a congested zone by incorporating a length dimension. Obstacle configuration is represented by terms such as area blockage ratio, average obstacle diameter, and number of rows of obstacles. In virtually all of the models, fuel reactivity is incorporated using the laminar burning velocity of the subject fuel. The scaling techniques use empirical data to determine various coefficients in the correlations. As a result, the correlations differ depending on the equations developed and data set used to determine coefficients. Correlations developed by Leeds, TNO and Shell are presented below.

6.2.7.   University of Leeds Correlation

Phylaktou and Andrews at the University of Leeds (Phylaktou and Andrews, 1993; Phylaktou, 1993; Phylaktou and Andrews, 1994; Phylaktou and Andrews, 1995) analyzed a large body of data from explosion experiments with obstacles and produced the following correlation for the ratio of turbulent to laminar burning velocities:

$$\frac{S_T}{S_l} - 1 = 0.67 \left(\frac{u'}{S_l}\right)^{0.47} R_l^{0.31} Le^{-0.46} \left(\frac{\upsilon}{\upsilon_a}\right)^{0.95} \qquad \text{(Eq. 6.4)}$$

where:

u' =   the rms turbulence velocity

ν =   the kinematic viscosity of the flammable mixture

ν_a =   the viscosity of air at standard temperature and pressure

Le =   the Lewis number, which is the ratio of thermal diffusivity to mass diffusivity for the reacting mixture

R_l =   the turbulent Reynolds number, defined as:

$$R_l = \frac{u'l}{\upsilon}$$

where l is the integral length scale of the turbulent flow. These researchers also performed a sequence of laboratory-scale experiments consisting of explosions in a confined pipe geometry containing an orifice plate of varying blockage ratio to determine the relationship between obstacle geometry and induced turbulence and flame acceleration.

Extending their findings to repeated arrays of obstructions, the following correlation for overpressure was determined:

$$P \propto \left[e^{(6.24 \cdot ABR)} D^{0.62} \right] \left[E^2 S_l^{1.06} \right] \left[(\alpha E S_l)^{1.56} \right] \qquad \text{(Eq. 6.5)}$$

where:

ABR =   the area blockage ratio

D =   the characteristic diameter of the obstruction (e.g. tube diameter)

E =   the expansion factor

S_l =   the laminar flame speed and α is the flame self-acceleration factor

α =   accounts for the increase in flame speed due to flame instabilities occurring during the initial stages of laminar propagation

A value for coefficient α is needed, which depends on both test rig geometry and mixture properties.

6.2.8.   TNO GAME Correlation

TNO, as part of the GAME project (Mercx, 1998), analyzed a large set of experimental vapor cloud explosion data collected during the European Union-sponsored MERGE and EMERGE studies. They developed a correlation for overpressure due to turbulent combustion in an obstacle laden environment based on a fully homogeneous obstacle configuration with length L_f defining the length available for flame travel, the combination (VBR*L_f/D) where VBR is the volume blockage ratio and D is the obstacle size. Fuel reactivity was represented by laminar burning velocity S_l of the fuel. Coefficients were developed from curve fit to the MERGE data, which for 3-D environments gives:

$$\Delta P_0 = 0.84 \left(\frac{VBR \cdot L_f}{D} \right)^{2.75} S_l^{2.7} D^{0.7} \qquad \text{(Eq. 6.6)}$$

For 2-D environments, the results of the CECFLOW (Visser, 1991) and DISCOE (van Wingerden, 1989) experiments were used to obtain a fit. The CECFLOW experiments were conducted in a wedge-shaped geometry whereas the DISCOE geometry consisted of concentric rings of cylindrical obstructions placed around the ignition point and confined by two parallel horizontal plates. The resulting 2-D correlation is:

$$\Delta P_0 = 3.38 \left(\frac{VBR \cdot L_f}{D} \right)^{2.25} S_I^{2.7} D^{0.7}$$

(Eq. 6.7)

6.2.9.   Shell CAM Correlation

The Congestion Assessment Method (CAM) method uses the concept of a "Severity Index", S, as a means of estimating source overpressure when extrapolating experimental data to plant geometries. (The term 'Severity Index' is unrelated to the TNO Severity Levels associated with the Multi-Energy method.) The relationship between this index and the source overpressures was determined by using the SCOPE 3 phenomenological model Puttock (2000) for a wide range of input conditions. That relationship was determined to be:

$$S = P_{max} \cdot \exp\left(0.4 \frac{P_{max}}{E^{1.08} - 1 - P_{max}} \right)$$

(Eq. 6.8)

where:

S =   the severity index (in bars)

P_{max} =   the source overpressure (in bars)

E =   the expansion ratio.

The severity index is calculated from the expression:

$$S = a_o \left(U_o (E - 1) \right)^{2.71} \left(n \cdot r_{pd} \cdot d \right)^{0.55} n^{a_i} \exp\left(a_2 b \right)$$

(Eq. 6.9)

where the empirically determined coefficients (for laminar burning velocity in m/s, length in m, pressure in bars) are shown in Table 6.9.

Table 6.9. CAM Coefficients

Coefficient	Without Roof	With Roof
a_0	3.9×10^{-5}	4.8×10^{-5}
a_1'	1.99	1.66
a_2	6.44	7.24

6.2.9.1.   Effect of Length-to-Width Ratio

Van Wingerden (1989) conducted experiments in the 4 m × 4 m × 1.6 cm 2-D configuration. For center ignition, when length-width ratio L/W =1 (i.e., 4 m × 4 m platform), side relief was not possible. That is, the flame had to travel 2 m to an edge in any direction. For $L/W \leq 2$, the L/W ratio appeared to be unimportant. In L/W = 2 tests (i.e., 4m × 2m platform), side relief is possible when the flame reached the closest edge of the array, or at a distance of 1 m from the ignition source. In tests using high L/W ratios, L/W = 3.25 and L/W = 4, side relief had a dramatic influence on flame propagation, with flame speeds remaining comparatively low, with a maximum flame speed at only 122 m/s or less. The subsequent study of this topic in the GAMES report (Mercx, 1998) suggests that there may be length-to-width aspect ratios above which explosion overpressures may be limited to that at the time of first venting; however, they concluded that this would not apply in all situations, and chose not to limit VCE energy based on aspect ratio in the Multi-energy Method.

6.2.9.2.   Effect of Edge Ignition

Most of the experiments summarized previously used center ignition. Van Wingerden (1989) conducted edge ignition experiments to examine this effect of ignition location using the same test setup as described in 6.2.9.1. The initial stages of edge ignition explosions are dominated by back relief, which is relief toward the open edge at the ignition source. Nevertheless, flame speeds attained in edge-ignited situations exceed those ignited in the center for those tests with closely spaced parallel plates. This was due to the fact that the flame has passed more rows of obstacles in edge-ignited cases. As a result, higher flame speeds were achieved in spite of the initial back relief. In the final parts of flame propagation in the tests with L/W = 1 and L/W = 2, a steady flame speed was reached between 635 m/s and 685 m/s for 7.5 % ethylene/air mixture. The pressure generated at the end of flame propagation was approximately 10 bar.

Van Wingerden and Zeeuwen (1985) also reported the ultimate flame velocity of 420 m/s from edge ignition exceeding the 175 m/s resulting from center ignition in the same configuration, i.e., a double plate configuration with horizontal cylindrical obstacles (large scale test with obstacle diameter $d = 0.1$ m, $h = 0.2$ d, the height between the two plates, $H = 3d$, $P = 2.25d$, using 7-8 % ethylene-air).

6.2.9.3.   Effect of Jet Ignition

To investigate the effect of turbulence generated by fuel discharge on combustion, several investigators conducted experiments to determine the intensity of pressure waves resulting from the rupture of vessels and pipes. Giesbrecht et al. (1981) published the results of a series of small-scale experiments in which vessels with release sizes ranging from 0.226 to 10,000 kg, and containing propylene under 40 to 60 bar pressure, were ruptured. After a pre-selected time lag, vapor clouds were ignited by exploding wires, and ensuing flame propagation and pressure effects were recorded. Flame speed was observed to be nearly constant, but increased with the scale of the experiment. The maximum pressure observed was found to be scale dependent, see Figure 6.26.

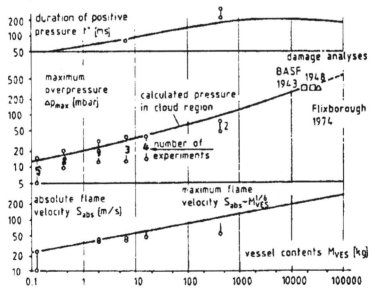

Figure 6.26. Flame velocity, peak overpressure, and overpressure duration in gas cloud explosions following vessels bursts. (Giesbrecht et al. 1981)

Battelle (Seifert and Giesbrecht 1986) and BASF (Stock 1987) each conducted studies on exploding fuel jets, the former on natural gas and hydrogen jets, and the latter on propane jets. The methane and hydrogen jet program covered subcritical outflow velocities of 140, 190, and 250 m/s, and orifice diameters of 10, 20, 50, and 100 mm. In the propane jet program, outflow conditions were supercritical with orifice diameters of 10, 20, 40, 60, and 80 mm. The jets were started and ignited after they had achieved steady-state conditions.   In the methane and hydrogen jet experiments, blast overpressure was measured at various distances from the cloud, and the propane jet experiments only produced measurements of in-cloud overpressures.

The findings of the studies were:
- In-cloud overpressure was dependent on outflow velocity, orifice diameter, and the fuel's laminar burning velocity. The effect of orifice diameter is shown in
- Figure 6.27.
- The maximum overpressure appeared to rise substantially when the jet was partially confined between 2-m-high parallel walls and obstructed by some 0.5-m-diameter obstacles.

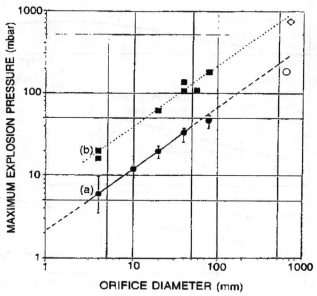

Figure 6.27. Maximum overpressure in vapor cloud explosions after critical-flow propane jet release dependent on orifice diameter:
(a) undisturbed jet; (b) jet into obstacles and confinement.

Several experiments with ethylene and hydrogen investigated the effects of jet ignition on flame propagation in an unconfined cloud, or on flame propagation in a cloud held between two or more walls (Figure 6.28). Such investigations were reported by Schildknecht and Geiger (1982), Schildknecht et al. (1984), Stock and Geiger (1984), and Schildknecht (1984). The jet was generated in a 0.5 × 0.5 × 1-m box provided with turbulence generators for enhancing internal flame speed. Maximum overpressures of 1.3 bar were observed following jet ignition of an ethylene-air cloud contained on three sides by a plastic bag. In a channel confined on three sides, maximum pressures reached 3.8 bar in ethylene-air mixtures. A transition to detonation occurred in hydrogen-air mixtures.

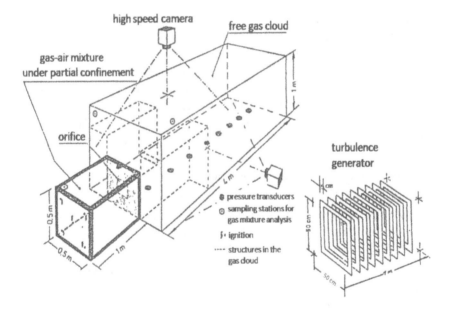

Figure 6.28. Experimental apparatus for investigating jet ignition of ethylene-air and hydrogen-air mixtures. (Schildknecht et al., 1984)

One experiment (Moen et al. 1985) revealed that jet ignition of a lean acetylene-air mixture (5.2% v/v) in a 4 m long, 2 m diameter bag can produce the transition to detonation.

A detailed study performed by McKay et al. (1989) revealed some of the conditions necessary for a turbulent jet to initiate a detonation directly. These experiments are covered in more detail in Section 6.3.1.

Pförtner (1985) reports experiments with hydrogen in a lane, 10 m long and 3 × 3 m in cross section, in which a fan was used to produce turbulence. In these experiments, a transition to detonation occurred at high fan speeds.

Experiments without additional turbulence produced flame speeds no higher than 54 m/s.

6.2.9.4.   Influence of Spacing between Congested Volumes

Van Wingerden (1989) conducted experiments to investigate to what extent an explosion in one congested area will influence the explosion in a separate congested area if both congested areas and the open area in between them are covered by the cloud of flammable mixtures. Here, the first congested area was called the "donor" and the second was called the "acceptor." The dimensions for both congested areas were 4 m × 2 m × 0.16 m, and the donor array was provided with a wall of symmetry. The obstacles were vertical cylinders placed in concentric circles with $ABR = 0.5$ in all tests. The donor array had a height $H = 2d$, pitch $P = 6d$, with center ignition. The acceptor array had a height $H = 2d$ and pitch $P = 3d$. The flammable mixture of 7-8 % ethylene-air was present in both donor and acceptor and the intervening area.

The distance, S, between the donor and the acceptor was varied. It is shown in Figure 6.29 that as soon as the flame left the donor array, the flame speed dropped due to the absence of confinement and obstruction. When the flame entered the acceptor, the flame started acceleration again.

For $S = 0.5$ m, the initial flame speed in the acceptor array was about 125 m/s, which resulted in a very high rate of flame acceleration and a very high ultimate flame speed. The initial flame speeds for $S = 1$m and $S = 2$m were about 75-100 m/s and 25 m/s, respectively, resulting in a longer distance of flame propagation in the acceptor before high flame speeds were achieved.

Both pressure-time histories measured within the array and blast pressure-time histories measured outside the congested areas exhibited two peaks; the first peak is due to the explosion in the donor, and the second peak is due to the explosion in the acceptor construction.

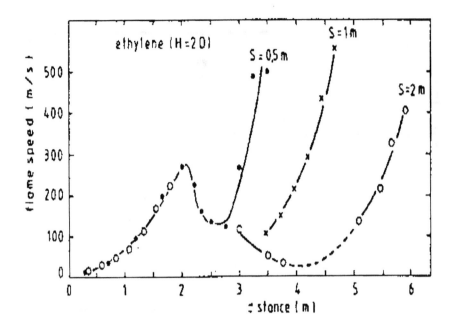

Figure 6.29. Effect of the gap between two congested areas on flame speeds.
(van Wingerden, 1989)

Mercx et al. (1992) performed experiments to investigate the gap effect using TNO medium and large scale set up. The medium scale set up is similar to that used by van Wingerden (1989) and the large scale is with a scaling factor of 6.35 to the medium scale, i.e., the dimensions of two parallel plates are 25.4 m × 12.7 m with a solid plate as the symmetric plane.

The results are shown in Figure 6.30. The data from medium scale experiments, represented by a dotted line, were performed with ethylene and a P = 6D donor array, and the gap between the two arrays, S = 0.5 and 2 m. The data from large scale experiments, represented by a solid line, were performed with propane and a P = 6D donor array. In order to compare medium and large scale data, the distance was non-dimensionalized, divided by the obstacle diameter D.

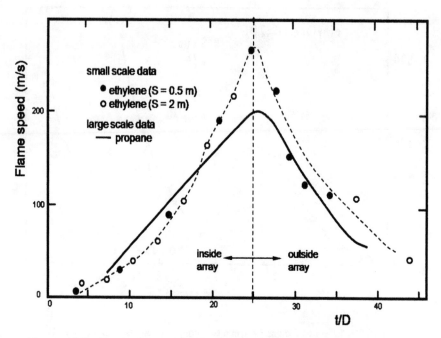

Figure 6.30. Comparison of flame propagation between two adjacent arrays in medium and large scale tests (Mercx 1992).

Harris and Wickens (1989) preformed tests which repeated obstacles (40% area blockage ratio) were located over the first 22.5 m length of the enclosure with the second half, also 22.5 m in length, left completely unobstructed. Following ignition at the end with obstacles, the flame accelerated through the repeated obstacles within the test rig, reaching a maximum flame speed at the end of the obstructed region. The flame rapidly decelerated immediately after the flame emerged from the obstructed region into the unobstructed part of the cloud. Typical results are illustrated in Figure 6.31 for cyclohexane-air and Figure 6.32 for natural gas-air. Within a few meters of the flame emerging from the obstructed region there is a very rapid reduction in flame speed, to a level which generates overpressures of less than 10 mbar.

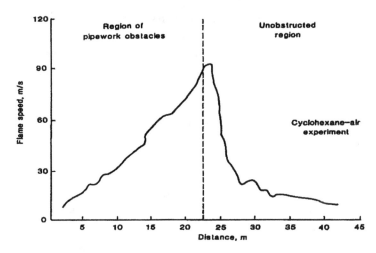

Figure 6.31 Flame speed/distance showing acceleration in the region of repeated obstacles and deceleration on emerging into the unobstructed region , cyclohexane-air experiment. (Harris and Wickens, 1989)

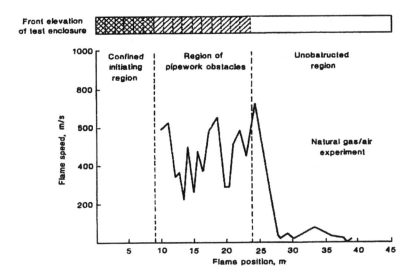

Figure 6.32. Flame speed/distance showing rapid deceleration on exit from a region containing repeated obstacles into an unobstructed region, natural gas-air. (Harris and Wickens, 1989)

Van den Berg and Mos (2005) conducted tests with stoichiometric ethylene-air mixtures to evaluate the separation distance needed between two congested volumes (donor and acceptor) to produce two separate blast waves, which they called the critical distance. A smaller separation distance produced a single blast wave. Half-cube shaped congested volumes of the same configuration as MERGE tests were used, measuring about 1.4 m length and width and 0.7m high. The program resulted in a limited number of concrete indications with respect to the critical distance. All tests were run in low overpressure range, and they were extrapolated to high pressure. They concluded that simple blast modeling methodologies tend to overestimate the blast effects from the acceptor explosion by more than an order of magnitude, particularly in the low overpressure range. Directional effects resulting in back venting from the acceptor volume were the cause of the discrepancies. At separation distances slightly larger than the critical distance, suppression of acceptor blast effects was documented.

6.3.   VAPOR CLOUD DETONATION THEORY AND RESEARCH

6.3.1.   Direct Initiation of Vapor Cloud Detonations

A vapor cloud detonation may be achieved through direct initiation or through deflagration to detonation transition (DDT). For direct initiation of detonation, a blast wave that is capable of maintaining its post-shock temperature above the mixture's auto-ignition temperature over some span of time is required (Lee and Ramamurthi 1976, Sichel 1977). The ignition source must have sufficient energy, which is called critical initiation energy, and is defined as the minimum source energy required for the direct initiation of an unconfined detonation in a detonable mixture. The critical initiation energy is also used to illustrate the detonation sensitivity of the given mixture. Directly measured data for critical initiation energy is limited due to experimental difficulties. Figure 6.33 presents data for critical initiation energy compiled by Bull, et al. (Bull, 1978).

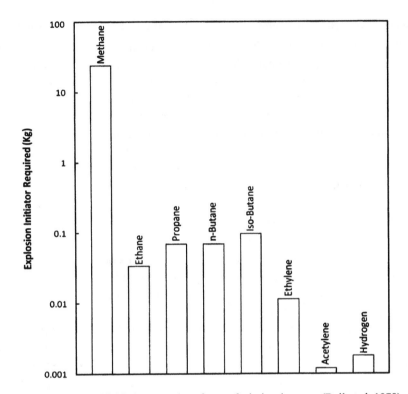

Figure 6.33. Critical initiation energies of some fuel-air mixtures. (Bull et al. 1978)

In Figure 6.33, all mixtures are in stoichiometric composition and the initiation energy is expressed by the charge weight of tetryl (a high explosive, similar to TNT in energy). It is shown that methane has the highest critical initiation energy among all hydrocarbons due to its particular molecular structure.

6.3.2.   Detonability of Commonly Used Fuels

6.3.2.1.   Dynamic detonation parameters

Lee et al. (1981) referred to the detonation cell size, critical tube diameter, critical initiation energy, and detonability limits as the "dynamic detonation parameters," which are dependent on the rate of the chemical reactions in a detonation wave. These parameters cannot be determined from chemistry alone and are correlated to the hydrodynamic structure of the detonation wave itself.

It has been demonstrated by numerous experiments that a detonation wave consists of a three-dimensional cellular structure formed by a series of intersecting transverse shock waves (see Figure 4.8). The detonation cell size, ë, which is the transverse wave spacing, has been measured. Detonation cell size varies with the fuel, oxidizer, and fuel-oxidizer ratio. The critical tube diameter, d_C, is defined as the minimum diameter of a detonation propagating in a tube to initiate a detonation in an unconfined vapor cloud outside the tube with the same composition as in the tube (confined planar detonation wave) for the transformation into a self-sustained spherical detonation wave. Mitrofanov and Soloukhin (1965) and Edwards et al. (1979) observed that $d_C \approx 13$ë for a circular tube and $d_C \approx 10$ë for a rectangular channel for all fuels.

Matsui and Lee (1978) measured the direct initiation energy of detonation for eight gaseous fuels (acetylene, ethylene, ethylene oxide, propene, methane, propane, and hydrogen) using a planar detonation from a linear tube for initiation. A dimensionless parameter, called the detonation hazard index, (D_H), is defined by the ratio of the minimum initiation energy of the fuel to that of an acetylene-oxygen mixture (acetylene $D_H = 1$). The magnitude of D_H is used for comparing the relative detonation sensitivity of the fuels. It was found that ethylene oxide-oxygen mixture requires about 10 times more direct initiation energy than acetylene. The olefins (ethylene, ethylene oxide, and propene) in oxygen having values of $D_H \cong 10^2$ are about 100 times more direct initiation energy than acetylene. The alkanes (methane and propane) in oxygen have values of $D_H \cong 10^3$, with the exception of methane, which requires particularly high direct initiation energy with a value of $D_H \cong 10^5$. Hydrogen-oxygen was found to be similar to normal alkanes with values of $D_H \cong 10^3$. Fuel-air mixtures in general have values of D_H about 10^6 times larger than the corresponding values for the same fuel with pure oxygen.

A fuel-air mixture is detonable only if its composition is between the detonability limits. The detonation limits for fuel-air mixtures are substantially narrower than their range of flammability (Benedick et al. 1970).

The measured critical initiation energy, E_C, detonation hazard index, D_H, and detonability limits for some hydrocarbons are presented in Table 6.10.

Table 6.10. Critical initiation energy and detonability for hydrocarbon fuels
(Matsui and Lee, 1978)

Mixtures	Fuel (vol. %)	E_C (joules)	D_H	Lower – Upper Detonation Limits (vol. % fuel)
C_2H_2/O_2	40	3.83×10^{-4}	1	18-59
C_2H_4O/O_2	40	1.2×10^{-2}	3.1×10^1	13.5-60.5
C_2H_4/O_2	33.3	7.2×10^{-2}	1.9×10^2	15-48
C_3H_6/O_2	25	2.03×10^{-1}	5.3×10^2	9-35
C_3H_8/O_2	22.2	5.77×10^{-1}	1.5×10^3	8-30.5
C_2H_6/O_2	28.6	1.07	2.8×10^3	12-38
CH_4/O_2	40	50.7	1.3×10^5	N/A
H_2/O_2	60	1.58	4.1×10^3	39-81
C_2H_2/air	12.5	1.29×10^2	3.4×10^5	6.7-21.4
C_2H_4O/air	12.3	7.62×10^3	2.0×10^7	5.3-18
C_2H_4/air	9.5	1.2×10^5	3.1×10^8	N/A
C_3H_6/air	6.6	7.55×10^5	2.0×10^9	N/A
C_3H_8/air	5.7	2.52×10^6	6.6×10^9	2.2-9.2
C_2H_6/air	7.7	5.09×10^6	1.3×10^{10}	N/A
CH_4/air	12.3	2.28×10^8	5.9×10^{11}	N/A
H_2/air	29.6	4.16×10^6	1.1×10^{10}	N/A

N/A = not available

6.3.2.2.   Detonation in non-homogeneous mixtures

The question of whether a non-homogeneous mixture can sustain a detonation wave is more relevant to the vapor cloud detonation problem than homogeneous mixtures since the composition of a vapor cloud dispersing in the atmosphere is, in general, far from homogeneous.

Experiments on the detonability of non-homogeneous mixtures are scarce. Two experiments reported in the literature may shed some light on this matter. Bull et al. (1981) investigated the transmission of detonation across an inert region in hydrocarbon-air mixtures under unconfined conditions. The transmission of a hydrocarbon-air detonation across an inert region in a tube was studied by Bjerketvedt and Sonju (1984), and Bjerketvedt, Sonju, and Moen (1986). Although apparatus for these experiments differed significantly, results are strikingly consistent. The experiments show that detonations in stoichiometric hydrocarbon-air mixtures are unable to cross a gap of pure air of approximately 0.2 m thickness. These results indicate that it is difficult for a detonation to maintain itself without contiguous fuel-air mixture. Data are lacking concerning propagation of a detonation in a contiguous heterogeneous mixture. Until such time as data are developed, a conservative assumption is that a detonation will propagate as long as the mixture is between detonability limits.

6.3.3.   Deflagration-to-Detonation Transition (DDT)

Detonation can occur in a vapor cloud through flame acceleration and transition from a deflagration. Very few accidents have been reported as vapor cloud detonations. The Port Hudson explosion in 1972, which involved a propane-air mixture within a valley, was the only one widely reported as a detonation. Initiation was suspected to have occurred inside a block building. Several other incidents with localized areas of high damage within the VCE combustion zone have been investigated where a relatively small area may have detonated, but they have not been published or investigators may not have recognized that a local detonation occurred.

Investigations by McKay et al. (1988) and Moen et al. (1989) showed that initiation of detonation in a fuel-air mixture by a burning, turbulent gas jet is possible, provided the jet is large enough. The diameter of the jet must exceed five times the critical tube diameter, which is approximately 65 times the cell size.

Experimental results are summarized and presented to correspond to the 3-dimensional (3-D), 2-dimensional (2-D) and one-dimensional (1-D) flame expansion geometries coupled with various congestion arrangements. Further, experimental results from various investigators are presented to illustrate the effects of several factors on flame acceleration.

6.3.3.1.   DDT in 1-D configurations

A large number of DDT events have been reported in 1-D experimental configurations. It has been demonstrated that flame propagation will eventually lead to detonation under 1-D flame expansion conditions (i.e., pipes, channels, etc.), even for low reactivity mixtures and without the presence of obstacles.

Introduction of obstacles into tubes greatly reduces the distance for transition to detonation.

Moen et al. (1982) reported high flame speeds and severe pressure buildup to 500 m/s and 4.0 bar for methane/air mixture, and 650 m/s and 13.9 bar for propane/air mixture in a tube measuring 2.5 m in diameter and 10 m in length with orifices. Lee et al. (1984) and Sherman et al. (1985) demonstrated that a hydrogen/air mixture could undergo a DDT at hydrogen concentrations as low as 13% (equivalence ratio of 0.36).

Equivalence ratio is the fuel-air ratio used in the test divided by stoichiometric fuel-air ratio. The 1-D experimental results are indicative of the effect of obstacles on flame acceleration, but the quantitative flame acceleration results are only applicable to 1-D configurations. Grossel (2002) discusses deflagration, detonation and DDT in pipes at length in his book on flame arresters.

6.3.3.2.   DDT in 2-D configurations

Mercx (1992) reported that a DDT in ethylene/air mixture occurred in all three large-scale TNO tests in which the mixture was 6.85% ethylene in air. The test rig was a 2-D platform configuration with dimensions of 25.4 m × 25.4 m × 1 m, and obstruction was provided by vertical tubes 0.5 m in diameter.

DDT events were realized in large-scale tests of 2-D configurations. For example, Mercx (1992) recorded DDT in 4.22% propane/air mixture in a large-scale 2-D configuration with dimensions of 25.4 × 25.4 × 1 m. Congestion was provided by vertical tubes of 0.5 m in diameter (d) with the pipe spacing (pitch P) P = 3d, area blockage ratio ABR = 50%, volume blockage ratio VBR = 14.5%, measured maximum flame speed = 575 m/s, and maximum overpressure = 55 bar.

6.3.3.3.   DDT in 3-D configurations

During the MERGE (Modeling and experimental Research into Gas Explosions) program (Mercx, 1995), it was reported that DDT occurred in 3-D configurations with ethylene-air mixtures, a pressure of around 3 bar, and with a pressure spike of 18 bar observed near the outer edge of the 4 m × 4 m × 2 m test rig in several tests. However, the MERGE test rig provided configurations up to highly congested environment, with 4.1 cm pipe obstacles oriented in three directions, which is rarely found together in industrial situations.

More recently, Thomas et al. (2003) reported the observation of DDT in ethylene-air mixtures with a wide range of fuel concentration, from lean (5.9% C_2H_4) to rich (9.3% C_2H_4). The test rig was also a 3-D configuration, 14.64 m long × 3.66 m wide × 1.83 m high (48 ft × 12 ft × 6 ft) and without any confining surfaces (other than the ground). An array of vertical circular tubes provided congestion with the pitch-to-obstacle diameter ratio of 4.5, and area and volume blockage ratios of 23% and 4.2%, respectively. These tests clearly demonstrate that high reactivity fuels can undergo a DDT in moderate levels of congestion even in the absence of any confinement at a scale that is applicable in actual process plants.

The results of the above experiments are summarized in Table 6.11.

Table 6.11. DDT in ethylene/air mixtures

Reference	Fuel vol.%	Obstacle parameters			Flame Speed V_f (max)	Over- pressure Δp (max)
		P*	ABR %	VBR %	m/s	bar
Mercx 1992, TNO 2D 25.4×25.4×1m, vertical tubes d = 0.5m	6.85	6d	50	8	1323	-----
	6.85	3d	30	8.7	-----	20.5
	6.85	3d	20	5.8	342	11.4
Mercx 1995, MERGE 3D 4×4×2m, tubes in 3 directions d = 4.1cm	6.8	4.65d	35	10	-----	10.8
	6.8	3.25d	45	20	-----	13.3
Thomas 2003, ERC 14.6×3.7×1.8m, vertical tubes d = 5.08cm	7.3	4.5d	23	4.2	1768	>6.9
	5.9	4.5d	23	4.2	>340	>6.9
	9.3	4.5d	23	4.2	>340	>6.9

P = Pitch;　ABR = Area Blockage Ratio;　VBR = Volume Blockage Ratio

* Pitch P is shown as multiples of the obstacle diameter d.

The above test results indicate that in a high reactivity fuel-air mixture, DDT can occur in unconfined situations with medium to high levels of congestion.

6.3.4.   Blast Effects Produced by Vapor Cloud Detonations

Although the detonation mode is unlikely for accidental vapor cloud explosions, the evaluation of blast effects produced by a detonating cloud is still of practical importance. First, in very rare situations, a detonation may be realized by DDT or high energy initiation. Second, detonation data are useful for validation of VCE models. Third, detonation gives the worst-case blast effects, which may be used for upper-bound consequence analysis. History indicates that accidental VCEs are typically deflagrations rather than detonations.

6.3.4.1.   Determination of detonation parameters

Detonation parameters (pressure, temperature, density, and propagation velocity at the wave front) are uniquely determined once initial thermodynamic conditions of the cloud are specified. This is because the detonation wave propagates at supersonic speed into the un-reacted mixture, which has not been disturbed by the wave propagation.

Standard computer codes, such as Gordon and McBride (1976, 1996), and Cowperthwaite and Zwister (1973, 1996), are available for the calculation of detonation parameters for any fuel-oxidizer mixture under user-specified conditions. It has been demonstrated that the results of calculations agree well with experimental measurements. Detonation properties for some fuel-air mixtures are presented in Table 6.12.

Table 6.12. Detonation properties for some stoichiometric fuel-air mixtures (McBride, 1996)

Fuel	Formula	Fuel Vol. %	Peak Detonation Pressure (bar)	Peak Detonation Temp. (K)	Detonation Velocity (m/s)
Methane	CH_4	9.5	17.19	2781	1804
Ethane	C_2H_6	5.66	17.98	2815	1803
Propane	C_3H_8	4.03	18.06	2822	1800
Butane	C_4H_{10}	3.13	18.11	2825	1798
Heptane	C_7H_{16}	1.87	17.77	2796	1784
Decane	$C_{10}H_{22}$	0.91	17.76	2794	1782
Ethylene	C_2H_4	6.54	18.36	2926	1825
Propylene	C_3H_6	4.46	18.30	2889	1811
Butylene	C_3H_6	3.38	18.30	2877	1806
Acetylene	C_2H_2	7.75	19.12	3112	1867
Propyne	C_3H_4	4.99	18.78	2999	1834
Butyne	C_4H_6	3.68	18.76	2967	1828
Butadiene	C_4H_6	3.68	18.44	2928	1812
Hydrogen	H_2	29.59	15.84	2951	1968
Benzene	C_6H_6	2.72	17.41	2840	1766
Ethanol	C_2H_5OH	6.54	17.68	2735	1773
Butanol	$C_4H_{10}O$	3.38	18.26	2819	1796
Propylene Oxide	C_3H_6O	4.99	18.43	2886	1810

6.3.4.2.   Experimental measurements of blast effects from VCE detonations

There have been a number of blast measurements made from detonating premixed gas mixtures in spherical free air or hemispherical ground configurations. Brossard, et al. (1984) collected previous data and conducted many more such tests, with gas volumes ranging from 5×10^{-4} m³ up to 1.45×10^{4} m³. Smooth curve fits for positive phase and positive plus negative phase properties are presented in Figure 6.34 and Figure 6.35.

A hemispherical charge on the ground with radius r_0 was assumed to have the same energy as a sphere with radius r_0 in free air, due to ground reflection (i.e. ground acting as a perfect reflector for hemispherical ground configurations, effectively doubling the energy). The source energy was always:

$$E = \frac{4\pi r_0^{3}}{3} \rho_f E_f$$

(Eq. 6.10)

where ρ_f and E_f are density and heat of combustion, respectively, for the specific gas mixture. The dimensionless distance is defined as:

$$\overline{R} = \frac{R}{\left(E / p_0\right)^{1/3}}$$

(Eq. 6.11)

Note that in these figures, impulse and duration parameters are not rendered non-dimensional, even though they are plotted versus non-dimensional distance. Units for dimensional parameters are shown in the figures.

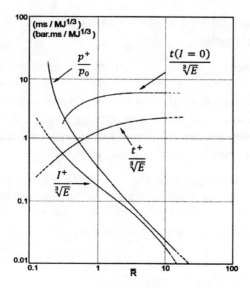

Figure 6.34. Positive phase characteristics from VCE detonations.
(Brossard et al. 1983)

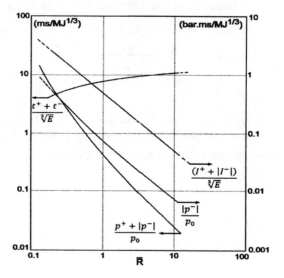

Figure 6.35. Total amplitude of characteristics from VCE detonations.
(Brossard et al. 1983)

Brossard (1983) gives the least squares fits to positive phase blast detonation properties as the following:

Positive phase scaled overpressure $\Delta p+/p_0 i$
$$\ln(\Delta p+/p_0) = -0.9126 - 1.5058\,(\ln\bar{R}) + 0.1675\,(\ln\bar{R})^2 - 0.0320\,(\ln\bar{R})^3$$
$$(Eq.\ 6.12)$$

Positive phase scaled duration $t+/E^{1/3}$
$$\ln(t^+/E^{1/3}) = +0.2500 + 0.5038\,(\ln\bar{R}) - 0.1118\,(\ln\bar{R})^2$$
$$(Eq.\ 6.13)$$

Positive phase scaled impulse $i^+/E^{1/3}$
$$\ln(i^+/E^{1/3}) = -1.5666 - 0.8978\,(\ln\bar{R}) - 0.0096\,(\ln\bar{R})^2 - 0.0323\,(\ln\bar{R})^3$$
$$(Eq.\ 6.14)$$

The above equations are valid for $0.3 \le \bar{R} \le 12$.

Based on several decades of experimental and analytical research in Russia and in other countries, Dorofeev (1995) summarized blast effects from spherical detonations and presented the following empirical equations:

For gaseous detonations:

$$\bar{P}^+ = 0.34/\bar{R}^{4/3} + 0.062/\bar{R}^2 + 0.033/\bar{R}^3 \qquad (Eq.\ 6.15)$$

$$\bar{I}^+ = 0.0353/\bar{R}^{0.968} \qquad (Eq.\ 6.16)$$

where:

$$\bar{R} = R/(E/p_0)^{1/3}, \ 0.21 < \bar{R} < 3.77$$

$$\bar{P}^+ = \Delta p^+/p_0$$

$$\bar{I}^+ = I^+ a_0/p_0^{2/3}/E^{1/3}$$

The comparison of Dorofeev empirical equations with the results from other authors are shown Figure 6.36 and in

Figure 6.37 for positive overpressure and positive impulse, respectively.

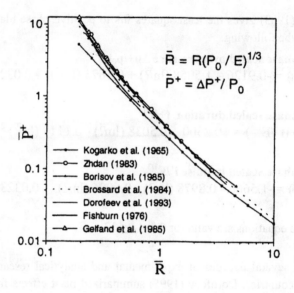

Figure 6.36. Positive overpressure versus distance for gaseous detonations.

(Dorofeev, 1995)

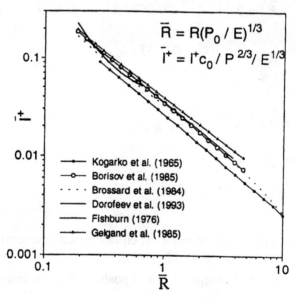

Figure 6.37. Positive impulse versus distance (c_0 is the same as a_0) (Dorofeev, 1995)

In general, blast effects from detonations of heterogeneous mixtures of vapor and droplets are less severe than that from homogeneous gaseous mixtures due to the incomplete combustion of fuel drops, and parameters of the blast waves depend on the fuel type, drop size, vapor fraction, etc. Dorofeev et al. (1995) suggested an approximation based on numerous experiments with air sprays of motor fuel. Figure 6.38 and Figure 6.39 present measurements of overpressure and impulse for near-stoichiometric mixtures. Equations for heterogeneous mist detonation are as follows:

$$\overline{P}^+ = 0.125/\overline{R} + 0.137/\overline{R}^2 + 0.023/\overline{R}^3 \qquad \text{(Eq. 6.17)}$$

$$\overline{I}^+ = 0.022/\overline{R} \qquad \text{(Eq. 6.18)}$$
$$0.3 < \overline{R} < 2.0$$

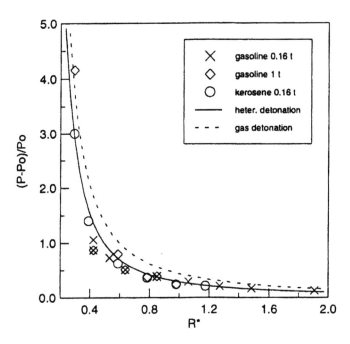

Figure 6.38. Positive overpressure versus distance for heterogeneous detonations. (Dorofeev, 1995) (R* is the same as $\overline{R}$)

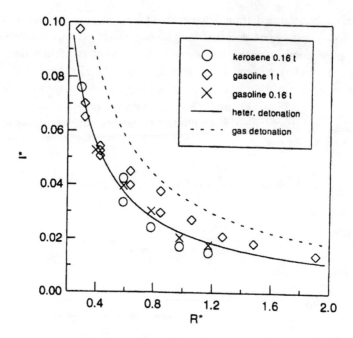

Figure 6.39. Positive impulse versus distance for heterogeneous detonations. (Dorofeev, 1995) (R* same as $\overline{R}$, I* is the same as $\overline{I}$)

6.4.   VCE PREDICTION METHODS

Blast prediction methods for vapor cloud explosions may be grouped into three categories: TNT equivalence method, numerical simulations, and blast curve methods.

The TNT equivalence method represents a VCE as a TNT explosion of equivalent explosion energy. The method is simple since the only parameter needed to represent the VCE is explosion energy, from which the TNT charge weight is calculated. TNT blast curves are then used to determine blast loads. The TNT equivalence method was developed before VCE blast curves and other methods were available. Besides the simplicity of this method, TNT blast curves are well documented. However, case studies and experimental data have revealed that blast loads predicted using an equivalent TNT charge are not representative of a VCE, overestimating pressure close to the explosion source and underestimating in the far field (Baker, 1991). It is now well understood that blast effects from

VCEs are determined not only by the explosion energy, but more importantly, by the combustion rate. Therefore, use of the TNT equivalence method is not recommended, but due to historical significance and continued use in the insurance sector, it is discussed briefly in this chapter.

The analytical methods are based on analytical solutions of the basic gas dynamic equations (conservation equations of mass, momentum, and energy). These partial differential equations can only be solved analytically by introducing some simplifications: the acoustic method, which is based on linearization of the conservation equations; and the self-similar method, which is based on the similarity principle.

The concept of self-similarity results directly from dimensional analysis, which is often the first step in treating a physical problem. Using dimensional analysis, the non-steady blast waves can be regarded as functions of just one independent variable (originally, the one-dimensional partial differential equations have two independent variables: a time variable and a space variable). Thus, the partial differential equations can be transferred into ordinary differential equations. Then, accurate solutions can be obtained by solving the ordinary differential equations. Due to the limitations brought by the assumptions, use of the analytical approach is applicable only to low flame speed VCEs (e.g., less than Mach 0.4).

Although numerical solutions of partial differential equations can be more easily obtained through advances in computational and numerical techniques in recent years, analytical solutions still play a role because they are obtained by directly solving the conservation equations, which provide the benchmark to calibrate numerical results. Analytical solutions seldom have use in chemical processing plant hazard assessments and are not discussed further in this chapter.

Numerical methods solve the gas dynamics of VCEs. Further development of numerical methods, however, is largely determined by advancements in computational fluid dynamics and computing technology. Consequently, the nature of published methods range from very simple methods capable of simulating one-dimensional, nonreactive, zero-viscosity flows to highly sophisticated methods capable of simulating the multidimensional process of premixed combustion in detail. In this section, these methods will be reviewed in increasing order of complexity.

Blast curve methods are based on VCE-specific blast curves developed using one-dimensional numerical calculations. Blast curves were developed for a range of flame speeds from slow speed deflagrations to detonations. Selection of the appropriate blast curve is based on experimental results for similar geometries and fuel reactivity. Blast curve methods are in widespread use in industry, and have been automated in a number of software packages.

6.4.1.   TNT Equivalency Method

For many years, the military has investigated the destructive potential of high explosives (Robinson 1944, Schardin 1954, Glasstone and Dolan 1977, and Jarrett 1968). Therefore, relating the explosion severity of an accidental VCE to an equivalent TNT charge was an understandable approach before VCE-specific methods were developed. Damage patterns observed in many historical VCE incidents have been related to equivalent TNT-charge weights.

The need to quantify the potential explosion severity of fuels arose long before the mechanisms of blast generation in vapor cloud explosions were fully understood. The use of TNT-equivalency methods for blast prediction purposes is quite simple. The available explosion energy in a vapor cloud is converted into an equivalent charge weight of TNT with the following formula:

$$W_{TNT} = \alpha_e \frac{W_f H_f}{H_{TNT}} = \alpha_m W_f \qquad \text{(Eq. 6.19)}$$

where:

W_f	=	the mass of fuel involved	(kg)
W_{TNT}	=	equivalent mass of TNT or yield	(kg)
H_f	=	heat of combustion of the fuel	(J/kg)
H_{TNT}	=	TNT heat of detonation	(J/kg)
α_e	=	TNT equivalency based on energy	
α_m	=	TNT equivalency based on mass	
units	=	any self-consistent unit set	

TNT equivalency is also called equivalency factor, yield factor, efficiency, or efficiency factor.

If the equivalent mass of TNT is known, the blast characteristics, in terms of the peak side-on overpressure of the blast wave, can be derived for varying distances from the explosion. This is done using a chart containing a scaled, graphical representation of experimental data. Presented in Figure 6.40 are side-on parameters for TNT hemispherical surface bursts from Kingery and Bulmash (1984) and adapted by Lees (1996) to SI units. The surface burst blast curves inherently account for ground reflection. Free-air TNT blast curves are also available Lees (1996) and other sources that do not take into account ground reflection. See Section 4.5.3 for a discussion of usage of free-air versus surface burst blast curves.

TNT-equivalency methods are the simplest means of modeling vapor cloud explosions. TNT equivalency can be regarded as a conversion factor by which the available heat of combustion can be converted into blast energy. In one sense, TNT equivalency expresses the efficiency of the conversion process of chemical energy (heat of combustion) into mechanical energy (blast).

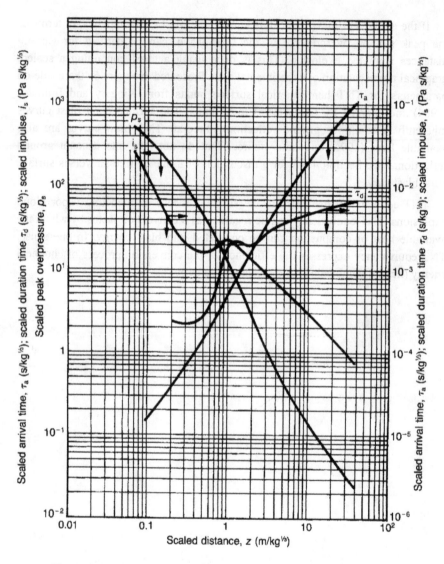

Figure 6.40. Side-on blast parameters for a TNT hemispherical surface burst.
(Lees, 1996 after Kingery and Bulmash, 1984)

The theoretical maximum efficiency of conversion of heat of combustion into blast for a stoichiometric, hydrocarbon-air detonation is equal to approximately 40%. If the blast energy of TNT is equal to the energy brought into the air as blast by a TNT detonation, a TNT equivalency of approximately 40% would be the theoretical upper limit for a gas explosion process under atmospheric conditions. However, the initial stages of shock propagation in the immediate vicinity of a detonating TNT charge are characterized by a high dissipation rate of energy. If this loss of energy is taken into account, the TNT equivalency for a gas detonation at lower blast overpressure levels is expected to be substantially higher than 40%.

Accidental vapor cloud explosions do not involve the full amount of available fuel. Therefore, practical values for TNT equivalencies of vapor cloud explosions are much lower than the theoretical upper limit. Reported values for TNT equivalency, deduced from the damage observed in many vapor cloud explosion incidents, range from a fraction of one percent up to some tens of percent (Gugan 1978 and Pritchard 1989). For most major vapor cloud explosion incidents, however, TNT equivalencies have been deduced to range from 1% to 10%, based on the heat of combustion of the full quantity of fuel released. Apparently, only a small part of the total available combustion energy is generally involved in actual explosive combustion.

Over the years, many authors, companies, and authorities have developed their own procedures and recommendations on the application of TNT equivalency method. Some of the differences in these procedures include the following:

- *The portion of fuel that should be included in the calculation:* The total amount released; the amount flashed; the amount flashed times an atomization factor; or the flammable portion of the cloud after accounting for dispersion over time.
- *The value of TNT equivalency:* A value based on an average deduced from observations in major incidents; or a safe and conservative value (whether or not dependent on the presence of partial confinement/obstruction and nature of the fuel).
- *The TNT blast data used:* A substantial scatter in the experimental data on high-explosive blast can be observed which is due to differences in experimental setup. Although often referenced differently, most recommendations can be traced back to ground burst data by Kingery and Pannill (1964).
- *The energy of explosion of TNT:* Values range from 1800 to 2000 Btu/lb, which correspond to 4.19 to 4.65 MJ/kg. Baker (1983) provided a TNT heat of detonation of 4.52 MJ/kg, and Crowl (2003) used 4.60 MJ/kg.

Below are examples of some of the many different approaches used. TNT equivalencies given by the sources identified below are based upon averages deduced from damage observed in a limited number of major vapor cloud explosion incidents:

- *Brasie and Simpson (1968) and Brasie (1976):* 2%-5% of the heat of combustion of the quantity of fuel spilled.
- *The UK Health & Safety Executive (1979 and 1986):* 1-3% of the heat of combustion of the quantity of fuel present in the cloud.
- *Industrial Risk Insurers (1990):* 2% of the available energy of the quantity of fuel spilled.
- *Factory Mutual Research (1990):* 5%, 10%, and 15% of the heat of combustion of the quantity of fuel present in the cloud for Class I (relatively nonreactive materials such as propane, butane, and ordinary flammable liquids), Class II (moderately reactive materials such as ethylene, diethyl ether, and acrolein), and Class III (highly reactive materials such as acetylene), respectively.
- *British Gas (Harris and Wickens 1989):* 20% TNT equivalency of available energy in the entire flammable cloud, which represents the explosive potential of the portion of the cloud in a congested/obstructed region. If transition to detonation is possible with higher reactivity fuels, the approach might be inappropriate because of involvement of portions of the cloud outside of congestion.

These figures can be used for predictive purposes to extrapolate "average major incident conditions" to situations under study, provided the actual conditions under study correspond reasonably well with "average major incident conditions." Such a condition may be broadly described as a spill of some tens of tons of a hydrocarbon in an environment with local concentrations of obstructions and/or partial confinement, for example, the site of an "average" refinery or chemical plant with dense process equipment or the site of a railroad marshaling yard with a large number of closely parked rail cars. It must be emphasized that the TNT equivalencies listed above should not be used in situations in which "average major incident conditions" do not apply.

To demonstrate the general procedure in applying TNT-equivalency methods in this work, one of the many methods, namely, that recommended by the UK Health & Safety Executive (HSE 1979; HSE 1986), is followed.

6.4.1.1.   Determine Charge Mass

In the HSE method, the equivalent-charge mass of TNT is related to the total quantity of fuel in the cloud; it can be determined according to the following stepwise procedure:

Determine the flash fraction of fuel on the basis of actual thermodynamic data. Eq. (6.20) provides a method of estimating the flash fraction.

$$F = 1 - \exp\left[\frac{-Cp\Delta T}{L}\right]$$ (Eq. 6.20)

where:

F	=	flash fraction	(–)
Cp	=	mean specific heat	(kJ/kg/K)
ΔT	=	temperature difference between vessel temperature	(K)
		and boiling temperature at ambient pressure	
L	=	latent heat of vaporization	(kJ/kg)
exp	=	base of natural logarithm (2.7183)	(–)

The mass of fuel Wf in the cloud is equal to the flash fraction times the quantity of fuel released. To allow for spray and aerosol formation, the cloud inventory should be multiplied by 2. (The mass of fuel in the cloud cannot exceed the total quantity of fuel released.)

The equivalent-charge mass of TNT can now be calculated as follows:

$$W_{TNT} = \alpha_c \frac{W_f H_f}{H_{TNT}}$$ (Eq. 6.21)

where:

W_{TNT}	=	equivalent mass of TNT	(kg)
W_f	=	mass of fuel in the cloud	(kg)
H_f	=	heat of combustion fuel	(MJ/kg)
H_{TNT}	=	blast energy TNT = 4.68 MJ/kg	
α_e	=	TNT-equivalency/yield factor = 0.03	(–)

6.4.1.2.   Determine Blast Effects

In Figure 6.40, the side-on blast wave peak overpressure produced by a detonation of a TNT charge is graphically represented as dependent on the Hopkinson-scaled distance from the charge. The side-on blast peak overpressure at some real distance (R) of a charge of a given weight (W_{TNT}) is found by calculating:

$$\bar{R} = \frac{R}{W_{TNT}^{1/3}}$$
(Eq. 6.22)

where:

$\bar{R}$	=	Hopkinson-scaled distance	$(m/kg^{1/3})$
W_{TNT}	=	charge mass of TNT	(kg)
R	=	real distance from charge	(m)

If the scaled distance $\bar{R}$ is known, the corresponding peak side-on overpressure, side-on impulse, positive phase duration, and time of arrival can be read from the chart in Figure 6.40.

6.4.2.   VCE Blast Curve Methods

There are three basic elements of blast curve methods: the blast curves, determination of blast strength or severity, and energy definition. First, there must be a set of blast curves as the basic tool to predict blast parameters (overpressure, impulse, duration, etc.). For an ideal explosion source, e.g., high explosives, the relationship between blast parameter and distance can be presented as a single curve in energy scaled coordinates, because an ideal explosion source can be characterized by a single parameter: its explosion energy. However, a vapor cloud explosion is a non-ideal explosion source, which cannot be defined by source energy alone. Both the explosion energy and the energy release rate are required to define the explosion source. Consequently, the relationship between blast parameters and distance for VCE blast must be represented by a family of curves instead of a single curve in energy scaled coordinates.

The common feature of the various sets of VCE blast curves is that they are all based on the explosion of a gas charge and are results of one-dimensional numerical calculations. The use of the explosion of a gas charge to replace the detonation of a TNT charge has taken into account the non-ideality of VCE as an explosion source, and therefore has greatly improved the accuracy in VCE blast prediction. The one-dimensional assumption indicates the symmetrical representation of blast effects (spherical or hemispherical). Consequently, the directional effects, non-uniform confinement and congestion, inhomogeneous mixtures, ignition location, and other non-ideal situations in real world scenarios are not considered in a simple application of the blast curves.

VCE blast curve methods were first published in the late 1970s and have been in continuous development since. The two predominant sets of blast curves are the Baker-Strehlow-Tang (BST) and the TNO Multi-Energy Method. Baker and Strehlow first published their blast curves in 1983 (Baker, 1983). TNO developed a piston-driven shock model, presented by Pasman (1976) and Wiekema (1980) in the TNO Yellow Book. TNO superseded the piston model with the Multi-Energy Method.

Shell developed the Confinement Assessment Method (CAM) (Cates, 1991; Puttock, 1995; Puttock, 2001). CAM does not use a blast curve, per se. Rather, an initial blast pressure is determined at the edge of the explosion combustion volume. Pressure decays with distance using a decay law that closely resembles the decay predicted with a blast curve.

The second element of a blast curve method is the determination of initial blast strength. Since a family of blast curves is used to represent VCE blast effects, a single curve in the family must be selected for a given scenario. Therefore, each curve in a family is given a label. The BST curves are labeled by flame speed, while the TNO curves are labeled by explosion strength. Once the flame speed or the explosion strength is determined, the initial blast is determined and the particular curve in a family can be selected. The relation between blast parameter and distance can be determined just by following the blast decay curve. The CAM method uses fuel reactivity, confinement assessment, and geometric parameters to determine initial blast pressure in lieu of a blast curve.

The third element of a blast curve method is the definition of explosion energy. The energy term E must be defined in order to calculate energy scaled standoff $\bar{R}$ and other energy scaled blast parameters. The energy term represents the heat of combustion corresponding to the portion of a cloud that contributes to the blast wave. In the BST, Multi-Energy, and CAM methods, the energy term is defined by the volume of the portion of cloud in a partially confined and/or congested region multiplied by the heat of combustion for the fuel per unit volume of mixture. For a large release in a small process unit, the flammable cloud may be larger than the process unit, and only the cloud volume in congestion would be considered for a deflagration. Conversely, a small release in a large process unit may result in the entire flammable cloud residing in the process unit. Regardless of scenario, the amount of fuel involved in a VCE will be less than the total amount of fuel released due to fuel outside the flammability limits at the time of the explosion, oxygen availability within the congested volume, and flammable fuel-air mixture that is outside of congestion.

6.4.3.   TNO Multi-Energy Method

The Multi-Energy Method (Van den Berg 1985; Van den Berg et al. 1987) recognized that only a portion of the cloud, not the entire volume of a vapor cloud, contributed to the blast effects. It has been demonstrated from damage patterns observed in major accidental vapor cloud explosions and experimental measurements that there is hardly any correlation between the total quantity of combustion energy in a flammable vapor cloud and the VCE blast effects.

Van den Berg (1985) and Van den Berg et al. (1987) stated that both the scale and strength of a blast are unrelated to total fuel quantity present in a cloud. Blast parameters are determined by the size and nature of partially confined and obstructed regions within the cloud. Reactivity of the fuel-air mixture is of an additional influence. Underlying considerations in the Multi-Energy method for VCE blast modeling are:

- A fuel-air cloud originating from an open air, accidental release is very unlikely to generate an overpressure.
- Damaging blast waves are produced by the portion of a vapor cloud in partially confined and/or obstructed environments, and the remaining portion of a cloud containing a flammable vapor-air mixture will burn out slowly without contributing significantly to blast.
- The Multi-Energy concept defines a vapor cloud explosion as a number of sub-explosions corresponding to the various sources of the blast in the cloud.

The basic tool used in the Multi-Energy method is blast curves that present blast parameters as a function of distance, as shown in Figure 6.41. In addition, the Multi-Energy method blast curves depict blast wave shape, with solid lines representing high strength shock waves and dashed lines representing low strength pressure waves that may steepen to shock waves at farther distances. In the development of the Multi-Energy method blast curves, steady flame-speed gas explosions were numerically simulated (Van den Berg, 1980).

Figure 6.41 represents the blast characteristics of a hemispherical fuel-air charge of radius R_0 on the ground surface, derived for a fuel-air mixture with a heat of combustion of 3.5×10^6 J/m^3. The figure presents only side-on peak overpressure (ΔP_s) and the positive-phase blast-wave duration ($\bar{t}_+$) as a function of scaled distance from the blast center ($\bar{R}$). The blast parameters are fully nondimensionalized according to Sachs scaling, which uses charge combustion energy (E), ambient atmosphere, pressure (P_0), and ambient speed of sound (c_0) in calculation of scaled standoff ($\bar{R}$), scaled overpressure, (ΔP_s), and scaled duration ($\bar{t}_+$). The Sachs method of scaling takes into account the influence of atmospheric conditions such as altitude. Moreover, Sachs scaling allows the blast parameters to be read in any consistent set of units.

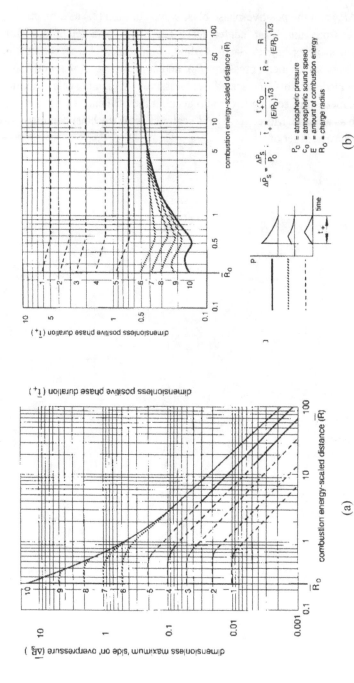

Figure 6.41. Multi-energy method positive-phase side-on blast overpressure and duration curves.

Initial blast strength in Figure 6.41 is represented by a number ranging from 1 (very low strength) up to 10 (detonative strength). In addition, this figure gives a rough indication of the blast-wave shape, which corresponds to the characteristic behavior of a gas-explosion blast. Pressure waves, produced by fuel-air charges of low strength, result in an acoustic overpressure decay behavior and a constant positive-phase duration. On the other hand, shock waves in the vicinity of a charge of high initial strength exhibit more rapid overpressure decay and a substantial increase in positive-phase duration with distance. Eventually, the high-strength blast decays to behavior approximating acoustic in the far field. Another significant feature is that, at a distance larger than about 10 charge radii from the center, a fuel-air charge blast is more or less independent of initial strength for values of 6 (strong deflagration) and above. As can be seen in Figure 6.41, blast curves 6, 7, 8 and 9 converge to the number 10 (detonation) curve.

In the application of the multi-energy concept, a particular VCE hazard is determined primarily by the plant geometry into which the flammable cloud disperses. An explosion hazard assessment requires an evaluation of the potential for generating blast from the plant geometry in the flammable cloud.

The procedure for applying the Multi-Energy method to model vapor cloud explosion blast as extracted from TNO guidance documents (van den Bosch, 2005) can be divided into the following steps:

Step 1. Apply constraints

Assume that blast is a deflagration. (The basis for this assumption is that an unconfined vapor cloud detonation is extremely unlikely.). Recall that the Multi-Energy Method is a simplification of reality that does not take directional effects due to inhomogeneities or deviations from point-symmetry into account.

Step 2. Determine cloud size

Determine the flammable mass of vapor in a cloud that could be formed after an accidental release. The volume of the flammable vapor cloud at stoichiometric concentration, c_s, is given by $V_c = Q_{ex}/(\rho \cdot c_s)$, where Q_{ex} is the flammable mass and ρ is the gas density.

Step 3. Identify potential blast sources

Potential sources of strong blast include:
- Extended spatial configuration of objects such as process equipment in chemical plants or refineries and stacks of crates or pallets;

- Spaces between extended parallel planes, for example, those beneath closely parked cars in parking lots, and open buildings, for example, multistory parking garages;
- Spaces within tube-like structures, for example, tunnels, bridges, corridors, sewage systems, culverts;
- An intensely turbulent fuel-air mixture in a jet resulting from release at high pressure.

The remainder of the fuel-air mixture in the cloud is assumed to produce a blast of minor strength. For a blast strength of 10 (detonation), the energy is defined by the volume within obstructed/confined areas.

Step 4. Define obstructed regions

The definition of obstructed regions is a multi-step process. The procedure is as follows:

Step 4a. Break down structures into basic geometrical shapes:
- Cylinders with length l_c and diameter d_c
- Boxes with sides of length b_1, b_2, and b_3, and
- Spheres with diameter d_s

Step 4b. Assume an ignition location. After ignition of a flammable cloud in a congested area the flame will travel outward so the orientation with respect to the flame propagation direction of each obstacle is known.

Step 4c. Determine obstacle orientation. Take the smallest dimension oriented in a plane perpendicular to the flame propagation direction to be D_1. Then:
- $D_1 = l_c$ or d_c for a cylinder
- $D_1 =$ smallest of b_1 and b_2, b_2 and b_3, or b_1 and b_3 for a box
- $D_1 = d_s$ for a sphere

Take the obstacle dimension parallel to the flame propagation direction to be D_2.

Step 4d. Build-up obstructed region. An obstacle A belongs in an obstructed region if the distance from the center of obstacle A to the center of any obstacle B already determined to be in the obstructed region is smaller than 10 times D_1 or 1.5 times D_2 of obstacle B. If the distance between the outer boundary of the obstructed region and the outer boundary of obstacle A is larger than 25 meters, then obstacle A does not belong in the obstructed region.

Step 4e. Define a box containing the obstructed region. The obstructed region is defined as a box that contains all of the obstacles determined to be in the obstructed region. This includes the space between a confining surface and an obstructed region if the distance between that surface and any obstacle in the obstructed region is less than 10 times D_1 or 1.5 times D_2 of that obstacle. The box should exclude parts of cylinders or boxes that clearly do not belong to the obstructed region, such as upper parts of chimneys, distillation columns (vertically oriented cylinders), or pipes (horizontally oriented cylinders) connecting process units that may each potentially be separate obstructed regions. These excluded parts may themselves form separate obstructed regions.

Step 4f. If appropriate, subdivide into multiple boxes. If the box defined in Step 4e contains free space that is not included in the obstructed region defined in Step 4d, subdivision into multiple directly adjacent boxes is acceptable for reducing the volume of the obstructed region.

Step 4g. Define additional obstructed regions if appropriate. If all obstacles present are not within the previously defined obstructed region, additional obstructed regions within the vapor cloud can be defined. If there are multiple obstructed regions, the region in which ignition occurs is called the "donor" region and other regions are called "acceptor" regions. The direction of flame propagation, which is required for the orientation of the obstacles in the acceptor region, depends on the orientation of the acceptor region with respect to the donor region.

If separate obstructed regions are located close to each other, they may be initiated nearly simultaneously. The coincidence of their blast effects in the far field is likely and their respective blasts should be superimposed. As of the most recent Yellow Book (van den Bosch, 2005), there is no clear guidance to the minimum distance between donor and acceptor regions at which they can be assumed to be separate explosions. The Yellow Book methodology given to build-up a congested region, given in Step 4d suggests that a Critical Separation Distance of 10 obstacle diameters or 25 meters can be used; however, van den Berg and Mos (2005) reported from the RIGOS testing that this is not always conservative, particularly at high explosion overpressures. The final report of the GAMES project (Mercx, 1998) provides the following *preliminary* guidance:

- The Critical Separation Distance around a potential blast source is equal to half of its linear dimension in each direction. If the distance between potential blast sources is larger, the sources should be modeled as separate blasts. If the distance is smaller, the two potential sources should be modeled as a single superimposed blast as suggested in Step 10 below.

The preliminary guidance from the RIGOS report (van den Berg and Mos, 2005) can be summarized as follows:

- Critical Separation Distance between obstructed regions equal to ½ of the donor dimension should be observed for any donor explosion overpressure of more than 100 kPa.
- Critical Separation Distance between obstructed regions equal to ¼ of the donor dimension should be observed for donor explosion overpressures of less than 10 kPa.
- Linear interpolation between these two values is proposed.
- Connecting obstacle configuration of sufficient cross-sectional area between two obstructed regions may substantially increase the Critical Separation Distance.

Step 4h. Determine the maximum part of the cloud V_{gr} that can be inside the obstructed regions. This is the sum of the free space calculated for each obstructed region. The free volume of an obstructed region, V_r, the volume of the box minus the volume occupied by the obstacles. If it is not possible to determine the volume occupied by the obstacles, assume that V_r is equal to the total volume of the box.

Step 4i. Calculate the volume V_0 of the unobstructed part of the vapor cloud according to: $V_0 = V_c - V_{gr}$.

Step 4j. Calculate the energy E [J] of each region, obstructed as well as unobstructed, by multiplying V_{gr} and V_0 by the combustion energy per unit volume. A typical value for the heat of combustion of an average stoichiometric hydrocarbon-air mixture is 3.5×10^6 J/m^3 (Harris 1983). Note that the flammable mixture may not fill an entire region. In this case, V_{gr} and V_0 can be reduced appropriately.

TNO guidance for selection of explosion energy and source strength proved to be very conservative ("The safe and conservative approach recommended was felt unsatisfactory.") (Mercx, 1998). TNO's simplified guidance recommended using full combustion energy in the turbulence generative conditions of the cloud and maximizing the source strength. Additional recommendations were that, if a user was not satisfied with the conservative approach because it leads to unacceptable overestimates of blast overpressure, he may refine his approach by seeking correlation with experimental data. Overestimates with the conservative approach and difficulty relating plant geometries to test configurations prompted development of additional guidance. The GAME project was undertaken (Eggen, 1998). Specific guidance was developed in the form of correlations. A relation was derived for a set of parameters describing the obstacle configuration and the fuel, which leads to an estimate of overpressure in the VCE. The overpressure estimate is used to determine an explosion strength (1 to 10).

The follow-up GAMES project investigated the applicability of the GAME guidance to realistic cases (Mercx, 1998). Emphasis in the GAMES project was on the determination of the Volume Blockage Ratio and Average Obstacle Diameter parameters. The main finding was that a safe approach in most situations is to apply the procedure of the Yellow Book for the determination of the volume of the obstructed region in combination with the hydraulic average obstacle diameter and a flame path length equal to the radius of a hemisphere with a volume equal to the volume of the obstructed region. Taking 100% of the full combustion energy originally present within the obstructed region assuming stoichiometry is conservative for low overpressures; the "efficiency" is less than 20% for source overpressures less than 0.5 bar. For clouds smaller than the obstructed region, GAMES recommends taking an efficiency factor of 100%.

Step 5. Estimate the source strength or class number for each region

A safe and most conservative estimate of the strength of the sources of strong blast can be made if a maximum strength of 10 is assumed. However, a source strength of 7 seems to more accurately represent actual experience. Furthermore, for scaled distances greater than 0.9 (side-on overpressures below about 0.5 bar), there is no difference in predicted overpressure for source strengths ranging from 7 to 10 since the blast curves converge to the 10 curve.

The blast resulting from the remaining unconfined and unobstructed parts of a cloud can be modeled by assuming a low initial strength. For extended and quiescent parts, assume minimum strength of 1. For more nonquiescent parts, which are in low-intensity turbulent motion, for instance, because of the momentum of a fuel release, assume a strength of 3.

If such an approach results in unacceptably high overpressures, a more accurate estimate of initial blast strength may be determined from the growing body of experimental data on gas explosions , or by performing an experiment tailored to the situation in question.

Another possibility is the application of numerical simulation by use of advanced computational fluid dynamic codes, such as FLACS (van Wingerden et al., 1993), or AutoReaGas (Van den Berg 1989), outlined in Section 6.4.6.

Definition of initial blast strength is a major research need for which work is ongoing. Further discussion is given below.

Step 6. Combination of obstructed regions

If more than one obstructed region has to be considered, define an additional blast source (obstructed region) by adding all of the energies of the separate blast sources together. Determine a center for this new blast source by considering the centers of each of the separate blast sources and their energies (e.g. determine an energy-weighted average location).

Step 7. Location of the unobstructed part of the vapor cloud

Determine the center of the unobstructed part of the vapor cloud. As in Step 6, an energy-weighted average location for the unobstructed regions can be calculated.

Step 8. Calculate equivalent radius

The blast from each source should be modeled as the blast from an equivalent hemispherical fuel-air charge of volume E/E_v m^3 (note that $E_v=3.5$ MJ/m3 is an average value applicable to most hydrocarbons at stoichiometric concentration). The radius, r_0, of each blast source can be determined from:

$$r_0 = \left(\frac{3E}{2\pi E_v} \right)^{1/3} \qquad \text{(Eq. 6.23)}$$

Step 9. Calculate blast parameters

Once the energy quantities E and the initial blast strengths of the individual blast sources are estimated, the Sachs-scaled blast side-on overpressure and positive-phase duration at some distance R from a blast source can be read from the blast charts in Figure 6.41. To calculate the Sachs-scaled distance:

$$\overline{R} = \frac{R}{(E/P_0)^{1/3}}$$

(Eq. 6.24)

where:

$\overline{R}$	=	Sachs-scale distance from charge	(–)
R	=	distance from charge	(m)
E	=	charge combustion energy	(J)
P_0	=	ambient pressure	(Pa)

The blast side-on overpressure and positive-phase duration can be calculated from the Sachs-scaled quantities:

$$P_s = \Delta\overline{P}_s \cdot P_0$$

And

$$t_+ = \overline{t}_+ \left[\frac{(E/P_0)^{1/3}}{c_0} \right]$$

(Eq. 6.25)

where:

P_s	=	side-on blast overpressure	(Pa)
$\Delta\overline{P}_s$	=	Sachs-scaled side-on blast overpressure	(–)
P_0	=	ambient pressure	(Pa)
t_+	=	positive-phase duration	(s)
$\overline{t}_+$	=	Sachs-scaled positive-phase duration	(–)

E =   combustion energy of the explosion source   (J)

a_0 = is ambient speed of sound   (m/s)

Note: any consistent set of units can be used

Step 10. Handling of multiple obstructed regions

If separate blast sources are located close to one another, they may be initiated nearly simultaneously and their effect in the far-field may be cumulative. The combustion energy for each of these sources should be summed. The blast parameters at a given stand-off distance determined for the blast source as defined in Step 6 should be used. If the distance between two obstructed regions exceeds the Critical Separation Distance (see Step 4g), they should be treated as independent blast sources.

Step 11. Construct blast history at a specific location

The blast history at a given location consists of the blast parameters and blast shape determined for the unobstructed region superposed onto the blast parameters and shape determined for the obstructed region.

For low ignition energy and an open (3-D) environment, the GAME relation is:

$$P_{max} = 0.84 \left(\frac{VBR * L_p}{D} \right)^{2.75} S_L^{2.7} D^{0.7} \qquad \text{(Eq. 6.26)}$$

whereas the relation for low ignition energy and confinement between parallel planes (2-D) is:

$$P_{max} = 3.38 \left(\frac{VBR * L_p}{D} \right)^{2.25} S_L^{2.7} D^{0.7} \qquad \text{(Eq. 6.27)}$$

where:

P_{max} =   overpressure in the VCE   (bar)

VBR =   the volume blockage ratio of the vapor cloud,
          i.e. the portion of volume occupied by obstacles   (–)

L_p =   flame path length, the longest distance for the point of
          ignition to an outer edge of the obstacle configuration   (m)

D =   the average obstacle diameter within the vapor cloud   (m)

S_L =   the theoretical laminar flame speed of the fuel   (m/s)

P_{max} from the above correlations is normalized by the ambient pressure to give the pressure at r_0 and source strength depicted in the Multi-Energy Method blast curves, Figure 6.41. Note that the average obstacle diameter can be calculated by several methods. Mercx (1998) found that the hydraulic average diameter in combination with a flame path length equal to the radius of a hemisphere with a volume equivalent to the volume of the obstructed regions provides a safe upper bound in most situations and scenarios.

The aspect ratio of the obstructed region of interest provides guidance for the choice of correlation above (Mercx, 1998). Take the length of the obstructed region to be L, the width W, and the height H. If a confining plane (other than the ground) is present, it is advisable to use the 2-D correlation if:

- L > 5H
- W > 5H
- The dimensions of the confining plane are of about the same size as L and W.

The MERGE data set (Mercx, 1995) was the basis for Eq. (6.26). The MERGE data was the upper bound for experimental data at that time. The MERGE data was obtained with a homogeneous configuration of three-directional tubular obstacles. Any lack of homogeneity in the obstacle distribution was found to justify the choice of a lower explosion overpressure (Mercx, 1998). Experimental data sets from Hjertager (1993) and Harrison and Eyre (1986, 1987) can be used as an indication of the overpressure to be expected in cases where tubular obstacles are predominately two- and one-directional, respectively. These data sets are shown in Figure 6.42 in comparison to the MERGE data. As can be seen in Figure 6.42, the Hjertager and Harrison/Eyer data are often an order of magnitude lower that MERGE for comparable scale and level of congestion.

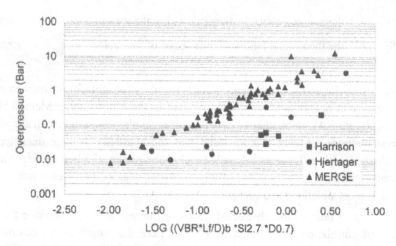

Figure 6.42. Observed overpressures from three datasets correlated to the parameter combination in the GAME relation. (Mercx, 2000)

6.4.4.   Baker-Strehlow-Tang (BST) Method

A one-dimensional numerical study was performed by Strehlow et al. (1979) to predict blast waves generated by constant velocity and accelerating flames propagating in a spherical geometry. This study resulted in the generation of plots of dimensionless overpressure and positive impulse as a function of energy-scaled distance from the cloud center. Baker et al. (1983) compared the Strehlow et al. (1979) curves to experimental data and developed the method for blast prediction to apply in research programs and accident investigations.

The numerical calculations for the original Strehlow curves were only carried out to a scaled Sach's distance of about two and extrapolated to ten due to the limitations on computing capabilities at that time. Tang et al. (1996, 1998) compared Strehlow curves with the results from other numerical calculations and experimental measurements and found that Strehlow supersonic curves decayed too slowly in the far field and departed significantly from other curves.

Tang and Baker improved the blast curves with advanced numerical modeling (Tang, 1996). These blast curves were named the "Baker-Strehlow-Tang" (BST) blast curves to distinguish them from the original Strehlow curves. Similar to the original Strehlow curves, the BST curves were based on one-dimensional numerical calculations in a Lagrangian coordinate system. However, several approaches were employed to improve the numerical results. One approach was the adoption of a shock recovery technique in which the peak pressure was corrected at each time step using fundamental shock relationships to prevent dissipation of blast pressure at each time step. Another approach was to optimize the numerical calculations by minimizing the influence of the artificial viscosity term on the pressure peak while still maintaining numerical stability. By optimizing numerical parameters and improving numerical techniques, BST curves are close to empirical results but still conservative in maintaining the slowest blast wave decay among numerical models.

BST curves are labeled by apparent flame speed, M_f, to be consistent with the observation in experimental measurements. A comprehensive set of blast curves were developed including positive and negative peak overpressures, positive and negative impulses, time durations for positive and negative phases, time of arrival for the wave front, and maximum particle velocity versus standoff distance.

The positive phase BST blast curves are presented in Figure 6.43 to

Figure 6.44, and negative phase parameters are presented in Figure 6.45 to

Figure 6.46. The blast parameters and distance are non-dimensionalized using Sach's scaling law as follows:

$$\overline{P} = \frac{p - p_0}{p_0} \qquad\qquad \overline{P}^- = \frac{|p - p_0|}{p_0}$$

$$\bar{i} = \frac{i a_0}{E^{1/3} p_0^{2/3}} \qquad\qquad \bar{i}^- = \frac{i^- a_0}{E^{1/3} p_0^{2/3}}$$

where:

p_0 =   atmospheric pressure

a_0 =   the acoustic velocity at ambient conditions

p =   the absolute peak pressure

R =   stand-off distance

E =   the explosion energy

i =   specific impulse for positive phase

i^- =   specific impulse for negative phase

Note: any consistent set of units can be used

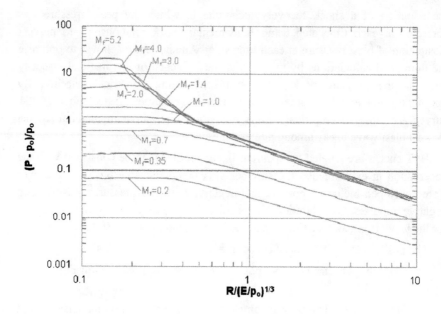

Figure 6.43. BST positive overpressure vs. distance for various flame speeds.

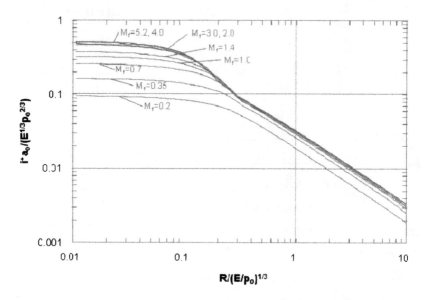

Figure 6.44. BST positive impulse vs. distance for various flame speeds.

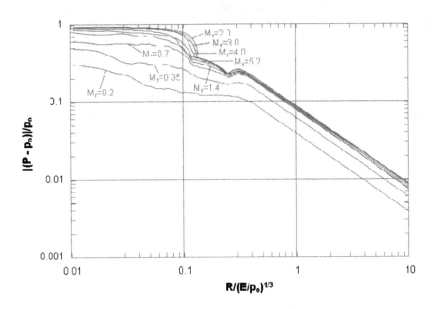

Figure 6.45. BST negative overpressure vs. distance for various flame speeds.

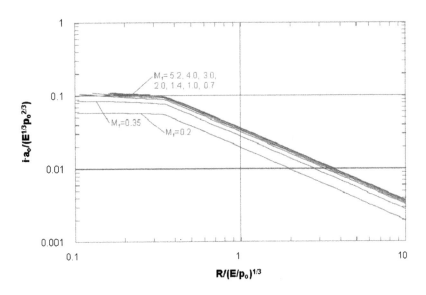

Figure 6.46. BST negative impulse vs. distance for various flame speeds.

BakerRisk developed guidance for determination of maximum flame speed based on empirical data as an objective means to select the appropriate blast curve using M_f. Initial guidance was provided by Baker (1994) and subsequently updated (Baker 1997, Pierorazio 2004). Experimentation is ongoing to improve the guidance on flame speed determination.

Pierorazio et al. (2004) reviewed the publications since 1995 and conducted a series of medium scale VCE tests and reported that the first edition of the BST flame table was not conservative for unconfined but congested configurations. Medium and large scale test data on unconfined but congested configurations became available subsequent to initial development of BST guidance. The BST flame correlation proposed by Pierorazio et al. (2004) is provided in Table 6.13. Combinations of confinement, congestion and reactivity for which there is potential for a DDT are designated as such in Table 6.13. In application of the BST method, the Mach 5.2 (detonation) blast curve would be selected for a DDT.

Table 6.13. BST flame speed correlations (flame speed Mach no. M_f)
(Pierorazio et al. 2004)

Confinement	Reactivity	Congestion		
		High	Medium	Low
2-D	High	DDT	DDT	0.59
	Medium	1.6	0.66	0.47
	Low	0.66	0.47	0.079
2.5-D	High	DDT	DDT	0.47
	Medium	1.0	0.55	0.29
	Low	0.50	0.35	0.053
3-D	High	DDT	DDT	0.36
	Medium	0.50	0.44	0.11
	Low	0.34	0.23	0.026

The BST flame table includes the 2.5-D confinement category introduced by Baker et al. (1997) and updated by Pierorazio (2004) to be used when blockage partially prevents a flame front from expanding in one dimension (Figure 6.47) such as piperacks with tightly packed pipes, partial coverage with intermediate solid platforms and lightweight roofs or frangible panels that dislodge with relatively low overpressure. As described, the 2.5-D values are obtained by taking a simple average between the 2-D and 3-D confinement values for the same congestion and fuel reactivity.

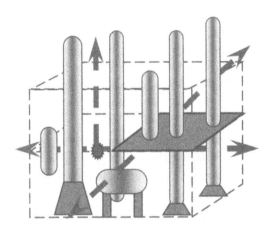

Figure 6.47. Quasi two dimensional (2.5-D) flame expansion geometry.
(courtesy of FM Global)

Earlier versions of the BST flame speed correlation (Baker, 1994; Baker 1997) included 1-D confinement. The 1-D entries have been deleted since the maximum flame speed achieved in pipe-like configuration is a function of the length-to-diameter ratio in addition to pipe geometry (elbows, tees, etc.), fuel reactivity, and internal congestion level. It should be noted that 1-D confinement has a very strong effect on the feedback mechanism for flame acceleration. As a result, many fuels are able to achieve DDT in a long tube even under low congestion. Therefore, it is recommended that 1-D geometries be treated as special cases with consideration of length-to-diameter ratio, tube geometry, congestion, and fuel reactivity. The possibility of DDT should be evaluated.

The energy term E must be defined to use the BST method. The energy term represents the heat released by that portion of the cloud contributing to the blast wave. Vapor cloud discharge/dispersion can be modeled to determine overall flammable cloud size and the portion of the cloud in congested zones can then be estimated. In situations when the flammable cloud does not fill the congested volume, multiple discharge orientations may be needed to determine the highest blast loads that may be applied to receptors (buildings, process equipment, etc.). A simplifying assumption that may be made for deflagrations as an upper bound case is that the congested volume is filled with a stoichiometric fuel-air mixture. A DDT may involve fuel that is outside of congestion, but guidance has not been published for DDT energy determination.

Once the energy has been calculated, it must be multiplied by a ground reflection factor (i.e., hemispherical expansion factor), because BST curves are based on a spherical explosion model. The ground reflection factor is generally 2 for vapor clouds that are in contact with the ground. If a vapor release is elevated and does not disperse to ground level, a factor between 1 and 2 must be selected, since most vapor cloud explosions are relatively close to the ground, and a factor of 2.0 is appropriate.

The BST method was developed for external VCEs. Internal VCEs, including vented deflagrations, are not addressed in the published BST method. All other factors being equal, an enclosed process will develop a high source blast pressure than the same process without external enclosure surfaces. Application of external VCE models to internal VCEs can lead to underestimate of the source blast pressure unless adjustments are made to account for increased severity.

6.4.4.1.   BST Procedure

The BST methodology requires selection of the maximum flame speed based on the combined effects of confinement, congestion and fuel reactivity. Confinement is rated as 3-D, 2.5-D or 2-D as discussed above.

Congestion

Congestion refers to the obstacles that obstruct the passage of the flame front enough to create turbulence and increase flame speed without preventing expansion. Obstacle density is classified as Low, Medium and High, as shown in Table 6.14. These categories are based on the area blockage ratio (ABR), pitch, and number of "layers" of the obstacles. Congestion is the most difficult parameter to assess. In general for realistic plant geometries, a higher area blockage ratio results in higher turbulence intensity and therefore higher flame speeds and overpressures.

Table 6.14. Congestion description for the BST method

Type	Obstacle Blockage Ratio per Plane	Pitch for Obstacle Layers	Geometry
Low	Less than 10%	One or two layers of obstacles	
Medium	Between 10% and 40%	Two to three layers of obstacles	
High	Greater than 40%	Three or more fairly closely spaced obstacles	

Definition of the level of congestion within a plant requires an inspection of the plant. It is necessary to walk through the congested area. While at the center of the area, inspect for:
- Shortest path(s) to the edge of congestion (all directions),
- Number of obstacles to the closest edge of congestion,
- Number of "layers" of congestion, and
- Approximate area blockage ratio.

Low congestion can be defined as ≤ 10% ABR and pitch ≥ 8D where D is the diameter of obstacles. As a simple rule, if it is easy to walk through an area relatively unimpeded and there are only one or two "layers" of obstacles, it can be considered low congestion.

Medium congestion can be defined as between 10-40% ABR and a pitch of 4D - 8D. When walking through an area of medium congestion, it is cumbersome to do so and often requires taking an indirect path.

High congestion can be defined as having more than 40% ABR and a pitch of less than 4D. It is not possible to walk through an area of high congestion since there is insufficient space to pass between obstacles and successive layers prevent transit through the area. Moreover, repeated layers of closely spaced obstacles block line of sight from one edge of the congested zone to the opposite side.

Low and high congestion are most easily rated. Everything in between should be medium congestion. If it is difficult to determine what congestion rating to select for an area, it is safest to assume the higher of the two categories in consideration (low to medium, or medium to high).

Fuel Reactivity

The fuel reactivity is a term used to describe the propensity of a flame to accelerate in a VCE for a given fuel. The reactivity ratings of low, medium, and high developed by TNO (Zeeuwen, 1978) were adopted for the original BST method, and has since been revised. Factors determining fuel reactivity are laminar (fundamental) burning velocity and fuel expansion (Taylor, 1988). Generally, medium reactivity single component fuels have burning velocities between 45 and 75 cm/s. Low and high reactivity fuels are usually less than 45 and greater than 75 cm/s, respectively (Baker 1997, Pierorazio 2004).

At the time the BST method was published (Baker, 1994), it was recognized that mixtures had a flame speed proportional to the composition. Mixtures containing 3% of a high reactivity fuel were conservatively to be treated as high reactivity. Puttock (1995) has since shown that 40% hydrogen in a hydrogen-propane mixture will have a burning velocity of 80 cm/s, and thus be high reactivity. Baker (1997) proposed that the burning velocities of mixtures involving fuels of different reactivity categories can be calculated using Le Chatelier's principle, shown in Eq. (6.28).

$$V_B = \frac{100}{\dfrac{x_1}{V_{B1}} + \dfrac{x_2}{V_{B2}} + \dfrac{x_3}{V_{B3}} + \ldots} \qquad \text{(Eq. 6.28)}$$

where:

V_B = laminar burning velocity of the mixture of fuels   (cm/s)

x_1, x_2, x_3 = mole % of each component in the mixture

V_{B1}, V_{B2}, V_{B3} = burning velocity of each component in the mixture   (cm/s)

For a mixture involving hydrogen (high reactivity) with propane (medium reactivity), the above equation can be solved for the minimum mole percent hydrogen to result in a burning velocity of 75 cm/s. The resulting hydrogen concentration (x_1) is 45 mole percent hydrogen, which is in reasonable agreement with Puttock's experiments.

Puttock also presented data showing that 10% hydrogen in methane (burning velocity = 40 cm/s, low reactivity) will behave like propane (burning velocity = 46 cm/s, medium reactivity). Calculating the resultant burning velocity of this mixture using the above equation results in a V_B = 44 cm/s. This is close to the propane burning velocity and would qualify as a medium reactivity according to the guidelines.

As previously mentioned, there are factors other than burning velocity to determine fuel reactivity, which causes overlap between some low and medium reactivity fuels' burning velocities around 45 cm/s. For instance, carbon monoxide is a low reactivity fuel with a burning velocity of 46 cm/s, while n-butane is a medium reactivity fuel with a burning velocity of 45 cm/s. The conservative approach is to take the mixture as low reactivity if the burning velocity is less than 40 cm/s or if all components in the mixture are low reactivity (regardless of burning velocity), high reactivity for burning velocities greater than 75 cm/s, and medium for all other cases.

Examples of fuels that are low reactivity include methane, carbon monoxide, and ammonia. High reactivity fuels include ethylene, ethylene oxide, propylene oxide, acetylene, and hydrogen.

Potential Explosion Sites (PES)

A Potential Explosion Site (PES) is defined as an area of congestion and/or confinement into which a flammable vapor cloud is introduced and ignited (Baker, 1997). Each PES is evaluated based on its confinement and congestion. A PES may contain several zones with different congestion and confinement categories. In this case, the zones are averaged into a single area of congestion and confinement.

Zones of congestion/confinement may be either adjoining, or very close to one another, such that they should be evaluated as one composite PES for a given explosion scenario. That rationale for this treatment is that a flammable cloud covering these zones would produce a single blast wave since the flame front would not de-accelerate between the zones. Conversely, zones of congestion/confinement separated sufficiently by uncongested/unconfined (i.e., "empty") space can be considered as separate PESs. The empty space would allow sufficient flame deceleration such that two separate blast waves (one for each congested zone) would result. Zones with less than 15 ft (4.6 m) of empty space between them are treated as a single PES within the BST method. For cases where the edges of the respective zones are filled with repeated obstacles that encroach uniformly to the zone borders, a minimum of 30 ft (9.1 m) between zones is generally recommended to consider the zones to form separate PESs. Situations where the zone borders do not have high concentrations of repeated obstacles or where a significant portion of the zone border is not congested favor the 15 ft separation, whereas high concentrations of repeated obstacles occupying a large portion of the zone borders favors the 30 ft separation.

Blast Loads from Multiple PESs and Ignition Sources

Two methods for combining blast pulses from multiple sources into an equivalent triangular pulse were presented by Baker (1997). The first method (Method 1) is to use the greater of the two peak pressures, sum their impulse, and compute duration of the equivalent triangular load. The second method (Method 2) is to use the duration of the most significant pulse, sum their impulses, and solve for an effective peak pressure for the equivalent triangle.

Method 1 is generally conservative; however, it is not conservative for either elastic or slightly plastic systems and when the blast duration (t_d, not t_d') is approximately equal to or less than the natural frequency. In cases where the blast duration is greater than the natural frequency, which is typical of larger VCEs, or in any case with significant plastic deformation, Method 1 tends to be conservative. Considering that the accepted practice in both Government and industry is to allow significant plastic deformation of ductile building components subjected to accidental explosions, Method 1 provides conservative results yet avoids the overly conservative predictions associated with Method 2. If an existing building is predicted to have an elastic or slightly plastic response using Method 1, a slightly greater actual response during an accidental explosion is normally considered acceptable for ductile systems.

Method 2 is always conservative, and is very conservative in the pressure sensitive regime (load durations much greater than the natural frequency of the building or component). This is especially true for systems that are elastic or only slightly plastic (maximum deflection 1.5 times greater than elastic limit deflection or less).

Method 1 is recommended for siting analyses (maximum of individual predicted peak pressures, enhanced duration by conserving impulse) and Method 2 considered on a case-by-case basis during detailed siting analysis or the design of building upgrades. Buildings can be analyzed for a multi-pulse loading if sufficient information is available to allow accurate prediction of the load histories.

BST Analysis Procedure

Step 1. Define Inputs

BST method estimates the blast pressure loading at a given distance from the VCE explosion source as a function of the following input parameters:

> Congested Height, L_h
>
> Congested Width, L_w
>
> Congested Depth, L_d
>
> Confinement Level (3-D, 2.5-D, 2-D)
>
> Congestion Level (low, medium, high)

Material Properties

> Fuel reactivity level (low, medium, high)
>
> Heat of Combustion, H_c
>
> Fuel molecular weight, MW
>
> Speed of Sound, a_o
>
> Atmospheric pressure, P_o
>
> Gas Constant, R_g

Step 2. Estimate VCE Energy

Step 2a. Determine the stoichiometric concentration of the fuel by balancing the chemical reaction for complete combustion.

For example, with propane:

$$C_3H_8 + 5(O_2 + 3.76N_2) \rightarrow 3CO_2 + 4H_2O + 18.8N_2$$

Stoichiometric concentration:

$$\eta = \frac{moles\ fuel}{moles\ fuel\ and\ air}$$

(Eq. 6.29)

Step 2b. Determine the moles of gas in the combustion volume

$$n = \frac{P \cdot V}{R_r \cdot T}$$

(Eq. 6.30)

where:

P	=	ambient pressure	(Pa)
V	=	the volume of congestion with a flammable fuel-air mixture	(m^3)
R	=	the universal gas constant	(kJ/kmol·K)
T	=	the temperature	(K)

The volume of the flammable cloud in congestion is appropriate for a deflagration, but may not be representative of a DDT scenario.

Step 2c. Determine the combustion energy, E

$$E = 2n\eta (fuel\ molecular\ weight)(Heat\ of\ combustion)$$

(Eq. 6.31)

The factor of 2 doubles the energy term to account for ground reflection.

Step 3. Define flame speed

The appropriate flame speed can be located in Table 6.13 for the predetermined congestion/confinement level and the fuel reactivity.

Step 4. Calculate scaled standoff distance

The following equation is used to calculate the scaled standoff distance,

$$\bar{R} = R \cdot \left(\frac{P_o}{E} \right)^{1/3}$$

(Eq. 6.32)

where $\bar{R}$ is the scaled standoff distance and R is the required distance.

Using BST blast curves, the scaled pressure, $\bar{P}_{so}$ and impulse, $\bar{i}_{so}$ can be determined for a given $\bar{R}$.

Step 5. Calculate Actual Side-on Pressure

$$P_{so} = \bar{P}_{so} \cdot P_o$$

(Eq. 6.33)

Step 6. Calculate Actual Side-on Impulse

$$i_{so} = \frac{\bar{i}_{so} \cdot (E)^{1/3} \cdot (P_o)^{2/3}}{a_o}$$

(Eq. 6.34)

6.4.5.   Congestion Assessment Method

6.4.5.1.   Overview of the Congestion Assessment Method

The Congestion Assessment Method (CAM) was initially developed by Cates (1991, and updated by Puttock (1995, 2001) as a simplified method of estimating the source overpressure, P_{max}, of a vapor cloud, and the blast pressure pulse as a function of distance from the source. The most recent version is known as CAM 2. For the remainder of the chapter, the method will be referred to as simply the CAM method.

VCE overpressures are empirically predicted from data from experiments by Shell, TNO, and BG Technology (Snowden, 1999; Mercx, 1993). The tests considered vapor cloud explosions of methane, propane, and ethane in half cubes of symmetrical congestion. The 2m to 9m half cubes were made up of meshes of cylinders oriented in all three coordinate directions and intersecting. The cylinder diameters and spacing were varied in the tests. Further test results (Snowden, 1999; van Wingerden, 1988; Visser, 1991) were used to develop predictive models for congestion with a roof.

The CAM method estimates the source overpressure (the pressure generated in a congested plant area on the ground) as a function of the fuel and the congestion characteristics. The congestion is assessed as a series of rows of obstructions in the two horizontal directions and the vertical, for each of which an *area* blockage is estimated.

Overpressures beyond the congested area are predicted as a function of the source overpressure, the congested volume, and the distance from the congestion.

6.4.5.2.   CAM Inputs

CAM uses two fuel properties to characterize the influence of the reactivity of the fuel on the blast loads: the laminar burning velocity, U_0, in m/s; and the ratio of the density of the unburned fuel to the burned fuel (expansion ratio), E. These are combined into a "fuel factor" F. The fuel factor for several common fuels are found in Table 6.15. For other fuels, the fuel factor can be calculated from U_0, the laminar burning velocity (in m/s) and E, the expansion ratio using:

$$F = \frac{\left(U_0\left(E-1\right)\right)^{2.71}\Big|_{Fuel}}{\left(U_0\left(E-1\right)\right)^{2.71}\Big|_{Propane}} \qquad \text{(Eq. 6.35)}$$

Table 6.15. CAM Fuel Factor F and Expansion Ratio E for Common Fuels

Fuel	Fuel factor, F	Expansion Ratio, E	Laminar Burning Velocity U_0 (m/s)
Methane	0.6	7.75	0.448
Propane	1.0	8.23	0.464
Pentane	1.0	8.34	
Methanol	1.0	8.22	
Ethylene (ethene)	3	8.33	0.735

The second set of input parameters pertain to the congestion characteristics, and are calculated along the length (x direction), width (y direction) and height (z direction) of the plant:

- Number of rows of obstacles, $2n_x$, $2n_y$, n_z
- The average area blockage of the obstacle rows in the three directions: b_x, b_y, b_z. (Note that sharp-edged obstacles, e.g. beams, have a bigger effect than pipes, and their blockage should be increased by a factor of 1.6.)
- The length, width and height of the congestion: $2L_x$, $2L_y$, L_z

Note that the length is defined as $2L_x$ so that the same calculation can be used for horizontal and vertical.

A real plant, with obstacles of many different scales, generates a much more complex flame surface than rows of uniform cylinders. The result can be faster flame acceleration and therefore higher overpressures. The "complexity factor", f_c, for typical industrial plant is 4.0.

In addition, the congested volume, V_o (m^3), must be defined. The source volume recommended for use in the CAM model is defined by the width, length and height of the congested volume plus an additional 2 meters beyond the last obstacle in each direction i.e. $(2L_x+4)(2L_y+4)(L_z+2)$.

6.4.5.3.   Severity Index

CAM uses the concept of a "Severity Index", S, as a means of maintaining a realistic source overpressure when extrapolating experimental data to real world conditions.   Note that the term 'Severity Index' is unrelated to the TNO Severity Levels associated with the Multi-Energy method. The relationship between this index and the source overpressures was determined by using the SCOPE 3 phenomenological model (Puttock, 2000) for a wide range of input conditions. That relationship was determined to be:

$$S = P_{max} \cdot \exp\left(0.4 \frac{P_{max}}{E^{1.08} - 1 - P_{max}}\right) \qquad \text{(Eq. 6.36)}$$

where S is the severity index (in bars), E is expansion ratio, and P_{max} is the source overpressure (in bars). Figure 6.48 is used to calculate source overpressure once the severity index is determined.

6.4.5.4.   CAM Solution Method

After the inputs identified above have been specified, the CAM predictions are made through the following five steps.

Step 1. Determine a Severity Index as a function of congestion parameters

This needs to be done for the x and y directions, and the z direction if there is no roof. In the following equations, i denotes x, y or z successively. S_x, S_y and S_z are calculated from:

$$S_i = 0.89 \times 10^{-3} f_c FL_i^{0.55} n_i^{1.99} \exp(6.44b) \quad \text{without roof} \qquad \text{(Eq. 6.37)}$$

$$S_i = 1.1 \times 10^{-3} f_c FL_i^{0.55} n_i^{1.66} \exp(7.24b_i) \quad \text{with roof} \qquad \text{(Eq. 6.38)}$$

where lengths L_i are in m, pressure S_t is in bars.

Step 2. Average the severity index for the various directions

$$S = (S_x + S_y + S_z)/3 \quad \text{without roof} \qquad \text{(Eq. 6.39)}$$

$$S = (S_x + S_y)/2 \quad \text{with roof} \qquad \text{(Eq. 6.40)}$$

Step 3. Determine the Source Overpressure, P_{max} as a function of the Severity Index

It is possible to solve for the peak or source overpressure, P_{max} from the severity index and the relationship shown in Eq. (6.36) by iteration. Alternatively, this book's authors suggest solving the for P_{max} by letting

$$X = E^{1.08} - 1 \qquad \text{(Eq. 6.41)}$$

then dividing the severity index by this value to get S/X. Next use Figure 6.48 below to obtain the value of P/X. Lastly, multiply the P/X by X to get the estimated source pressure, Pmax.

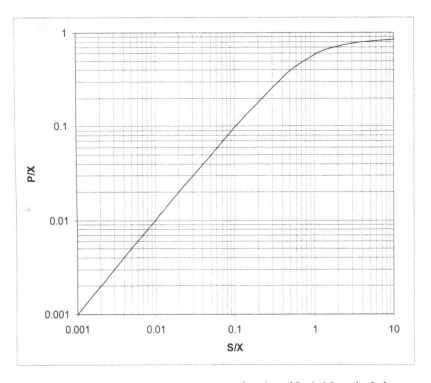

Figure 6.48. Scaled source overpressure as a function of Scaled Severity Index.

Step 4. Calculate the effective cloud radius

CAM calculates an effective radius R_o from the total congested volume by:

$$R_o = \left(\frac{3V_o}{2\pi} \right)^{1/3}$$ (Eq. 6.42)

Step 5. Calculate the free-field pressure at some distance

The predicted overpressure P (bar) at a given distance r (m), from the *edge* of the congested area can be read from Figure 6.49.

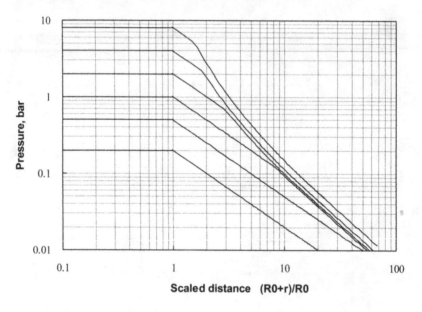

Figure 6.49. CAMS pressure decay as a function of distance $(R_0+r)/R_0$ for $P_{max} = 0.2, 0.5$, 1, 2 4 and 8 bar (contours bottom to top).

Step 6. Estimate pulse duration and shape

The duration of the pulse at this distance is found by defining a dimensionless distance parameter, d_f, defined as

$$d_f = \left(\frac{r}{R_o} \right) \left(\frac{P_{max}}{P_o} \right)^2 \qquad \text{(Eq. 6.43)}$$

where P_o is the atmospheric pressure (in bars).

The positive phase duration t_d (s) is estimated by

$$t_d = C \frac{R_o}{\sqrt{P'_o / \rho_{air}}} \qquad \text{(Eq. 6.44)}$$

where P'_0 is the source overpressure in Pa, ρ_{air} is the density of air (about 1.2 kg/m^3), and

$$C= 0.65 \text{ for } d_f<5 \qquad \text{(Eq. 6.45)}$$
$$C=0.65(d_f+10)/15 \text{ for } 5<d_f<20$$
$$C=1.3 \text{ for } 20<d_f$$

6.4.5.5. Other Considerations for CAM

Adjustments in CAM used to account for the following considerations were provided by Puttock (2001):

- One wall – if a wall is present a reflection factor can be attributed.
- Long, narrow areas – Lateral venting in a long plant which is small in the other dimensions may limit flame acceleration. An effective plant length is defined to account for these situations.
- Partial fill – defines a way to calculate the congested volume to be filled.

6.4.6. Numerical Methods

6.4.6.1. Computational Fluid Dynamics (CFD) Models

As discussed in Section 6.2, after ignition the expanding, unburned gas flow will interact with obstacles and/or confinement, generating a turbulent flow field ahead of the flame. Turbulent combustion enhances the energy release rate and flow velocity, causing the turbulence ahead of the flame to further increase. This strong positive feedback mechanism causes flame acceleration and in turn, elevated vapor cloud explosion blast loads. Computational Fluid Dynamics (CFD) models of vapor cloud explosions have focused on this turbulent combustion flame acceleration mechanism. The availability of faster and more powerful computers, in combination with improved numerical techniques for computing transient, turbulent flow has made it possible to simulate the process of turbulent, premixed combustion in a gas explosion in more detail. A number of CFD codes are available for gas explosion modeling. A review of some existing models for gas explosion modeling can be found in literature (Hjertager et al., 1996; Lea & Ledin, 2002).

As of the date of publication of this edition, CFD models were incapable of modeling a DDT. Situations involving high reactivity fuels that may DDT should be addressed using a blast curve method.

The sections below provide a brief description of the CFD method, the representation of geometric details within current CFD VCE codes, and a discussion of the turbulence and combustion sub-models used by such codes. Validation of CFD codes against test data is then discussed, followed by a summary of CFD VCE codes and an example of CFD modeling of a gas explosion for an onshore facility.

6.4.6.2.   Description of CFD Methods

CFD models are based on the fundamental equations governing turbulent and reacting flows (e.g., the explosion processes). These models start with a computational domain that covers the physical space of interest. The computational domain is discretized either with structured or unstructured meshes, yielding individual control volumes sufficiently small to capture the nature of a transient flow field associated with turbulent combustion and the resulting blast waves. The governing equations are then discretized for each control volume using appropriate numerical schemes. CFD models solve the discretized equations numerically by an iterative method for all the control volumes to obtain flow field properties. Relative to the empirical and phenomenological explosion models, CFD models are capable of providing a more detailed description of an explosion over a wider range of conditions and geometries. In addition to the blast loads (overpressure and impulse/positive duration), CFD codes can provide gas velocities, temperatures, densities, drag loads, turbulence parameters, species concentration, and other computed parameters.

It should be noted that the blast curve methods (BST and Multi-Energy Method) presented in the preceding sections are based on CFD models. Both methods were developed using one-dimensional, inviscid flow equations. The development of the BST blast curves utilized a model with a prescribed energy addition rate (i.e., a prescribed flame speed). The Multi-Energy Method used a constant rate of volumetric expansion to simulate the combustion zone. The fundamental feedback mechanism in turbulent combustion from flow interaction with obstacles and confining surfaces was not directly modeled.

Three major numerical treatments need to be efficiently implemented within a CFD model. These are:

- Conservation equations for mass, momentum and energy (as formulated in the Navier-Stokes equations which govern the transient flow),
- Geometry representation, and
- Physical sub-models to treat turbulence and combustion.

Three-dimensional compressible, transient flows governed by Navier-Stokes equations have been numerically investigated for more than a half a century. As a result, numerous robust numerical methods have been developed and are widely and successfully employed. Godunov's scheme (Godunov, 1959), the FCT method by Boris and Book (1976), and the TVD scheme by Harten (1983) are among the well-known numerical methods. The state-of-the-art numerical methods for compressible, transient flows are of 2nd order or higher accuracy for both space and time, and are sufficient for gas explosion simulations. The following sections discuss the geometry representation and sub-models used by existing codes.

6.4.6.3.   Geometry Representation

A typical onshore process unit is composed of thousands of objects (i.e., pipes, process equipment, structural elements, buildings, etc.) which cover a wide range of dimensions. Objects with characteristic dimensions down to roughly 1 inch (2 to 3 cm) must be represented within the geometric model to accurately simulate a vapor cloud explosion. While modeling the geometry associated with a basic test rig does not represent a significant challenge, the representation of such a large number of separate objects within the computational domain is a major issue for CFD VCE simulations of actual process units.

The geometry can be approximated using both resolved and unresolved geometrical objects. Resolved objects are treated explicitly and all object surfaces are meshed. Unresolved objects are represented using porosities and distributed resistances (PDRs), an approach first proposed within this context by Patankar & Spalding (1974). The PDR approach was extended by Sha et al. (1982) to include more advanced turbulent modeling. Objects such as structural elements and pipes are represented as area porosities (the opposite of blockages) on control volume (CV) faces, and are represented as volume porosities within the CV interior. CV surfaces and volumes are each either fully open, fully blocked, or partially blocked. The porosity for partially blocked surfaces or volumes is defined as the fraction of the area/volume that is available for fluid flow. The resulting porosity model is used to calculate flow resistance terms, turbulence source terms from small sub-grid objects, and flame speed enhancement arising from flame folding in the sub-grid wake. Numerically and practically, an object will be resolved if it is larger than the characteristic dimensions of the CV used in the computational domain (i.e., if it is "large"), and will be unresolved if it is smaller than the CV characteristic dimensions (i.e., if it is "small").

The PDR concept was developed as a compromise between the need to characterize the geometric details and the need to have the code run in a reasonable amount of time. It is not practical to use resolved geometrical objects in the computational domain for a typical process unit, where thousands of separate objects covering a wide range of dimensions must be included in the model. The PDR method makes it feasible to simulate a process unit VCE. The PDR method can be used for unresolved objects and complicated geometries using either a structured or unstructured meshing scheme.

There are a few dedicated CFD models which have been developed for VCE simulations, such as EXSIM (Hjertager et al. 1996), FLACS (van Wingerden et al., 1993) and AutoReaGas (Mercx et al. 2000). They are all based on the PDR method for representation of the geometry. These are the codes that have been used by industry up to the current time. They are capable of evaluating a VCE with run times of a few hours to a few days. FLACS and AutoReaGas are commercially available at the time of publication. The remainder of this section deals with these CFD VCE simulation codes.

6.4.6.4.   Summary of CFD VCE Codes

All of the aforementioned CFD VCE codes (EXSIM, FLACS and AutoReaGas) utilize turbulence and combustion models implemented via the PDR method (i.e. unresolved geometrical objects) in one form or another. These models incorporate a number of coefficients and parameters that have been set, or tuned, through comparison with a suite of VCE test data. The suite of test data employed for these efforts is fairly broad and covers a wide range of congestion levels and patterns, degrees of confinement, fuel mixture compositions and length scales. It should be recognized, however, that a VCE is an extremely complex process. Hence, the use of these codes outside their range of validation is subject to considerable uncertainty. Even for conditions within their range of validation, the CFD VCE codes discussed above can generate significantly different results. As a result of these limitations, it is recommended that the user of a CFD VCE code have a very strong background of VCE phenomena. The user should be aware of the expected behavior (e.g., the expected blast load range) in advance of a calculation and hence be able to determine if the calculated result is reasonable and in keeping with the existing body of VCE test data. In addition, the user should be well-trained in the use of the specific CFD code of interest.

Another class of CFD VCE codes has been employed to predict VCE blast loads. They are BWTI (Blast Wave Target Interaction) (Geng and Thomas, 2007a, 2007b) and CEBAM Computational Explosion & Blast Assessment Model) (Clutter, 2001).   These codes are listed in Table 6.16 in comparison to AutoReaGas and FLACS. Unlike AutoReaGas and FLACS that solve the Navier-Stokes (N-S) equations and model detailed turbulent combustions, both BWTI and CEBAM use a simplified combustion model in which flame speed is prescribed by the user, and modeled as an energy wave (Luckritz, 1977). The flame speed within congested areas is determined based on the BST flame speed table (Baker at al. 1998). The flame Mach No. in the flame speed table depends on the dimensionality of flame expansion (i.e. confinement), fuel reactivity and congestion levels.

The advantage of this class of CFD VCE codes (BWTI, CEBAM) is that the use of the flame speed table eliminates the effort of the turbulent combustion modeling. The resolved mesh formulation adopted in the codes allows irregular geometries to be represented. The main application area for these codes is engineering-level analyses involving scenarios where blast wave shielding and focusing are relevant, such that simplified methods that neglecting these effects would yield inaccurate results.

Table 6.16. CFD codes used to predict VCE blast loads

CFD Code	Governing Equations	Geometry Representation	Physical Sub-model	Blast Field of Interest
BWTI	Euler (N-S)	Resolved, Unstructured Mesh	Prescribed Flame Speed Energy Wave	Near- to Far-Field (Interaction with Target Structures)
CEBAM	Euler (N-S)	Resolved, Structured Mesh	Prescribed Flame Speed Energy Wave	Near- to Far-Field (Interaction with Target Structures)
AutoReaGas	N-S	Unresolved, Structured Mesh	Turbulent Combustion	Within Cloud, Near-field
FLACS	N-S	Unresolved, Structured Mesh	Turbulent Combustion	Within Cloud, Near-field

6.4.6.5.   Physical Sub-Models to Treatment of Turbulence and Combustion

The physical sub-models (turbulence and combustion models) used in CFD codes can be summarized as follows. All the codes use finite-domain approximations to the governing equations including the effect of turbulence and the rate of combustion. Turbulence effects are considered by the kε model of Launder and Spalding (1974). The FLACS code uses a burning velocity controlled flame mode (Arntzen, 1998), where the burning velocity is expressed empirically based on the experimental data presented by Abdel-Gayed et al. (1987) and the Bray flame speed correlation (Bray, 1990). The EXSIM code uses the eddy dissipation combustion model proposed by Magnussen and Hjertager (1976). The turbulent combustion rate in AutoReaGas is modeled with the eddy breakup model (Spalding, 1977), the eddy dissipation model (Magnussen and Hjertager, 1976) and the Bray flame speed correlation (Bray, 1990).

6.4.6.6.   Validation of CFD Models

CFD VCE simulation codes have been increasingly used over the last decade to perform safety analyses, particularly within the oil and gas industry. There are several reasons for the increasing popularity of such codes. First, the computational resources now widely available make it possible to simulate a large scale VCE in a reasonable amount of time. This makes the application of CFD VCE simulation codes tractable within the context of engineering projects.

Second, while medium- to large-scale VCE tests continue to provide valuable insights, they are extremely expensive compared with CFD analyses. CFD analyses also allow detailed parametric studies (e.g., effect of cloud size, cloud composition, ignition locations, mitigation schemes, etc.). However, a CFD code must be validated against a wide range of representative tests in order to provide the level of confidence required to allow its use for engineering design and safety analysis. The validation efforts associated with the two commercially available CFD VCE codes are discussed below.

The FLACS (FLame ACceleration Simulator) code was initially developed by the Christian Michelsen Research (CMR) Institute in Norway for the purpose of simulating gas explosions in offshore modules. The FLACS code has undergone continuous improvements over the last decade by CMR and GexCon, and has been validated against a wide range of explosion experiments including both simple idealized geometries and realistic geometries representative of process equipment. A description of explosion tests and results for a range of the geometries can be found in Bjerketvedt et al. (1997) and Arntzen (1998).

The AutoReaGas code was jointly developed by TNO and Century Dynamics and has been extensively validated against many experiments on several key projects, including MERGE and Blast and Fire Joint Industry (JIP) Phase 2 projects that involved full scale tests (Selby and Burgan, 1998).

Figure 6.50 through Figure 6.52 show a FLACS code validation calculation performed by GexCon (Foisselon, Hansen and van Wingerden, 1998) against the BFETS tests carried out by British Gas (now Advantica) at Spadeadam (U.K.). Figure 6.50 shows the FLACS model of the test module and external target distribution. The external targets were arranged along 0°, 45° and 90° directions to examine the directional effects. Figure 6.51 shows a comparison of the measured internal pressure histories and those predicted by FLACS for both the center and end ignitions, indicating a good agreement in terms of the positive phase rise time, positive overpressure, and positive and negative phase durations. Figure 6.52 compares the FLACS predictions and BFETS data for the decay of overpressure with distance external to the rig. The numerical results and test data are in a good agreement and both show the expected trends with respect to the decay of blast overpressure with distance and directionality effects.

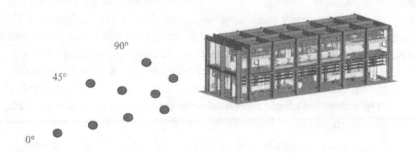

Figure 6.50. BFETS FLACS model and target distribution.

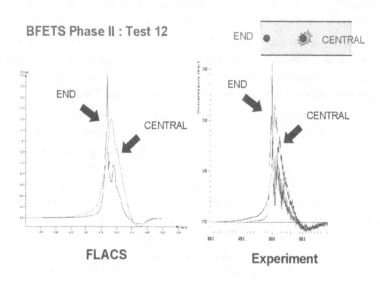

Figure 6.51. Comparison of FLACS results and experimental data
(internal pressure histories).

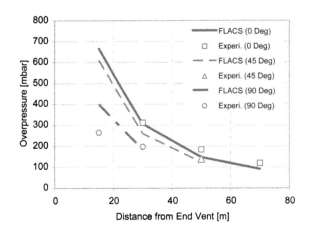

Figure 6.52. Comparison of FLACS results and experimental data (external).

All of these CFD VCE codes utilize turbulence and combustion models implemented via the PDR method in one form or another. These models incorporate a number of coefficients and parameters that have been set, or tuned, through comparison with a suite of VCE test data. The suite of test data employed for these efforts is fairly broad and covers a wide range of congestion levels and patterns, degrees of confinement, fuel mixture compositions and length scales.

6.4.6.7.   Example of CFD Simulation of an Onshore Installation Gas Explosion

This section provides an illustrative example of a CFD simulation of a gas explosion for an onshore process unit. The FLACS code was used for this example (Arntzen, 1998). Figure 6.53 shows the FLACS model of the process unit geometry. Four buildings were placed within the model, one on each side and end of the unit, as shown in the figure; the standoff distance from the unit center is the same for each building. The ignition location is at the southeast corner of the unit. Figure 6.54 shows an example of a flame contour.

Elevation view (without surrounding buildings)

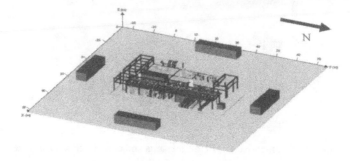

Northeast view (looking southwest)

Figure 6.53. FLACS model of an onshore installation.

East View (Looking West)

Figure 6.54. Flame front contour.

Pressure contours at selected times are shown in Figure 6.55. The north-running flame focuses the blast wave towards the building located north of the process unit, with the wave reaching this building at about 1.15 seconds. The buildings on the other sides of the process unit are exposed to a reduced blast load. This near-field blast load asymmetry cannot be predicted to the same degree using simplified blast-curve methods.

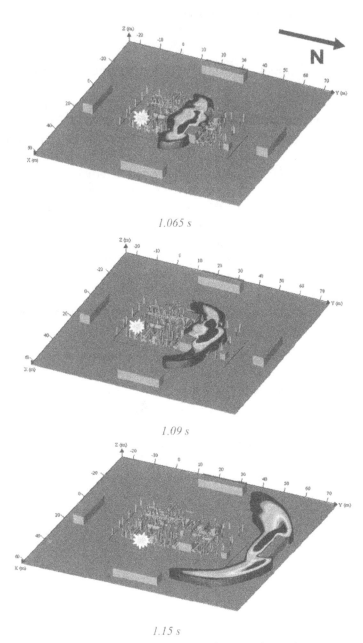

Figure 6.55. Pressure contours at selected times (northeast view).

6.5.   SAMPLE PROBLEMS

6.5.1.   Sample Problem – TNT Equivalence Method

Conventional TNT-equivalency methods state a proportional relationship between the total quantity of flammable material released or present in the cloud (whether or not mixed within flammability limits) and an equivalent weight of TNT expressing the cloud's explosion severity. The value of the proportionality factor - called TNT equivalency, yield factor, or efficiency factor - is directly deduced from damage patterns observed in a large number of major vapor cloud explosion incidents.   Over the years, many authorities and companies have developed their own practices for estimating the quantity of flammable material in a cloud, as well as for prescribing values for equivalency, or yield factor. Hence, a survey of the literature reveals a variety of methods.

6.5.1.1.   Vapor Cloud Explosion Hazard Assessment of a Storage Site

Problem. A storage site consists of three propane storage spheres (indicated as F9110, F9120, and F9130) and a 50-m diameter butane storage tank (indicated as F9210) on an open site, Figure 6.56. To diminish inflow of heat from the soil, the butane storage tank is placed 1 m above the earth's surface on a concrete pylon array. A parking lot with space for 100 cars is situated next to the tank farm. An accidental release of 20,000 kg of propane is postulated in this environment. The propane is released from a 0.1-m-diameter leak in the unloading line of sphere F9120. The propane is released at about 8 bars overpressure and mixes with air in a high-velocity jet.

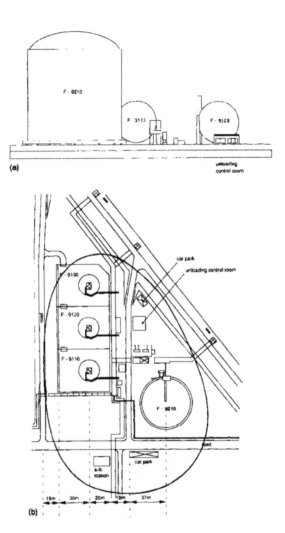

Figure 6.56. (a) View of a storage tank farm for liquefied hydrocarbons.
(b) Plot plan of the tank farm.

Quantify the explosive potential of a vapor cloud which results from the postulated propane release, and calculate the potential blast effects. Because it is dense, the flammable propane-air cloud spreads in a thin layer and covers a substantial area, including the tank farm and parking lot. An overview of the tank farm site is given by the map in Figure 6.56 (b).

Data:

heat of combustion for propane = 46.3 MJ/kg

mean specific heat for liquid propane = 2.41 kJ/kg/K

latent heat for propane = 410 kJ/kg

boiling temperature of propane at ambient pressure = 231 K

ambient temperature = 293 K

heat of combustion for TNT = 4.18 MJ/kg

Step 1). Determine charge weight

The HSE TNT-equivalency method expresses the potential explosion severity of a vapor cloud as one, single, equivalent TNT-charge located in the cloud's center. The equivalent charge weight of TNT is proportionally related to the quantity of fuel in the cloud and can be determined according to the following stepwise procedure:

- Determine the flash fraction of the fuel on the basis of actual thermodynamic data, using Eq. (6.46):

$$F = 1 - \exp\left[\frac{C_p \Delta T}{L}\right] = 1 - 2.718\left[-\frac{(2.41 kJ/kg/K)(293K - 231K)}{410 kJ/kg}\right] = 0.31$$

(Eq. 6.46)

- The weight of fuel in the cloud is equal to the flash fraction times the quantity of fuel released. To allow for spray and aerosol formation, the cloud inventory should be multiplied by 2. (The weight of fuel in the cloud may not exceed the total quantity of fuel released.) Consequently, the cloud inventory equals:

$W_f = 2 \times 0.31 \times 20,000$ kg = 12,400 kg of propane.

- The equivalent charge weight of TNT can now be calculated using Eq. (6.47) as follows:

$$W_{TNT} = \alpha_e \frac{W_f H_f}{H_{TNT}} = 0.03 \times \frac{12,400 \, kg \times 46.3 \, MJ/kg}{4.18 \, MJ/kg} = 4130 \, kg$$

(Eq. 6.47)

Note that the α is a factor used to express only a fraction, in this case 3%, of the vapor cloud energy used in defining an "equivalent" TNT charge.

Step 2). Define Blast effects

Once the equivalent charge weight of TNT in kilograms has been determined, the side-on peak overpressure of the blast wave at some distance R from the charge can be found with Eq. (6.48):

$$\overline{R} = \frac{R}{W_{TNT}^{1/3}} = \frac{R}{(4130\,kg_{TNT})^{1/3}} = \frac{R}{16\,kg_{TNT}^{1/3}} \qquad (\text{Eq. 6.48})$$

Once the Hopkinson-scaled distance from the charge is known, the corresponding side-on peak overpressure can be read from the chart in Figure 6.40. Table 6.17 gives results for several distances.

Table 6.17. Side-on peak overpressure for several distances from charge

Distance from Charge (m)	Scaled Distance from Charge (m/kg$^{1/3}$)	Side-on Peak Overpressure (bar)
50	3.24	0.68
100	6.48	0.21
200	12.95	0.084
500	32.38	0.025
1000	64.77	0.013

6.5.1.2.   Vapor Cloud Explosion as a Consequence of Pipe Rupture at a Chemical Plant

Problem. Large quantities of ignitable materials are stored and processed in process industries, often under high pressures and at high temperatures. Such activities pose vapor cloud explosion hazards. This was demonstrated, for example, by the events that occurred on June 1, 1974, at the Nypro Ltd. plant at Flixborough, UK, whose layout is shown by the plot plan in Figure 6.57.

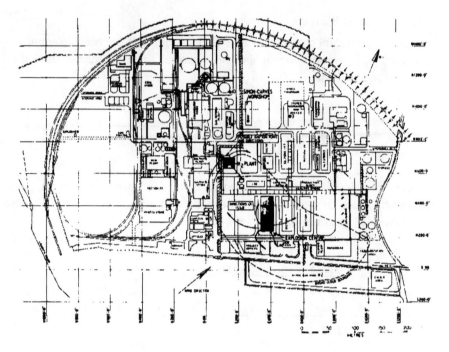

Figure 6.57. Plot plan of Nypro Ltd. plant at Flixborough, UK.

Data published by Sadee et al. (1976/1977), Gugan (1978), and Roberts and Pritchard (1982) serve as starting points for this sample problem. Because a pipe between two reactor vessels in the oxidation plant (see plan) ruptured, a large amount of cyclohexane was released within some tens of seconds at high pressure (10 bars) and temperature (423 K). The material quickly mixed with air, thus resulting in a large vapor cloud covering a substantial part of the plant area. In addition to the oxidation plant and the caprolactam plant, indicated in Figure 6.57 as Section 7 and 27 to the right of the explosion center, the cloud covered a large, more-or-less open area toward the hydrogen plant. The flammable cloud found an ignition source, probably somewhere in the hydrogen plant. The fire flashed back to the gas leak where, in between the densely spaced process equipment of the oxidation plant and the caprolactam plant, it found conditions under which intense and explosive combustion developed. The consequences were devastating. Twenty-eight people were killed, and dozens were injured. The plant was totally destroyed. Windows were damaged for several miles. Reconstruct the explosion severity and blast effects of the vapor cloud explosion on the basis of the available data.

The exact amount of cyclohexane released is unclear, but it escaped from a system consisting of five reactor vessels containing a total quantity of 250,000 kg (Gugan 1978). However, a complete discharge is unlikely. If an almost complete discharge of the two vessels adjacent to the ruptured pipe is assumed, a total quantity of 100,000 kg of cyclohexane would have been released.

Data

heat of combustion cyclohexane = 46.7 MJ/kg

mean specific heat liquid cyclohexane = 1.8 kJ/kg/K

latent heat cyclohexane = 674 kJ/kg

process temperature in reactor vessels = 423 K

boiling temperature at ambient pressure = 353 K

Step 1). Determine charge weight.

If conventional TNT-equivalency methods are applied, the potential explosion severity of a vapor cloud is expressed as one single, equivalent-TNT charge located at the cloud's center. The equivalent-charge weight of TNT is proportionally related to the fuel quantity within the cloud and can be determined according to the following stepwise procedure:

- Determine the flash fraction of fuel on the basis of actual thermo-dynamic data. The flash fraction for cyclohexane at 423 K can be calculated from Eq. (6.49):

$$F = 1 - \exp\left(\frac{-C_p \Delta T}{L}\right) = 1 - 2.718^{((1.8\,kJ\,/\,kgK)(423K - 353K)\,/\,674\,kJ\,/\,kg)} = 0.17$$

(Eq. 6.49)

- The weight of fuel in the cloud is equal to the flash fraction times the quantity of fuel released. To allow for spray and aerosol formation, the cloud inventory should be multiplied by 2. (The weight of fuel in the cloud may not exceed the total quantity of fuel released.)

If a release of 100,000 kg of cyclohexane is assumed, the weight of fuel in the cloud equals:

$$W_f = 2 \times 0.17 \times 100,000 \text{ kg} = 34,000 \text{ kg}$$

This rather speculative figure is in reasonable agreement with Sadèe et al.'s (1976/1977) estimate.

- The equivalent charge weight of TNT can now be calculated from Eq. (6.50)as follows:

$$W_{TNT} = \alpha_c \frac{W_f H_f}{H_{TNT}} = 0.03 \times \frac{34,000 kg \times 46.7 \, MJ/kg}{4.68 \, MJ/kg} = 10,178 kg$$

(Eq. 6.50)

Step 2). Define blast effects

Once the equivalent charge weight of TNT in kilograms is known, the side-on peak overpressure of the blast wave at some distance R from the charge can be found by calculating the Hopkinson-scaled distance using Eq. (6.51):

$$\bar{R} = \frac{R}{W_{TNT}^{1/3}} = \frac{R}{(10,178 kg)^{1/3}} = \frac{R}{21.7 \, kg_{TNT}^{1/3}}$$

(Eq. 6.51)

Once the Hopkinson-scaled distance from the charge is known, the corresponding side-on peak overpressure can be read from the chart in

Figure 6.44, which includes values provided in Table 6.18 for several distances.

Table 6.18. Side-On peak overpressure for several distances from charge expressing explosion severity of the Flixborough vapor cloud explosion.

Distance from Charge (m)	Scaled Distance from Charge (m/kg$^{1/3}$)	Side-on Peak Overpressure (bar)
50	2.3	1.2
100	4.6	0.39
200	9.2	0.13
500	23	0.04
1000	46	0.018
2000	92	0.010

6.5.2.   Sample Problem - Multi-Energy Method

6.5.2.1.   Multi-Energy Method – Assumed Initial Strength

The Multi-Energy Method treats vapor cloud explosions as a number of subexplosions, and recognizes that the explosive potential of a vapor cloud is primarily determined by the blast-generative properties of the environment in which the vapor is released and disperses. Therefore, the following steps to determining blast strength and effects apply:

Step 1. Identify potential blast sources.

Data provided by the literature (Sadèe et al. 1976/ 1977; Gugan 1978; Robert and Pritchard 1982) identified potential blast sources. The plot plan in Figure 6.57 shows that the cloud covered a substantial area: the oxidation and caprolactam plants (indicated in Figure 6.57) and also the more-or-less open area toward the hydrogen plant.

The process units had elements of partial confinement: densely spaced process equipment mounted in open buildings consisting of parallel concrete floors. The areas covering the cyclohexane oxidation and caprolactam plants should be considered sources of strong blasts. No significant contribution to the blasts should be expected from the rest of the cloud, because it is unconfined and unobstructed.

Step 2. Determine the scale of equivalent fuel-air charges.

Consider each blast source separately. Assume that the entire volume of fuel-air mixture present in each cloud portion identified as a source of strong blast contributes to the blast. The blast originating from each source is modeled as though it were from a hemispherical fuel-air charge. The combustion energy contributing to each respective charge is found by assuming a stoichiometric composition and by multiplying the volume of each source by the fuel's heat of combustion, 3.5 MJ/m^3.

The scale of the charge representing the potential explosion severity of the single source of the strong blast identified is determined by calculation of the quantity of combustion energy of flammable mixture within the partially confined volume. In this case, it is the volume of space between parallel concrete floors and obstructed by the densely spaced equipment in both the hexane oxidation plant and the caprolactam plant. On the basis of the scale of the plan in Figure 6.57, an approximate estimate of the partially confined or obstructed volume V of vapor can be made:

$$V = 100 \text{ m} \times 50 \text{ m} \times 10 \text{ m} = 5 \times 10^4 \text{ m}^3.$$

This volume corresponds with a quantity of combustion energy of:

$$E = 50{,}000 \text{ m}^3 \times 3.5 \text{ MJ/m}^3 = 175{,}000 \text{ MJ}.$$

Consequently, the potential explosion severity of the rest of the cloud, covering a more-or-less open area, can be expressed as a fuel-air charge of:

$$34{,}000 \text{ m}^3 \times 46.7 \text{ MJ/m}^3 - 175{,}000 \text{ MJ} = 1{,}412{,}800 \text{ MJ}.$$

Step 3. Determine the initial strengths of the charges.

A quick and simple approach to estimation of initial strengths of the charges expressing the potential explosion severity within the vapor cloud is to use the following safe and conservative approach:

- The fuel-air charge expressing the explosion severity of the source of strong blast is assumed to be of strength number 10.
- The fuel-air charge expressing the explosion severity within the rest of the vapor cloud is assumed to be of strength number 2.

Thus, the potential explosion severity of the vapor cloud can be expressed as two equivalent fuel-air charges whose characteristics and locations are listed in Table 6.19.

Once equivalent charges expressing the vapor cloud's potential explosion severity are known, both in scale and strength, corresponding blast effects can then be determined.

Table 6.19. Characteristics and locations of fuel-air charges expressing potential explosion severity of the Flixborough vapor cloud

	Combustion Energy E (MJ)	Strength (Number)	Location
Equipment (charge I)	175,000	10	Center of equipment
Rest of the cloud (charge II)	1,412,800	2	Center of cloud

Step 4. Calculate blast effects.

The side-on peak overpressures and positive-phase durations of blast waves produced by the respective charges for any selected distance, R, can be found by calculating separately for each charge:

$$\bar{R} = \frac{R}{(E/P_0)^{1/3}}$$

where:

$\bar{R}$	=	nondimensionalized distance from charge	(–)
R	=	distance from charge	(m)
E	=	charge combustion energy	(J)
P_0	=	ambient pressure = 101,325 Pa	

Calculate the properties of the blast produced at a distance of 1,000 m from each of the two charges. The nondimensionalized distance equals:

Charge I:
$$\bar{R} = \frac{1000\,m}{(175,000 \times 10^6\,J/101,325\,Pa)^{1/3}} = 8.3$$

Charge II:
$$\bar{R} = \frac{1000\,m}{(1,412,800 \times 10^6\,J/101,325\,Pa)^{1/3}} = 4.2$$

Once the nondimensionalized distances from each charge are known, the corresponding nondimensionalized blast parameters can be read from the charts in Figure 6.41(a) and (b). The nondimensionalized blast parameters read are tabulated in Table 6.20.

Table 6.20. Nondimensionalized blast parameters at 1,000m distance from two charges, read from charts in Figure 6.40

	R (m)	E (MJ)	Strength Number	$\bar{R}$	$\Delta\bar{P}_s$	$\bar{t}_+$
Charge I	1,000	175,000	10	8.3	0.028	0.45
Charge II	1,000	1,412,800	2	4.2	0.0032	3.0

The nondimensionalized side-on peak overpressures and positive-phase durations read from the tables can be converted into real values for side-on peak overpressures and positive-phase durations as follows:

Charge I:

$$\Delta P_s = \Delta \bar{P}_s \times P_0 = 0.028 \times 101,325 = 2,837 \text{ Pa} = 0.028 \text{ bar}$$

$$t_+ = \bar{t}_+ \frac{(E/P_0)^{1/3}}{c_0} = 0.45s \times \frac{(175,000 \times 10^6 \, J / 101,325 \, Pa)^{1/3}}{340 \, m/s} = 0.159s$$

Charge II:

$$\Delta P_s = \Delta \bar{P}_s \times P_0 = 0.0032 \times 101,325 = 324 \text{ Pa} = 0.003 \text{ bar}$$

$$t_+ = \bar{t}_+ \frac{(E/P_0)^{1/3}}{c_0} = 3s \times \frac{(1,412,800 \times 10^6 \, J / 101,325 \, Pa)^{1/3}}{340 \, m/s} = 2.1s$$

where:

ΔP_s	=	side-on peak overpressure	(Pa)
$\Delta \bar{P}_s$	=	nondimensionalized side-on peak overpressure	(–)
P_0	=	ambient pressure = 101,325 Pa	
t_+	=	positive-phase duration	(s)
$\bar{t}_+$	=	nondimensionalized positive-phase duration	(–)
E	=	charge combustion energy	(J)
c_0	=	ambient speed of sound = 340 m/s	

This operation can be repeated for any desired distance. Results for selected distances are given in Table 6.21 and Table 6.22.

Table 6.21. Side-on peak overpressure and positive-phase duration of blast produced by Charge I (E = 175,000 MJ, strength number 10)

R (m)	$\overline{R}$		ΔPs (bar)	$\overline{t}_+$	t+ (s)
50	0.41	3.4	3.45	0.15	0.053
100	0.83	0.70	0.71	0.19	0.067
200	1.67	0.21	0.21	0.29	0.102
500	4.17	0.065	0.066	0.40	0.141
1000	8.34	0.028	0.028	0.45	0.159
2000	16.67	0.013	0.013	0.49	0.173
5000	41.68	0.0050	0.005	0.53	0.187

Table 6.22. Side-on peak overpressure and positive-phase duration of blast produced by Charge II (E = 1,412,800 MJ, strength number 2)

R (m)	$\overline{R}$	$\Delta\overline{P}_s$	ΔPs (bar)	$\overline{t}_+$	t+ (s)
100	0.42	0.020	0.020	3.3	2.3
200	0.83	0.016	0.016	3.0	2.1
500	2.08	0.0065	0.007	3.0	2.1
1000	4.15	0.0032	0.003	3.0	2.1
2000	8.31	0.0016	0.002	3.0	2.1

6.5.2.2.   TNO Multi-Energy Method – Initial Strength per GAME

Problem. Propane has spilled underneath a large LPG storage tank. The tank has a diameter of 20 meters and is 10 meters tall. To minimize heat transfer from the ground, the tank is supported by a uniform array of 267 cylindrical concrete piers. Each pier is 1 meter tall and has a diameter of 0.3 meters. Assume that the entire volume underneath the tank is filled with a stoichiometric mixture of propane and air. Determine the overpressure and impulse at a distance of 50 meters resulting from the explosion of this mixture. Employ the GAMES methodology to determine the source strength of the blast.

- Case a: Assume ignition at the center of the cloud.
- Case b: Assume a flame travel distance equal to the radius of a hemisphere with the same volume as the congested region consistent with the GAMES recommended approach.

Solution:

Step 1. Calculate VBR. The diameter of congested region, D_{tank}, is 20 m and its height, H, is 1 m. Therefore, the volume of the congested region underneath the tank is

$$V_{region} = \frac{\pi}{4} D^2_{tank} H = \frac{\pi}{4} (20m)^2 (1m) = 314.2\ m^3$$

Each pier has a diameter, D, of 0.3 m and a height, H, of 1 m. There are a total of 267 piers under the tank. The obstructed volume occupied by the piers is

$$V_{obstructed} = 267 \cdot \frac{\pi}{4} D^2 H = \frac{\pi}{4} (267)(0.3m)^2 (1m) = 18.9\ m^3$$

The volume blockage ratio, VBR, is defined as the ratio of the obstructed volume to the total volume. Therefore, the VBR is

$$VBR = \frac{V_{obstructed}}{V_{region}} = 0.06$$

Step 2. Determine source pressure and blast load. The source pressure of this VCE is estimated from the 2D GAMES correlation. The fuel specified is propane, which has a laminar burning velocity of $S_L = 0.46$ m/s. The characteristic obstacle diameter is the diameter of each pier, D=0.3 m. The vapor cloud is confined between the ground and the bottom of the LPG storage tank, an example of 2D confinement.

Case a)   If we assume center ignition then L_p=10 m, since the distance from the center of the cloud to its edge is equal to the tank radius. The source strength is determined to be:

$$P_0 = 3.38 \left(\frac{VBR \cdot L_p}{D} \right)^{2.25} S_L^{2.7} D^{0.7}$$

$$= 3.38 \left[\frac{(0.06)(10\,m)}{(0.3m)} \right]^{2.25} (0.46\,m/s)^{2.7} (0.3m)^{0.7} = 0.85\ bar$$

From Figure 6.41, this corresponds to a strength number between 6 and 7. To assure conservatism, we will use a strength number of 7. The heat of combustion, H_c, of a stoichiometric propane-air mixture is 3.5 MJ/m^3, so the energy of the blast source is:

$$E = H_c \left(V_{region} - V_{obstructed} \right)$$
$$= (3.5\,MJ/m^3)(314.2m^3 - 18.9m^3) = 1033.6\ \text{MJ}.$$

This means that 50 meters corresponds to a scaled distance of:

$$\overline{R} = \frac{R}{(E/P_a)^{1/3}} = \frac{50\,m}{(1033.6 x10^6\ J/101{,}325\,Pa)^{1/3}} = 2.3$$

The resulting side-on overpressure at the scaled distance corresponding to 50 m is:

$$P = 13.3\ \text{kPa}$$

The duration is determined from Figure 6.41 and is equal to t_p=23.8 ms. From the side-on overpressure and duration, the impulse at a distance of 50 m is given by:

$$i = \frac{1}{2} \cdot P \cdot t_p$$
$$= \frac{1}{2}(13{,}300Pa)(23.8ms) = 158.3\ Pa \cdot s$$

Case b) The radius of a hemisphere having the same volume as the congested region is given by:

$$R = \left(\frac{3V_{region}}{2\pi}\right)^{1/3}$$

$$= \left(\frac{3(314.2m^3)}{18.9m^3}\right)^{1/3} = 5.3 \; m$$

Using this value for L_p in the 2D GAMES correlation yields a source strength of:

$$P_0 = 3.38\left(\frac{VBR \cdot L_p}{D}\right)^{2.25} S_L^{\,2.7} D^{0.7}$$

$$= 3.38\left(\frac{(0.06)(5.3m)}{(0.3m)}\right)^{2.25} (0.46 \, m/s)^{2.7}(0.3m)^{0.7} = 0.2 \; bar.$$

From Figure 6.41 this corresponds to a strength number of 5. Since P_0 is less than 0.5 bar, an energy efficiency factor of 20% is applied. Therefore, the energy of the blast source is taken to be:

$$E = 0.2(1033.6 \; MJ) = 206.7 \; MJ.$$

This means that 50 meters now corresponds to a scaled distance of

$$\overline{R} = \frac{R}{(E/P_a)^{1/3}} = \frac{50m}{(206.7x10^6 \, J/101325Pa)} = 3.9.$$

The resulting side-on overpressure at this scaled distance is

$$P = 3.0 \; kPa$$

The duration is determined from Figure 6.41 and is equal to t_p=24.3 ms. From the side-on overpressure and duration, the impulse at a distance of 50 m is given by:

$$i = \frac{1}{2} \cdot P \cdot t_p = \frac{1}{2}(3000 Pa)(24.3ms) = 37\ Pa \cdot s$$

6.5.3.   BST Sample Problem

Propane has spilled in a congested process unit measuring 15m long by 13m wide by 7m high. The process area has no confining surfaces. Congestion level in the unit is medium. Assume that the entire congested volume is filled with a stoichiometric mixture of propane and air. Determine the blast overpressure and impulse as a function of distance from the VCE.

Step 1. Define inputs

The congested volume is:

V = 7m × 15m × 13m = 1,365m³

The confinement level is 3-D due to the absence of confining surfaces. Congestion level was given as Medium. Fuel reactivity of propane is Medium. Other constants needed for the calculations are shown in Table 6.23.

Table 6.23. Constants used in the BST sample problem

Heat of Combustion (Propane) :	4.6E+07 J
Gas Constant, R_g :	8.3 kJ/kmol·K
Molecular Weight (Propane) :	44 kg/kmole
Speed of Sound (a_o) :	330 m/s
Atmospheric Pressure :	101325 Pa

Step 2. Estimate VCE energy

Step 2a. Determine the stoichiometric concentration of the fuel by balancing the chemical reaction for complete combustion:

For propane,

$$C_3H_8 + 5(O_2 + 3.76N_2) \rightarrow 3CO_2 + 4H_2O + 18.8N_2$$

Stoichiometric concentration:

$$\eta = \frac{moles\,fuel}{moles\,fuel\,and\,air}$$

$$\eta = \frac{1}{(1+5\cdot4.76)} = 0.04$$

Step 2b. Determine the moles of gas in the combustion volume (assuming the volume is completely filled with a stoichiometric propane/air mixture):

$$n = \frac{P\cdot V}{R_r \cdot T} = \frac{(101325\,Pa)(1365m^3)}{(8.3x10^3\,J/kmoleK)(300K)} = 55.8\,kmole$$

Step 2c. Determine the combustion energy, E

$$E = 2n\eta(fuel\ molecular\ weight\ )(Heat\ of\ combustion\ )$$

$$E = 2(55.8\,kmol)(0.04)(44kg/kmol)(4.6x10^7\,J)$$

$$E = 2.1x10^{10}\,J$$

Note: energy was doubled to account for ground reflection

Step 3. Define flame speed

According to Table 6.13, $M_f = 0.28$

Step 4. Calculate scaled standoff distance

Required standoff distance = 10 m

$$\overline{R} = R \cdot \left(\frac{P_o}{E}\right)^{1/3} = (10 \ m) \cdot \left(\frac{101325 \ N/m^2}{2.1x10^{10} \ J}\right)^{1/3} = 0.17$$

Step 5. Calculate actual side-on pressure

From the BST blast curves at $\overline{R} = 0.17$

$$\overline{P}_{so} = 0.14$$

$$P_{so} = \overline{P}_{so} \cdot P_o = (0.14) \cdot (101325 \ Pa) = 1.5x10^4 \ Pa \ (gauge)$$

Step 6. Calculate actual side-on impulse

From the BST blast curves at $\overline{R} = 0.17$

$$\overline{i}_{so} = 0.10$$

$$i_{so} = \frac{\overline{i}_{so} \cdot (E)^{1/3} \cdot (P_o)^{2/3}}{a_o}$$

$$i_{so} = \frac{(0.10) \cdot (2.01x10^{10} \ J)^{1/3} \cdot \left(101325 \ \frac{N}{m^2}\right)^{2/3}}{330 \ m/s}.$$

$$i_{so} = 1720 \ Pa \cdot s$$

Table 6.24 provides side-on and impulse results for a range of standoff distances.

Table 6.24. Blast overpressure and impulse for different standoff distances using the BST method

Standoff R (m)	Scaled Standoff $\overline{R}$	Scaled Pressure $\overline{P}_{so}$	Overpressure P (Pa)	Scaled Impulse, $\overline{i}_{so}$	Impulse, i (Pa*s)
10	0.17	0.14	15,000	0.10	1720
25	0.43	0.12	12,000	0.052	890
50	0.86	0.06	6,500	0.027	470
100	1.7	0.03	3,300	0.014	230

6.5.4.   CAM Example Problem

Estimate the free-field pressure and duration for a propane vapor cloud explosion with the following conditions:

Congested height = 6 m

Congested width = 12 m

Congested length = 12 m

Number of rows of obstacles through height = 3

Number of rows of obstacles through width = 6

Number of rows of obstacles through length = 6

Area blockage ratio, $b_x=b_y=b_z = 0.5$

Thus, $L_x=L_y=L_z =6m$ and $n_x=n_y=n_z =3$.

Congested Volume, $V_o= (2L_x+4)(2L_y+4)(L_z+2) = 2048 \text{ m}^3$

From Table 6.15, for propane,

   Fuel factor $F = 1$; $X = 8.74$

Solution:

Step 1. Determine Severity Index as a function of congestion parameters

   For the case of congestion without a roof from Eq. (6.15),

$$S_x=0.89 \times 10^{-3} \ f_c FL_x^{0.55} n_x^{1.99} \ \exp(6.44 b_x)$$

$$S_x=(0.89 \times 10^{-3}) \cdot 4.0 \cdot 1 \cdot (6m)^{0.55} \cdot 3^{1.99} \exp(6.44 \cdot 0.5) = 2.13$$

   In this particular case, S_y and S_z are equal to S_x, so $S = 2.13$.

   Using the coefficients for congestion with a roof gives $S = 2.73$

Step 2. Determine the Source Overpressure, P_{max} as a function of the Severity Index

$S/X = 2.13/8.74 = 0.243$ for congestion without a roof

$S/X = 2.73/8.74 = 0.312$ for congestion with a roof

From Figure 6.48,

$P_{max}/X = 0.217$ for congestion without a roof

$P_{max}/X = 0.269$ for congestion with a roof

Then,

$P_{max} = P/X \cdot X = 0.217 \cdot 8.74 = 1.90 \, bar$ for congestion without a roof

$P_{max} = P/X \cdot X = 0.269 \cdot 8.74 = 2.35 bar$ for congestion with a roof

Step 3. Calculate the Initial Cloud Radius

$$R_o = \left(\frac{3V_o}{2\pi} \right)^{1/3}$$

$$R_o = \left(\frac{3 \cdot 2048 m^3}{2\pi} \right)^{1/3} = 9.9 \, m$$

Step 4. Calculate pressure at some distance, r, from the congested area

At r = 50 m,

$$\frac{R_o + r}{R_o} = 6$$

Interpolating between the $P_{max}= 2$ bar and $P_{max}= 3$ bar curves in Figure 6.48 gives:

P = 0.21 bar for no roof.

Further calculations are shown in Table 6.25.

Table 6.25. Predicted blast loads using the CAM method

	P_{max} (bar)		d_f		C		t_d (ms)		i_s (bar-ms)	
$r(m)$	w/o roof	w/ roof	w/o roof	w/ roof	w/o roof	w/ roof	w/o roof	w/ roof	w/o roof	w/ roof
1	1.73	2.14	0.35	0.19	0.65	0.65	16	15	13,8	16.5
5	1.27	1.57	1.76	0.95	0.65	0.65	16	15	10.2	11.8
10	0.95	1.18	3.52	1.89	0.65	0.65	16	15	7.6	8,9
25	0.50	0.52	8.79	4.73	0.65	0.65	16	15	4.0	3.9
50	0.21	0.21	17.6	9.47	0.81	0.65	20	15	2.1	1.6
75	0.12	0.13	26.4	14.2	1.20	0.84	30	19	1.8	1.2
100	0.09	0.09	35.2	18.9	1.30	1.05	33	24	1.5	1.1

Step 5. Estimate pulse duration and impulse

For $r = 50$ m, dimensionless distance parameter, d_f is:

$$d_f = \left(\frac{r}{R_o}\right)\left(\frac{P_o}{P_\infty}\right)^2$$

$$= \left(\frac{50m}{10.0m}\right)\left(\frac{1.9bar}{1.013bar}\right)^2 = 17.6$$

According to Eq. (6.45), C= 1.2

$P_0' = 10,000$ Pa/bar × 1.9 bar = 190,000 Pa

According to Eq. (6.44), positive phase duration is:

$$t_d = C \frac{R_o}{\sqrt{P_o' / \rho_{air}}}$$

$$t_d = 1.2 \frac{10.0m}{\sqrt{190000Pa / 1.2kg/m^3}} = 0.030s$$

Impulse can be calculated assuming a triangular wave form (instantaneous pressure rise, linear decay to ambient) for which:

$$i = \frac{1}{2} P_{max} t_d$$

$$i = \frac{1}{2} (0.21\ bar)(0.020\ s) = 0.0021\ bar \cdot s = 2.1\ bar \cdot ms$$

7. PRESSURE VESSEL BURSTS

Pressure vessel burst (PVB), as the name implies, is a type of explosion that involves burst of a pressure vessel containing gas at elevated pressure. The term "pressure vessel" in PVB is not necessarily synonymous with the definition of pressure vessel used in ASME code; rather, any vessel or enclosure that can build significant pressure before bursting can generate a PVB. Upon burst, the sudden expansion of a compressed gas generates a blast wave that propagates outward from the source. The shell of the vessel along with attached external appurtenances is thrown, creating a fragment hazard. It is not necessary for the vessel contents to be flammable or contain reactive chemicals. PVBs occur with inert gases or mixtures as well as flammable or reactive materials.

This chapter includes only PVBs, which involve only the release of energy from the compressed gas contents. Flashing of the superheated liquid upon vessel failure can also contribute to the explosion energy, but that is a separate type of explosion called a Boiling Liquid Expanding Vapor Explosion (BLEVE). BLEVEs are addressed separately in Chapter 8. PVBs may involve a vessel that contains liquids that do not change phase when the vessel depressurizes upon bursting.

Mechanisms of PVBs are discussed first. Experimental and analytical research results are presented, followed by prediction methods for airblast parameters, and lastly, fragment velocity and throw prediction methods.

Most pressure vessels in service today are fabricated from metal. However, composite pressure vessels are becoming more common, most often when lighter weight is desirable. Examples of widespread use of composite pressure vessels include portable breathing air cylinders. Composite transportation fuel cylinders are used for hydrogen, liquefied petroleum gas (LPG), and propane. Composite vessels may have different failure patterns from metal vessels, which could affect hazard predictions. Prediction methods specifically for composite pressure vessels have not been developed. Airblast prediction methods in this chapter will provide conservative estimates for all types of vessel construction. Fragment velocity and throw prediction methods will need to consider the vessel failure pattern and number of fragments, which may differ from metal vessels.

7.1.   MECHANISM OF A PVB

7.1.1.   Accident Scenarios

Prior to the 1900s, a PVB was a frequent phenomenon. The frequency of PVBs led the American Society of Mechanical Engineers (ASME) to the creation standards for the design and manufacture of the boilers and pressure vessels, which helped to drastically reduce the frequency of PVB accidents. Today, the ASME boiler and pressure vessel code is used worldwide and has been adopted as law in many jurisdictions.

ASME code-compliant vessels have a design ultimate failure pressure that has a significant safety margin above expected working pressures. In fact, ASME code-compliant vessels should not exceed yield stress of the vessel at the maximum allowable working pressure (MAWP). As a result, ultimate failure pressure for an ASME code-complaint vessel is typically of 3 to 4 times design MAWP, depending on the version or section of the code in use. Corrosion and fabrication allowances may further increase the ultimate failure pressure, in some instances up to about 5 times MAWP. European codes design margins are typically less than U.S., such a 2.4 times MAWP for carbon steels without corrosion and fabrication allowances.

American Petroleum Institute (API) Standard 521 (API, 2008) requires that pressure relief valves (PRVs) on pressure vessels achieve rated flow at 1.21 times MAWP in a fire exposure case. If mechanical integrity of a vessel is not compromised, pressure generation does not exceed flow capacity of the PRV, and the PRV performs properly, the PRV should prevent vessel failure. Cases of loss of mechanical integrity can result in failures at or below the PRV relief pressure. Situations in which pressure generation exceeds relief capacity or the relief valve is compromised can lead to failure above PRV relief pressure, possibly approaching ultimate failure pressure.

In spite of the significant design safety factors and pressure relief devices, accidents involving pressure vessels can still occur for a variety of reasons. Accident scenarios can involve one or more of the following causes of failure:

- Loss of mechanical integrity – the pressure vessel's attached connections are compromised in some fashion; as a consequence, the stresses in a pressure-carrying component can exceed the ultimate capacity. Failures due to loss of mechanical integrity may occur at normal working pressure. Examples in this category include:

- o Temperature drop below the metal transition (cold temperature embrittlement)
- o Stress corrosion cracking
- o Corrosion and/or erosion
- o Fatigue
- Pressurization beyond ultimate failure pressure – pressure goes beyond the ultimate capacity without a coinciding reduction in ultimate failure pressure due to other factors, with scenarios considering the rate of pressurization versus the ability to relieve pressure through safety devices. Examples of this type of failure include:
  - o Connection of a high pressure supply source to a vessel not designed for that pressure
  - o Compressor or pump that fails to cut off
  - o Chemical reaction inside vessel
  - o Rapid Phase Transition (RPT) internal to the vessel (e.g., water introduced into a hot oil tank)
  - o Combustion internal to the vessel
  - o Pressure relief device(s) fail to operate, or are blocked or plugged
  - o Vessel is moved into an incorrect orientation (e.g., tank car roll over), and the PRV is submerged
- External agencies – an external source that compromises mechanical integrity, such as:
  - o Impact - an object hits the vessel with sufficient energy to compromise the shell, a vessel is dropped, or a vessel in motion (i.e., transportation vessel) suffers an impact.
  - o Cutting
  - o Welding
  - o Stress imparted to the vessel or attached appurtenances from external sources such as lifting, jacking, prying, settling of supports, natural event, etc.
  - o Overheating due to fire or other external mechanism causing loss of strength due to high vessel temperature
- Fabrication, transportation and installation – Flaws, weaknesses and deficiencies induced during fabrication, such as post weld heat treatment

and transportation induced damage, causing the installed vessel to not correspond sufficiently with the design intent.

7.1.2.   Damage Factors

Damage to the surroundings from a PVB may be caused by blast pressure, fragments, and thermal effects. Blast effects provide the primary source of damage in a PVB. Fragmentation typically causes damage relatively close to the vessel. Unlike BLEVEs, thermal effects for PVBs are usually minor to negligible. This chapter will address blast and fragmentation effects.

The energy of a PVB explosion resides in the compressed gases in the vessel just prior to burst. Liquids (not in suppressed boiling) and/or solids may occupy volume inside a pressure vessel, which reduces the volume of compressed gas. (Liquids in suppressed boiling can contribute to a BLEVE, which is addressed in Chapter 8). The volume of liquids and solids in pressure vessels may vary during operations, and the explosion energy is greatest for a given pressure when liquid and solid levels are lowest.

The energy in the compressed gas is also affected by the temperature of the gas. Pressure vessels are often operated at elevated temperature, or a runaway reaction may increase the temperature above the normal operation range. For a given vessel volume, the explosion energy varies inversely with temperature due to the reduction in gas density at elevated temperature.

7.1.3.   Phenomena

For simplicity, a PVB will be depicted by an idealized case of a massless, spherical vessel burst. Upon burst, flow is uniformly outward, which can be studied as a one-dimensional problem. Prior to vessel burst, the massless vessel wall separates the high pressure region (pressurized gas inside the vessel) and low pressure region (surrounding air).

When the vessel bursts, a shock wave propagates into the surrounding air. Pressure and temperature increase very rapidly across the shock front. Concurrently, expansion waves propagate into the high pressure vessel. Between the shock wave and the expansion wave there is a contact discontinuity that separates the gases initially inside the vessel from the surrounding air. Neglecting diffusion, the gases that were initially inside and outside the vessel do not mix, but are separated across the contact discontinuity sometimes referred to as the contact *surface.* The contact surface moves outward from the vessel until pressure of the gases from inside the vessel equalizes with the pressure of the surroundings.

Because vessel size is finite, the inward propagating expansion or rarefaction waves will collide at the center. The collision causes the rarefaction waves to reflect at the center and then the reflected rarefaction waves propagate outward. The outward propagating rarefaction waves will catch up with and reflect from the contact surface. Secondary and tertiary shocks are formed by the successive implosions and expansions of shock waves and rarefaction waves from the contact surface and vessel center, but these shocks are weak and are not considered in the predictive models.

Another noticeable characteristic of blast waves from vessel bursts is the sizable negative phase when compared with the positive phase. The relatively large negative phase is typical for explosion sources with low energy density, which is also observed in vapor cloud explosions with low flame speeds.

7.1.4.   Factors that Reduce Available Explosion Energy

In previous sections, it has been assumed that all of the energy within a pressure vessel is available to drive the blast wave. In fact, energy must be spent to rupture the vessel, propel its fragments, and accelerate the gases. In some cases, the vessel expands in a ductile manner before bursting, thus absorbing additional energy. Should a vessel also contain liquid or solids, a fraction of the available energy will be spent in propulsion of the contents.

Vessel Rupture. The energy needed to tear the vessel wall is a relatively low proportion of the total energy, and can be neglected in the calculation of the explosion energy. For a steel vessel, rupture energy is on the order of 1 to 10 kJ, that is, less than 1% of the energy of a small explosion.

Fragments. As will be explained in Section 7.5.2.1, 20% and 50% of available explosion energy may be transformed into kinetic energy of fragments and liquid or solid contents (Baum, 1984).

Vessel Contents. Solids and liquids in the vessel may absorb a portion of the energy if propelled from the vessel by the compressed gas. An example is packing material within a reactor. Analysis of propulsion of vessel contents is beyond the scope of this edition and requires consideration of the specific circumstances. Propulsion of vessel contents can be neglected to produce a conservative estimate of explosion energy.

7.2.   SCALING LAWS USED IN PVB ANALYSES

Airblast data for PVBs are presented as scaled parameters developed from similitude analysis (Baker, 1973). The most common form of scaling used for PVBs is Sach's scaling, also called energy scaling. All of the scaled parameters are dimensionless, which allows any set of self-consistent units to be applied. The scaled parameters used in this chapter are:

Scaled standoff distance:

$$\overline{R} = \frac{R\,p_o^{1/3}}{E^{1/3}}$$
(Eq. 7.1)

Scaled side-on overpressure:

$$\overline{P} = \frac{P_{so}}{p_o}$$
(Eq. 7.2)

Scaled side-on specific impulse:

$$\overline{i} = \frac{i\,a_0}{p_o^{2/3}E^{1/3}}$$
(Eq. 7.3)

Scaled duration:

$$\overline{t} = \frac{ta_o p_o^{1/3}}{\gamma_o^{1/2}E^{1/3}}$$
(Eq. 7.4)

where:

p_0 $=$ atmospheric pressure

a_o $=$ speed of sound in ambient air

E $=$ explosion energy

γ_1 $=$ the specific heat ratio of the vessel gas

γ_o $=$ the specific heat ratio of surrounding air

p_o $=$ ambient pressure

P_{so} $=$ peak side-on overpressure

R =   standoff distance from the center of the vessel

t =   duration

Other parameters used throughout the chapter include:

Burst pressure ratio: $\dfrac{P_1}{P_o}$ (Eq. 7.5)

Temperature ratio: $\dfrac{T}{T_o}$ (Eq. 7.6)

where:

P_1 =   vessel absolute pressure

T =   vessel absolute temperature

T_o =   ambient absolute temperature

7.3.   BLAST EEFFECTS OF PRESSURE-VESSEL BURSTS

This section addresses the blast effects of PVBs. The most important blast-wave parameters are peak overpressure, P_s, and positive specific impulse, i_s, as shown in

Figure 7.1. The deep negative phase and second shock are clearly visible in this figure.

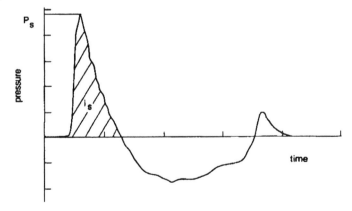

Figure 7.1. Pressure-time history of a blast wave from a PVB
(Esparza and Baker 1977a).

The strength and shape of a blast wave produced by a PVB depend on many factors, including type of fluid released, energy it can produce in expansion, rate of energy release, shape of the vessel, type of rupture, and character of the surroundings (i.e., the presence of wave-reflecting surfaces and ambient air pressure). In the following subsections, a selection of the theoretical and experimental work on PVBs will be reviewed. Attention will first be focused on an idealized situation: a spherical, massless vessel filled with ideal gas and located high above the ground. Increasingly realistic situations will be discussed in subsequent subsections.

7.3.1. Free-Air Bursts of Gas-Filled, Massless, Spherical Pressure Vessels

The pressure vessel under consideration here is spherical and is located far from surfaces that might reflect the shock wave. Furthermore, it is assumed that the vessel will fracture into many massless fragments, that the energy required to rupture the vessel is negligible, and that the gas inside the vessel behaves as an ideal gas. The first consequence of these assumptions is that the blast wave is perfectly spherical, thus permitting the use of one-dimensional calculations. Second, all energy stored in the compressed gas is available to drive the blast wave. Certain equations can then be derived in combination with the assumption of ideal gas behavior.

7.3.1.1. Experimental Work

Few experiments measuring the blast from exploding, gas-filled pressure vessels have been reported in the open literature. One was performed by Boyer et al. (1958) in which they measured the overpressure produced by the burst of a small, glass sphere pressurized with gas.

Pittman (1972) performed five experiments with titanium-alloy pressure vessels which were pressurized with nitrogen until they burst. Two cylindrical tanks burst at approximately 40 bar, and three spherical tanks burst at approximately 550 bar. The volume of the tanks ranged from 0.0067 m^3 to 0.170 m^3. A few years later, Pittman (1976) reported on seven experiments with 0.028-m^3 steel spheres that were pressurized to extremely high pressures with argon until they burst. Nominal burst pressures ranged from 1000 bar to 3450 bar. Experiments were performed just above ground surface.

Finally, Esparza and Baker (1977a) conducted twenty small-scale tests in a manner similar to that of Boyer et al. (1958). They used glass spheres of 51 mm and 102 mm diameter, pressurized with either air or argon, to overpressures ranging from 12.2 bar to 53.5 bar. They recorded overpressures at various places and filmed the fragments. From these experiments, it was learned that when compared to the shock wave produced by a high explosive of equal energy, shock waves produced by bursting gas-filled vessels have lower overpressures near the source, longer positive-phase durations, much larger negative phases, and stronger secondary shocks.

Figure 7.1 depicts such a shock. Pittman (1976) also found that the blast can be highly directional, and that real gas effects must be dealt with at high pressures.

7.3.1.2.   Numerical Work

The results of experiments described above can be better understood when compared to the results of numerical and analytical studies. Numerical studies, in particular, provide real insight into the shock formation process. Chushkin and Shurshalov (1982) and Adamczyk (1976) provide comprehensive reviews of the many studies in this field. The majority of these studies were performed for military purposes and dealt with blast from nuclear explosions, high explosives, or fuel-air explosions (FAEs: detonations of unconfined vapor clouds). However, many investigators studied (as a limiting case of these detonations) blasts from volumes of high-pressure gas as well. Only the most significant contributions will be reviewed here.

Many numerical methods have been proposed for this problem, most of them finite-difference methods. Using a finite-difference technique, Brode (1955) analyzed the expansion of hot and cold air spheres with pressures of 2000 bar and 1210 bar. The detailed results allowed Brode to describe precisely the shock formation process and to explain the occurrence of a second shock.

Baker et al. (1975) supplemented and extended the work reported by Strehlow and Ricker (1976) to include many more cases. They used a finite-difference method with artificial viscosity to obtain blast parameters of spherical pressure vessel explosions. They calculated twenty-one cases, varying pressure ratio between vessel contents (gas) and surrounding atmosphere, temperature ratio, and ratio of the specific heats of the gases. They used ideal-gas equations of state. Their research was aimed at deriving a practical method to calculate blast parameters of bursting pressure vessels, so they synthesized the results into graphs presenting shock overpressure and impulse as a function of energy-scaled distance.

Adamczyk (1976) noted that overpressures equivalent to high explosives are usually attained only in the far field for high burst pressures. When low burst pressure is used, overpressure curves do not converge in the far field; hence, equivalence with high explosives may not be attained.

Guirao and Bach (1979) used the flux-corrected transport method (a finite-difference method) to calculate blasts from fuel-air explosions. Three of their calculations were of a volumetric explosion, that is, an explosion in which the unburned fuel-air mixture is instantaneously transformed into combustion gases. By this route, they obtained spheres whose pressure ratios (identical with temperature ratios) were 8.3 to 17.2, and whose ratios of specific heats were 1.136 to 1.26. Their calculations of shock overpressure compare well with those of Baker et al. (1975). In addition, they calculated the work done by the expanding contact surface between combustion products and their surroundings. They found that only 27% to 37% of the combustion energy was translated into work.

Tang et al. (1996) presented the results of a study on the blast effects generated by bursting vessels. The properties of the shock wave and the associated flow field were determined by solving the non-steady, non-linear, one-dimensional equations numerically by means of the total variation diminishing (TVD) scheme proposed by Harten (1983). In addition to positive overpressure and impulse blast waves, negative pressure and impulse, arrival time of the shock front, durations of positive and negative phases, and flow velocity were presented, which had not been published previously. Higher accuracy was achieved due to the employment of the high resolution shock capturing scheme, which prevented dissipation of overpressure in numerical computations and produced more realistic overpressure predictions in the far field.

Baker et al. (2003) reported that blast pressures from higher temperature gaseous vessel contents are greater than those from lower temperatures at close distances, but become close and even slightly lower than lower temperature bursts at far distances. Prior practices of assuming that blast curves are parallel resulted in over-prediction of blast loads from high temperature in the far fields. Baker and Tang (2003) presented and discussed the effect of vessel temperature at different pressure levels. A modification of the engineering method is presented in Baker (2003), which gives more accurate results and therefore avoids the over-prediction of blast loads from high temperature bursts.

7.3.1.3.   Analytical Work

Analytical work has been performed to address key parameters related to pressure vessel explosions, including explosion energy and initial shock strength.

Step 1. Evaluation of explosion energy

Explosion energy is a very important parameter; of all the variables, explosion energy has the greatest influence on rate of decay of blast pressure and impulse with distance and the magnitude of impulse. Thus, the destructive potential of a PVB is strongly influenced by explosion energy. However, the evaluation of PVB explosion energy may vary significantly among definitions based on different thermodynamic processes.

Regardless of the availability of various energy equations, it is imperative when using scaled blast curves that the energy equation upon which the blast curves were created be used for blast predictions. Substitution of a different energy equation can result in prediction errors.

a). Constant volume energy addition – Brode's definition

Brode's (1959) definition gives the upper limit of the energy release, which is based on a constant volume process. Thus, explosion energy is defined as the energy, $E_{ex,Br}$, that must be employed to pressurize the vessel gas from ambient pressure to the burst pressure; that is, the difference in internal energy between the two states.

Therefore $E_{ex,Br}$ is

$$E_{ex,Br} = \frac{(p_1 - p_0)V_1}{(\gamma_1 - 1)}$$

(Eq. 7.7)

where the subscript 1 refers to initial state prior to vessel burst and the subscript 0 refers to ambient conditions, and:

γ_1	=	ratio of constant pressure to constant volume specific heats of the gas in vessel,
p_1	=	pressure in vessel prior to burst (absolute),
p_0	=	ambient pressure,
V_1	=	vessel volume.

b). Constant pressure energy addition

Another extreme is the definition based on a constant pressure process, which is the lower limit of explosion energy. This represents the energy release rate that is insufficient to form a blast wave.

$$E_{ConsP} = p_0 \left(V_2 - V_1 \right)$$

(Eq. 7.8)

where V_2 is the final volume occupied by the gas which was originally in the vessel.

c). Isentropic expansion

Strehlow and Adamczyk (1977) included an estimation based on isentropic expansion from initial burst pressure to atmospheric pressure.

For an ideal gas,

$$E_{ex,\, isen} = \frac{p_1 V_1 - p_0 V_2}{(\gamma_1 - 1)}$$

(Eq. 7.9)

For a low energy density explosion source, the expansion can be approximated as isentropic expansion with constant γ as, $pV^{\gamma} =$ constant. Therefore:

$$V_2 = V_1 \left[\frac{p_1}{p_0} \right]^{1/\gamma_1}$$

(Eq. 7.10)

Thus

$$E_{ex,isen} = \frac{p_1 V_1}{(\gamma_1 - 1)} \left[1 - \left(\frac{p_0}{p_1} \right)^{(\gamma_1 - 1)/\gamma_1} \right]$$

(Eq. 7.11)

d). Isothermal expansion

Smith and van Ness (1987) presented the isothermal case, which assumes the gas expands isothermally.

$$E_{ex,isoth} = R_g T_o \ln\left(\frac{p_1}{p_o} \right) = p_1 V_1 \ln\left(\frac{p_1}{p_o} \right)$$

Where R_g is the ideal gas constant and T_o is the ambient temperature.

e). Aslanov's definition

Aslanov and Golinskii (1989) give yet another definition of explosion energy. By taking into account the internal energy of the air that is displaced by the expanded gases, they proposed the following explosion energy:

$$E_{ex,AG} = \frac{p_1 V_1 - p_0 V_2}{\gamma_1 - 1} + \frac{p_0 (V_2 - V_1)}{\gamma_0 - 1}$$

(Eq.7.12)

where γ_0 is the ratio of specific heats of ambient air. When γ_1 is equal to γ_0, the above equation is equivalent to Brode's Eq. (7.7).

f). Thermodynamic availability - Crowl's definition

Crowl (1992) used batch thermodynamic availability to predict the maximum explosion energy of a gas contained within a vessel:

$$E_{ex,avail} = p_1 V_1 \left[\ln \left(\frac{p_1}{p_0} \right) - \left(1 - \frac{p_1}{p_0} \right) \right]$$

(Eq.7.13)

In summary, the constant pressure definition is not used because it is for the very slow process that generates no blast effects. At the other extreme, the isothermal expansion method overpredicts energy by neglecting waste heat. Brode's definition, Crowl's availability method, and isentropic expansion give realistic values. Brode's definition based on constant volume process has been used in most numerical calculations. Adamczyk and Strehlow (1977) showed blast yield must lie below the definition by isentropic expansion. The process in a real world scenario is the combination of constant pressure and isentropic expansion processes.

As mentioned above, it is imperative when using scaled blast curves that the energy equation upon which the blast curves were created be used for blast predictions. Substitution of a different energy equation will result in prediction errors.

Step 2. Determination of initial shock strength

When an idealized spherical vessel bursts, the air shock starts and has its maximum overpressure right at the contact surface between the vessel gas and the air. Since the flow is one-dimensional, the shock-tube relationship can be used to calculate the pressure jump across the shock front from the pressure ratio of vessel gas and surrounding air. The shock-tube relation is derived from the conditions at the contact surface and shock and expansion wave relations (Liepmann and Roshko, 1967).

The basic shock-tube equation is:

$$\frac{p_1}{p_0} = (\overline{P}_{so} + 1) \left[1 - \frac{(\gamma_1 - 1)(a_0 / a_1)\overline{P}_{so}}{\left[2\gamma_0 \left\{ 2\gamma_0 + (\gamma_0 + 1)\overline{P}_{so} \right\} \right]^{1/2}} \right]^{-2\gamma_1 / (\gamma_1 - 1)}$$

(Eq. 7.14)

where: $\overline{P}_{s0} = \dfrac{p_{so}}{p_o} - 1$

p_o = ambient pressure

p_1 = vessel absolute pressure

p_{so} = the initial shock absolute pressure just outside the vessel

a_o = the speed of sound in air outside the vessel

a_1 = the speed of sound in the gas inside the vessel

This gives the initial shock strength p_{so}/p_0 implicitly as a function of the vessel pressure ratio, p_1/p_0. Usually p_{so}/p_0 is unknown and one must solve the above equation by iteration. It is shown in the shock tube equation that the pressure in the vessel prior to burst is the predominant factor in determining the initial shock strength. However, the blast wave strength is also dependent on temperature and specific heat ratio of the vessel gas.

7.3.2.   Effects Due to Surface Bursts

In the previous subsection, an idealized configuration was studied. In this and following subsections, the influences of other factors will be discussed. When an explosion takes place at the ground surface or slightly above it, the shock wave produced by the explosion will reflect on the ground surface. The reflected wave overtakes the first (incident) wave and coalesces, and the combined wave is stronger than either of the individual waves. The resulting shock wave is similar to a shock wave that would be produced in free air by the original explosion together with its imaginary mirror image below the ground surface.

This subject has received little attention in the context of PVBs. Pittman (1976) studied it using a two-dimensional numerical code. However, his results were inconclusive because the number of cases he studied was small and because the grid he used was coarse. Based on experimental results with high explosives, Baker et al. (1975) recommends multiplying the explosion energy by 2 to take into account ground surface reflection when reading blast curves for free-air bursts.

Ground surface reflection is not to be confused with blast wave reflection from other surfaces, such as the wall of a building. Both types of reflection occur and both must be taken into account in blast predictions.

7.3.3.   Effects Due to Nonspherical Bursts

When a pressure vessel is not a sphere, or if the vessel does not fracture evenly, the resulting blast wave will be nonspherical. This is the case in many actual PVBs. Nonspherical geometry means that detailed calculations and experimental measurements become much more complicated, because the calculations and measurements must be made in two or three dimensions instead of one. Numerical calculations of bursts of pressurized-ellipsoid gas clouds were made by Raju and Strehlow (1984) and by Chushkin and Shurshalov (1982). Raju and Strehlow (1984) computed the expansion of a gas cloud corresponding to a constant volume combustion of a methane-air mixture ($p_1/p_0 = 8.9$, $\gamma_1 = 1.2$), demonstrating that the shape of the shock wave is almost elliptical near the vessel surface. Since strong shock waves travel faster than weak ones, the shape of the shock wave approached spherical in the far field. Shurshalov (Chushkin and Shurshalov 1982) performed a similar calculation; results confirmed those of Raju and Strehlow (1984). Shurshalov also found that the shock wave approaches a spherical shape more rapidly when the explosion source is higher pressure.

Even greater differences in shock pressure can be found when a pressure vessel does not burst evenly, but ruptures into two or three pieces. In that case, a jet emanates from the rupture, and the shock wave becomes highly directional. Pittman (1976) found experimental overpressures along the line of the jet to be greater, by a factor of four or more, than pressures along a line in the opposite direction from the jet.

Pipelines and tube reactors are special cases of nonspherical geometry. A straight pipeline in free air that ruptures over a long length produces a cylindrical blast wave in the near field. Rate of decay of blast pressure and impulse are less than for a sphere since expansion is 2-dimensional rather than 3-dimensional.

Likewise, a long tube reactor may produce a cylindrical blast wave in the near field. No blast curves have been produced for cases of pipeline and tube reactors; however, 2-dimensional CFD models that properly model shock waves can address this problem.

Geng, Baker and Thomas (2009) conducted a parametric study of the directional blast effect from cylindrical and elevated spherical PVBs using a CFD model, resulting in a group of adjustment factors for overpressure and impulse that will be discussed in Section 7.4.4.2. Figure 7.2 shows typical pressure contours at selected times of a blast field generated from a cylindrical pressure vessel with an aspect ratio of L/D=5 and an initial bursting pressure of $p_1/p_0 = 50$. L and D are cylinder length and diameter, respectively. The vertical axis in Figure 7.2 is the cylinder axis, and represents the axis of symmetry of the 2D simulations. Figure 7.2 only displays one quarter of the geometry. It should be noted that the pressure and dimension scales are changed for the different times to maximize resolution at the given time step. Dimensions of L and D were determined to ensure that the equivalent sphere (with the same volume as the cylinder) has characteristic length $r_0 = 1$. The scaled distance in the X-direction is x/r_0, and the scaled distance in the Y-direction is similarly scaled by r_0. Characteristic time is defined as the time it takes for an acoustic signal to propagate from the origin to r_0. Time in Figure 7.2 is scaled by characteristic time. All scaled parameters are dimensionless.

The first frame (t=0) in Figure 7.2 shows the initial cylindrical gas cloud before expansion. At time of t=2, an elliptical blast wave front is generated, with weaker shock waves along the long-axis (i.e., cylinder axis) and stronger shock waves along the short-axis (i.e., the direction normal to the cylinder axis). As the blast wave propagates outward further, the blast effects become more evenly distributed and begin to approach spherical. Note, however, that even in the last frame (t=14) shown in Figure 7.2, the compression and rarefaction waves are not completely uniform across all directions.

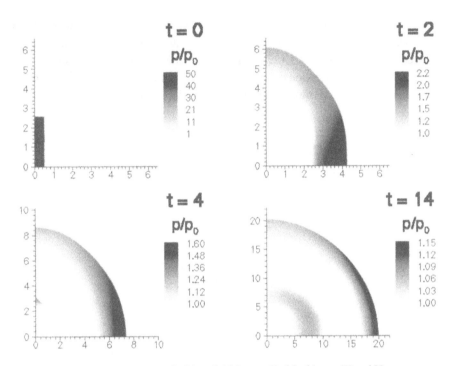

Figure 7.2. Pressure contours of a blast field for a cylindrical burst (X and Y axes are scaled distances based on characteristic distance r_o).

(Geng, 2009)

Figure 7.3 shows a sequence of typical pressure contours of a blast field produced by an elevated spherical gas cloud with an initial bursting pressure of $p_1/p_0 = 50$ (at t=0). The figure only displays one half of the geometry. The vertical axis in this figure is the axis of symmetry of the 2D simulations. The height to radius ratio of $H_S/R_S=2$ was examined. The pressure and dimension scales changed for the different times to maximize resolution at the given time step. The characteristic length is $R_S =1$. The scaled distance in the X-direction is x/R_S, and the scaled distance in the Y-direction is similarly scaled by R_S. Characteristic time is defined as the time it takes for an acoustic signal to propagate from the origin to R_S.

Time in Figure 7.3 is scaled by characteristic time. All scaled parameters are dimensionless. The waves emanating from the sphere are initially uniform; however, by the second frame (t=2) shown in the sequence in Figure 7.3, the waves propagating toward the ground plane have begun to reflect off of the ground. By the third frame of the sequence (t=4), the leading front of the reflected wave has intersected the initial leading wave along the ground plane. This interaction leads to non-uniform propagation of the compression and rarefaction waves. Noticeable directionality of the blast from this elevated sphere remains evident through the last frame (t=14) included in the sequence.

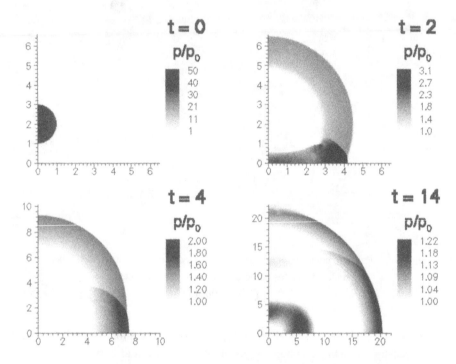

Figure 7.3. Pressure contours of a blast field for an elevated spherical burst (X and Y axes are scaled distances based on characteristic distance r_o).

(Geng, 2009)

Examples of scaled peak overpressure and positive impulse as a function of scaled distance are given in Figure 7.4 and Figure 7.5, respectively, for cylindrical bursts with an aspect ratio of L/D = 5 and an initial busting pressure $p_1/p_0=50$. Sach's scaling with the Brode energy equation was used in these two figures. For the sake of comparison, the blast curves from a bursting sphere with an equivalent volume (or energy) are also included in figures. For the cylindrical case, the blast from the PVB is distinctly non-uniform. Note that in the near field, the overpressure along the axis of symmetry (0 degrees) remains higher than for the spherical case. As distance increases, however, it drops to a value that is less than the corresponding overpressure for the sphere. Conversely, overpressures along the direction normal to cylinder axis (90 degrees) are initially lower than the corresponding values for the sphere but deteriorate less rapidly after $\overline{R}$ =0.1, remaining higher than the overpressures for the sphere out to $\overline{R}$ =10. Along the direction of 45 degrees, overpressure is similar to that for the sphere, in particular, in the far field ($\overline{R}$ >1). Similar trends are observed when comparing the radial distributions of impulse for the two cases.

Adjustment factors can be calculated for the cylinder and elevated sphere by taking the ratio of peak blast overpressure or impulse for the given geometry along a given direction to the corresponding value for the equivalent spherical PVB. A detailed discussion of the calculated adjustment factors is given in Section 7.4.4.2.

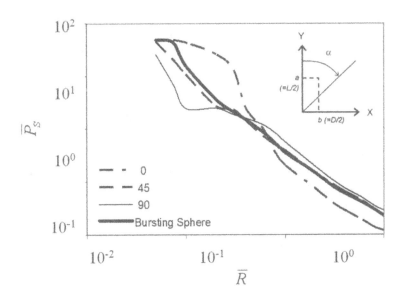

Figure 7.4. Surface burst scaled side-on overpressure generated by a cylindrical burst at angles of 0, 45 and 90° compared to a bursting sphere. (Geng, 2009)

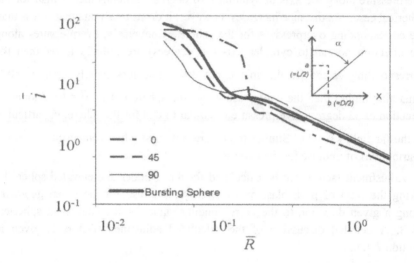

Figure 7.5. Surface burst scaled side-on impulse generated by a cylindrical burst at angles of 0, 45 and 90° compared to a bursting sphere. (Geng, 2009)

7.4.　　METHODS FOR PREDICTING BLAST EFFECTS FROM VESSEL BURSTS

Blast prediction methods for vessel bursts may be grouped into three categories according to their nature and complexity: the TNT equivalence method, blast curve methods, and numerical simulations. Attention will be focused on the blast curve method, which is the most widely used engineering method in practice.

The TNT equivalence method is not recommended. An explosive like TNT is a poor representation for a bursting vessel for most practical cases. The TNT equivalent method overpredicts PVB overpressure close in, and underpredicts in the far field. Impulse is underpredicted by the TNT equivalence at all distances except very near the vessel.

Numerical simulations based on computational fluid dynamics (CFD) have been developed. CFD modeling of a PVB is a fundamental approach that involves solving conservation equations in a grid system. Since the three-dimensional (3-D) numerical simulation can best represent complex, real world scenarios, it has the potential of providing higher accuracy and directly addressing the details of given scenario. As the result of intensive development programs, several CFD codes have been developed.

Experimental validation is critical for numerical models. Therefore, the application of any computer code should be limited to the type of problems for which the code has been validated by experiments. Also, a number of variables such as grid and time resolution can have significant effects on the results, requiring a high level of user expertise. Because of the complexity of the subject, detailed description of numerical simulation is beyond the scope of this text.

7.4.1.   Development of Blast Curves

One-dimensional numerical studies were carried out by researchers using different numerical techniques that resulted in several sets of blast curves. The numerical simulations accurately simulated the explosion source conditions (pressure, temperature, gas properties, etc.) and shock expansion. As a result, the PVB blast curves were a significant improvement over the TNT equivalence method.

7.4.1.1.   Baker-Strehlow Blast Curves

Overpressure and impulse versus distance curves for bursting spheres were developed by Strehlow and Ricker (1976). These curves were produced by means of one-dimensional numerical computations in Lagrangian coordinates and based on the von Neumann-Richtmyer finite difference scheme (von Neumann-Richtmyer, 1950).

Baker et al. (1975) supplemented and extended the work reported by Strehlow and Ricker (1976) to include many more cases of spherical PVBs; pressure ranged from 5 to 37,000 atmosphere, temperature ranged from 0.5 to 50 times the atmospheric absolute temperature, and the vessel gas ratio of specific heats was 1.2, 1.4, and 1.667 to investigate the dependence. The collaboration of Baker and Strehlow ultimately resulted in publication of revised PVB blast curves and methodology in Baker (1983).   The Baker-Strehlow curves have found international acceptance and are often referenced. They were adopted in the TNO Yellow Book (Yellow Book, 1979, 1997).

7.4.1.2.   Baker-Tang Blast Curves

Tang, Cao and Baker (1996) published a complete set of blast curves based on a systematic one-dimensional (spherical) numerical study (referred to as the Baker-Tang curves). Effects of the containing vessel and its fragments were ignored in this study; that is, all of the energy was put into the flowfield, rather than into the fragments as kinetic energy. Also, the surrounding gas was assumed to be air, and all fluids were perfect gases. The Baker-Tang PVB blast curves are spherical, free-air and do not include ground reflection. Pressure vessels on or near the ground are modeled by doubling the energy to account for ground reflection.

The spherical free-air geometry assumes that a spherical vessel fails abruptly and uniformly over its entire surface, creating a symmetrical shock wave. Non-uniform vessel failures, such as one end separating from a cylindrical pressure vessel, will produce non-symmetrical blast waves close to the vessel. Near field blast pressure could be higher or lower than predicted by the spherical free-air blast curves, and specialized CFD modeling may be needed. Section 7.3.3 addresses geometry effects from cylindrical PVBs and elevated spherical PVBs with uniform vessel failures.

The Baker-Tang curves were developed with a numerical model that reduced overpressure losses associated with artificial viscosity in prior models. As a result, far-field predictions are somewhat higher than the Baker-Strehlow curves in the far field. The Baker-Tang curves were also developed to simplify calculations by developing curves for specific burst pressures, eliminating the need for iterative solution of the shock tube equation and the graphical method of plotting the blast curve in Baker-Strehlow method.

The Baker-Tang PVB overpressure and impulse blast curves for both positive and negative phases are presented in Figure 7.6 through Firgure 7.9. The high explosive (HE) blast curve is shown for comparison. The scaled parameters and distance used in the figures are given in Section 7.2. Brode's energy was used in scaling all blast parameters.

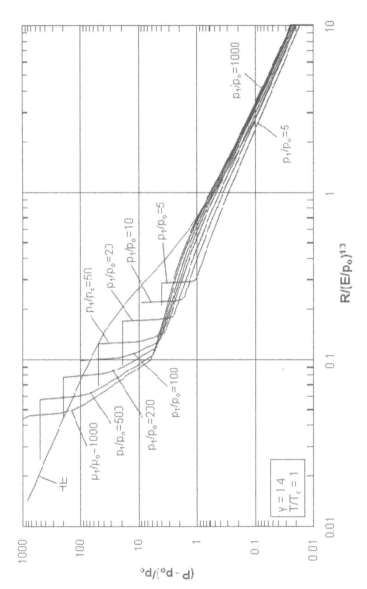

Figure 7.6. Positive overpressure curves for various vessel pressures. (Tang, et al. 1996)

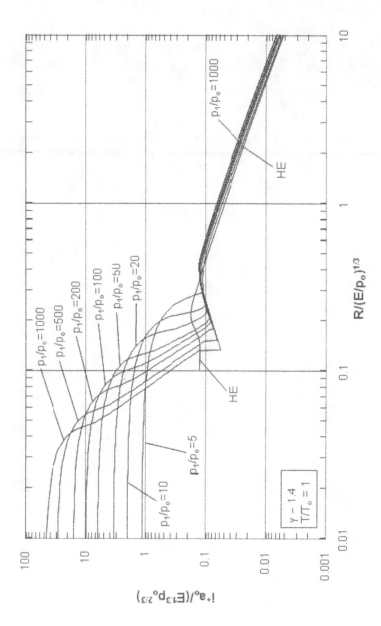

Figure 7.7. Negative pressure curves for various vessel pressures.

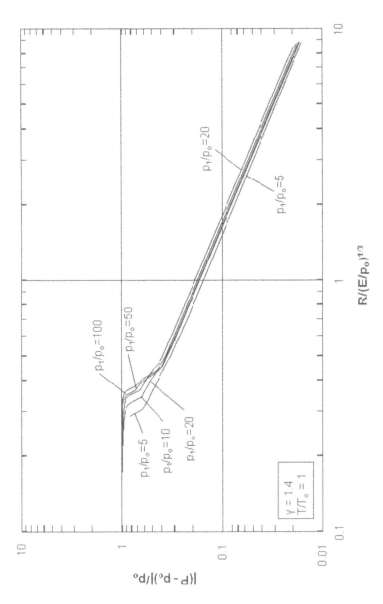

Figure 7.8. Positive impulse curves for various vessel pressures. (Tang, et al. 1996)

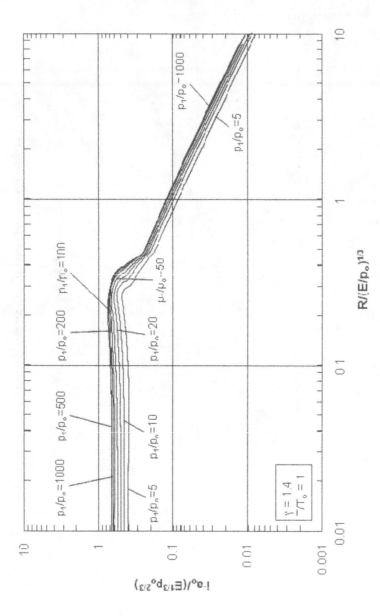

Figure 7.9. Negative impulse curves for various vessel pressures. (Tang, et al. 1996)

7.4.2.   Factors Influencing Blast Effects from Vessel Bursts

Tang and Baker (1996, 2004) carried out a parametric study that investigated the sensitivity of the pressure and impulse predictions to initial conditions. The results are summarized as follows.

The pressure in the vessel prior to bursting is a predominant factor in determining blast wave parameters. Higher initial pressure at a fixed vessel volume means greater total energy and higher energy density of the explosion source, producing a stronger blast wave. Unlike ideal explosion sources such as high explosives, the bursting vessel blast curves corresponding to various source energy levels do not merge into a single curve even when Sach's energy scaling is applied. This departure from ideal explosion source behavior is significant since it indicates that a smaller fraction of the source energy is transmitted into the blast wave for a source with a lower energy density.

Explosion energy varies directly with gas volume in a pressure vessel as can be seen in each of the energy equations in Section 7.3.1.3. Volume occupied by solids and liquids (not in suppressed boiling) inside the vessel can be deducted from the total vessel volume when calculating explosion energy A conservative estimate of explosion energy can be made by assuming the entire vessel is gas filled. The volume of attached piping is generally not included in the volume estimate since it typically contributes little to the explosion energy.

The effect of elevated vessel temperature needs to be taken into account only at close distances, $\overline{R} < 1.0$ for overpressure and $\overline{R} < 0.5$ for impulse. Blast pressure and impulse are slightly lower in the far field for high temperature sources. Therefore, it is reasonable and conservative to conclude that source temperature above ambient can be neglected for $\overline{R}$ greater than 1.0.

Source gases with higher values of γ produce greater impulses than with lower values of γ. The effect of γ on blast overpressures is not readily apparent in the energy scaled overpressure curves.

7.4.3.   Procedure for Calculating Blast Effects

The discussion of practical prediction methods will start with a massless sphere filled with ideal gas. Then adjustments will be introduced to account for the effect of fragmentation and real gas behavior and/or other factors on blast wave properties.

Step 1: Collect data

Collect the following data:

- the vessel's internal pressure (absolute), p_1
- the ambient pressure, p_0
- the vessel's volume of gas-filled space, V_1
- the ratio of specific heats of the gas, γ_1
- the distance from the center of the vessel to the "receptor," r,
- the shape of the vessel: spherical or cylindrical,
- speed of sound of the vessel gas, a_1,
- speed of sound of ambient air, a_0
- ratio of specific heats of the vessel gas, γ_1
- ratio of specific heats of ambient air, γ_0

Step 2: Calculate vessel-gas energy

The energy E of a compressed gas in the vessel is calculated using Brode's formula for free air explosions:

$$E = \frac{(p_1 - p_0)V_1}{\gamma_1 - 1}$$

(Eq. 7.15)

For ground burst, the energy is doubled to account for the ground reflection as:

$$E = \frac{2(p_1 - p_0)V_1}{\gamma_1 - 1}$$

(Eq. 7.16)

Step 3: Select blast curve

Calculate the vessel burst pressure ratio, p_1/p_0. Then locate the overpressure and impulse curves that correspond to the vessel burst pressure ratio in Figure 7.6 and Figure 7.7, respectively.

Step 4: Calculate Scaled Standoff Distance $\bar{R}$ of the receptor

Calculate the scaled distance of the receptor, $\bar{R}$, with:

$$\bar{R} = R\left[\frac{P_0^{1/3}}{E^{1/3}}\right]$$

(Eq. 7.17)

where R is the distance at which blast parameters are to be determined.

Step 5: Determine Scaled Positive Overpressure $\bar{P}_s$

To determine the scaled side-on overpressure $\bar{P}_s$, use the curve that corresponds to the vessel burst ratio p_1/p_0, and read $\bar{P}_s$ from Figure 7.6 for the appropriate $\bar{R}$.

Step 6: Determine Scaled Positive Impulse $\bar{i}_+$

Similarly, read $\bar{i}_+$ from Figure 7.7 for the appropriate $\bar{R}$.

Step 7: Adjust $\bar{P}_s$ and $\bar{i}_+$ for different initial conditions such as vessel temperature, and geometry

See Section 7.4.4 for various adjustments.

Step 8: Calculate P_s and i_s

Use Equations 7.2 and 7.3 rearranged to calculate side-on peak overpressure P_s and side-on impulse i_s from scaled side-on peak overpressure $\bar{P}_s$ and scaled side-on impulse $\bar{i}$:

$$P_s = \bar{P}_s p_o$$

(Eq 7.18)

$$i_s = \frac{\bar{i}p_o^{2/3}E^{1/3}}{a_o}$$

(Eq 7.19)

where a_o is speed of sound in ambient air. For sea-level average conditions, p_0 is approximately 101.3 kPa and a_0 is 340 m/s.

7.4.4.   Adjustments for Vessel Temperature and Geometry

7.4.4.1.   Adjustments for vessel temperature

Elevated temperature increases the positive overpressure near the vessel compared to the same vessel at ambient temperature, but the effect diminishes with distance away from the vessel. If the target is far enough from the vessel ($\bar{R}$ >1), adjustment for temperature effects is not necessary, and the use of ambient temperature   blast   curves   in Figure 7.6 through Figure 7.9 to represent the higher temperature ratios is adequate and conservative. For impulses, using the ambient temperature blast curves is always conservative for higher temperatures.

Baker and Tang (2004) provide overpressure and impulse blast curves for elevated temperatures at a vessel burst pressure ratio of 50. CFD models may also be used for specific pressure and temperature conditions.

7.4.4.2.   Adjustments for geometry effects

The procedure presented in Section 7.4.3 produces blast parameters applicable to a spherical blast wave, which is the case for a spherical vessel that abruptly breaks into many pieces. In practice, vessels are either spherical or cylindrical, and placed at some height above the ground. This influences blast parameters. To adjust for these geometry effects, some adjustment factors derived from experiments with high-explosive (HE) charges of various shapes were developed by Baker (1975). However, it was found by Geng (Geng, Baker and Thomas, 2009) that the adjustment factors based on HE experiments are not directly applicable to cylindrical or elevated spherical PVBs. Instead, Geng developed a group of adjustment factors specifically for PVBs.

These PVB adjustment curves are shown in Figure 7.10 for cylindrical PBVs in free air and in Figure 7.11 for elevated spherical PBVs. The bursting pressure ranges from 10 to 100 for both the cylindrical and elevated spherical PVBs. The length to diameter ratio falls between 2 and 10 for cylindrical PVBs, and the height to radius ratio falls between 1 and 5 for elevated spherical PVBs. $\bar{P}_s$ and $\bar{I}$ can be multiplied by adjustment factors derived from these curves.

There are two equivalent configurations for the cylindrical free air PVB: vertical and horizontal cylinders on ground, as shown in Figure 7.12. Considering the ground as a reflecting plane, the original length L of a vertical cylinder on the ground can be doubled ($L'=2L$) to achieve the same blast pressures without the ground present. The resultant cylinder is equivalent to a cylinder with an aspect ratio of L'/D ($=2L/D$) in free air. Similarly, if a horizontal cylindrical vessel with a diameter of $d = D/\sqrt{2}$ is placed directly on the ground, the ground can also be treated as a reflecting plane in terms of resultant blast field. The equivalent free air cylinder is the one with an aspect ratio of L/D ($= L/\sqrt{2}\,d$). Thus, the cylinder with an aspect ratio of L'/D in free air, the vertical cylinder with an aspect ratio of L/D on the ground, and the horizontal cylinder with an aspect ratio of L/d on the ground are equivalent. The adjustment factor curves given in Figure 7.10 can be applied to both the vertical and horizontal cylinders placed directly on the ground.

The blast wave from a cylindrical vessel is weakest along its axis (see Section 7.3.3). Thus, the blast field is asymmetrical for a vessel placed horizontally. The method will only provide maximum values for a horizontal tank's parameters along 90 degrees direction (normal to the cylinder axis).

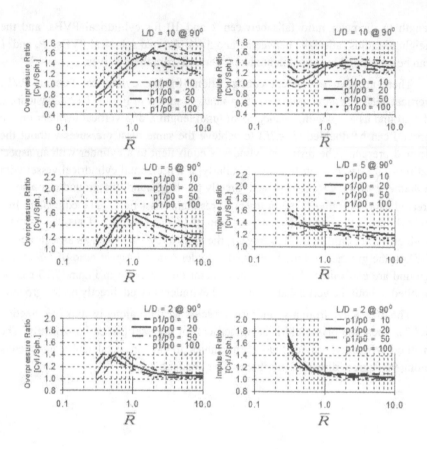

Figure 7.10. Adjustment factors for cylindrical free air PVBs compared to a spherical free air burst. (Geng, 2009)

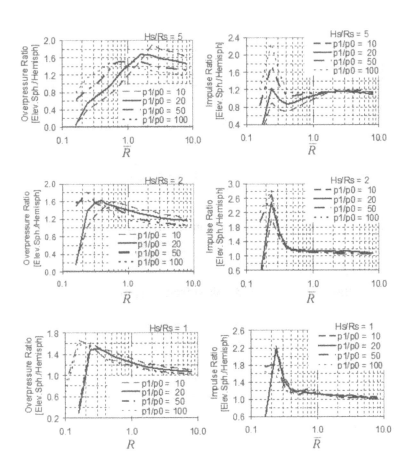

Figure 7.11. Adjustment factors for elevated spherical PVBs compared to a hemispherical surface burst. (Geng, 2009)

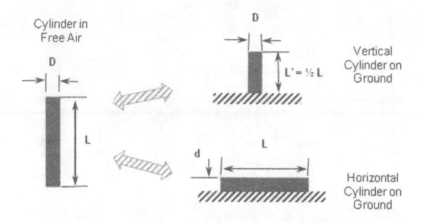

Figure 7.12. Equivalent surface burst cylindrical PVB geometries to a free air burst

7.4.5.   Sample Problem: Airblast from a Spherical Vessel

A spherical pressure vessel at grade is being pressure tested with nitrogen gas at a pressure 25% above maximum allowable working pressure (MAWP). If the vessel bursts during the test, a large storage tank, located 15 m from the vessel, and a control building, located 100 m from the vessel, might be endangered. What would be the side-on overpressure and impulse at these points?

The procedure:

Step 1: Collect data
- The ambient pressure p_o is 0.10 MPa.
- The MAWP of the vessel is 1.92 MPa, and the test pressure is 25% higher. Therefore, the absolute internal pressure p_1 is
    $p_1 = 1.25 \times 1.92$ MPa $+ 0.1$ MPa $= 2.5$ MPa (25 bar)
- The volume of the vessel, V_1, is 25 m^3.
- The ratio of specific heats of nitrogen, γ_1, is 1.40.
- The distance from the center of the vessel to the receptor, r, is 100 m for the control building and 15 m for the large storage tank.
- The vessel is spherical at ground level.

- The ratio of the speed of sound in the compressed nitrogen to the speed of sound in the ambient air, a_1/a_0, is approximately 1.
- The ratio of specific heats of the ambient air is 1.40.

Step 2: Calculate source energy

The energy of the compressed gas is calculated by:

$$E_{ex} = \frac{(p_1 - p_0)2V_1}{\gamma_1 - 1}$$

(Eq 7.20)

Substitution gives

$E_{ex} = (2.5 \times 10^6 Pa - 0.1 \times 10^6 Pa) \times 2 \times 25 m^3 / (1.4 - 1) = 300$ MJ

Step 3: Select the blast curve

To   select   the   proper   curve   in Figure 7.6, calculate the vessel burst pressure ratio p_1/p_0:

$$p_1/p_0 = 2.5 \text{ MPa} / 0.\,10 MPa = 25$$

Step 4: Calculate the scaled standoff $\bar{R}$ of the receptor

The scaled standoff $\bar{R}$ is calculated:

$$\bar{R} = R \left[\frac{p_0}{E_{ex}} \right]^{1/3}$$

(Eq. 7.21)

Substitution gives, for the control building:

$$\bar{R} = 100 \text{ m} \times (0.1 \times 10^6 \text{ Pa}/ 300 \times 10^6 J)^{1/3} = 6.9$$

And, for the large storage tank:

$$\bar{R} = 15 \text{m} \times (0.1 \times 10^6 \text{ Pa}/ 300 \times 10^6 J)^{1/3} = 1.04$$

Step 5: Determine $\bar{P}_s$

To determine the scaled side-on peak overpressure $\bar{P}_s$ at the large storage tank,

$\bar{P}_s$ is read from
Figure 7.6 at a scaled distance of $\bar{R} = 1.04$. When the curve for $P_1/P_0 = 25$ is followed from the starting point, a $\bar{P}_s$ of 0.4 is found.

To determine the scaled side-on peak overpressure $\bar{P}_s$ at control building, $\bar{P}_s$ is read from the same curve in
Figure 7.6 at a scaled distance of $\bar{R} = 6.9$. When the curve is followed from the starting point, a $\bar{P}_s$ of 0.03 is found.

Step 6: Determine $\bar{i}$

The scaled side-on impulse $\bar{i}$ at the tank is read from
Figure 7.7. For $\bar{R} = 1.04$, $\bar{i} = 0.05$.

The scaled side-on impulse $\bar{i}$ at the control building is read from
Figure 7.7, for $\bar{R} = 6.9$, $\bar{i} = 0.008$.

Step 7: Adjust P_s and I for geometry effects

The vessel is spherical at ground level; as a result, geometry will not have an effect on the blast loads at the tank or control building.

Step 8: Calculate P_s and i_s

Calculate the side-on peak overpressure P_s and the side-on impulse i_{so} from the scaled side-on peak overpressure $\bar{P}_s$ and the scaled side-on impulse $\bar{i}$:

$$P_s = \bar{P}_s p_o \qquad i_s = \frac{\bar{i} p_0^{2/3} E_{ex}^{1/3}}{a_0}$$

The ambient speed of sound a_0 is approximately 340 m/s.

Substitution gives, for the tank:
$$P_s = 0.4 \times 0.1 \times 10^6 Pa = 40 \text{ kPa (0.40 bar)}$$
$$i_s = [0.05 \times (0.1 \times 10^6 Pa)^{2/3} \times (300 \times 10^6 Pa)^{1/3}]/340 \text{ m/s} = 211 \text{ Pa-s.}$$

And, for the control building:
$$P_s = 0.03 \times 0.1 \times 10^6 = 3.0 \text{ kPa (0.03 bar)}$$
$$i_s = [0.008 \times (0.1 \times 10^6 Pa)^{2/3} \times (300 \times 10^6 Pa)^{1/3}]/340 \text{ m/s} = 34 \text{ Pa-s.}$$

7.5.   FRAGMENTS FROM A PVB

Case histories (see Chapter 3) demonstrate a variety of failure patterns for pressure vessels and throw distances for fragments. Failure patterns tend to be dominated by the direction of the initial crack. If the initial crack propagates circumferentially the vessel may split into two major pieces, either of which can become a rocket due to the thrust produced by expelled compressed gas. Throw distances of hundreds of meters or more have been documented.

Initial longitudinal cracks can produce failure patterns that range for a single lengthwise split with the vessel remaining in a single piece to multiple cracks that fragment the vessel into many pieces.

The size of fragments is affected by the rate of pressure rise in the vessel and the failure pressure relative to design pressure. Vessels that fail while at stable pressure at or near working pressure tend to generate only a few fragments. Conversely, vessels that experience a sudden increase in pressure and reach ultimate stress in much of the vessel surface will break into many pieces.

7.5.1.   Generation of Fragments from PVBs

In addition to blast waves, significant damage can also be caused by the impact of fragments or objects generated during PVBs. There are two types of fragments: primary fragments and secondary fragments. A primary fragment denotes a fragment from the vessel itself, it may be a part of the vessel wall, or a particle or object contained in the vessel before the burst. A secondary fragment denotes an object located near the explosion source and accelerated by the explosion to a velocity that can cause impact damage.

The primary fragments of vessel bursts are different in many aspects than those produced from high explosive bombs or shells. If the explosion source is high explosive, the container or casing usually ruptures into a very large number of small primary fragments, which can be projected at velocities up to several thousand meters per second. On the other hand, a vessel burst produces only a small number of fragments. In some cases, the vessel only splits into two or a few fragments. The velocities are usually much lower than those from high explosives, ranging only up to hundreds of meters per second.

Fragments from high explosives are usually small (about one gram) and of "chunk" geometry, i.e., all linear dimensions are of the same order of magnitude. Fragments from vessel bursts are often relatively large. Fragments can travel long distances, because large, half-vessel fragments can "rocket," and disk-shaped fragments can "frisbee." Frisbee fragments are flat, disk-shaped fragments that rotate edge-on during flight, and achieve a stable aerodynamic position that generates lift. They fly longer distances than tumbling fragments that have more drag. Frisbee-type fragments have a low probability of occurrence.

In many cases, pressurized gases in vessels do not behave as ideal gases. At very high pressures, van der Waals forces become important, that is, intermolecular forces and finite molecule size influence the gas behavior. Another nonideal situation is that in which, following the rupture of a vessel containing both gas and liquid, the liquid flashes.

Very little has been published covering such nonideal, but very realistic, situations. Two publications by Wiederman (1986a,b) treat nonideal gases. He uses a co-volume parameter, which is apparent in the Nobel-Abel equation of state of a nonideal gas, in order to quantify the influence on fragment velocity. The co-volume parameter is defined as the difference between a gas's initial-stage specific volume and its associated perfect gas value.

For a maximum value for the scaled pressure $p = 0.1$, a reduction in v_i of 10% was calculated when the co-volume parameter was applied to a sphere breaking in half. In general, fragment velocity is lower than that calculated in the ideal-gas case. Baum (1987) recommends that energy E be determined from thermodynamic data for the gas in question.

Wiederman (1986b) treats homogeneous, two-phase fluid states and some initially single-phase states that become homogeneous (single-state) during decompression. It was found that fragment hazards were somewhat more severe for a saturated-liquid state than for its corresponding gas-filled case. Maximum fragment velocities occurring during some limited experiments on liquid flashing could be calculated if 20% of the available energy, determined from thermodynamic data, was assumed to be kinetic energy (Baum 1987).

For vessels containing superheated liquids, the energy available for initial velocity can be determined by calculation of the energy contained in the gas plus the energy of the flashing liquid. This value can be refined by taking into account the released energy of the expanding, originally compressed liquid.

Fragment initial velocity calculations presented below were developed assuming ideal gas behavior.

7.5.2.   Initial Fragment Velocity for Ideal-Gas-Filled Vessels

7.5.2.1.   Initial velocity based on total kinetic energy

A theoretical upper limit of initial fragment velocity can be calculated if it is assumed that the total internal energy E of the vessel contents is translated into fragment kinetic energy. Two simple relations are obtained:

$$v_i = \left[\frac{2E_k}{M_C}\right]^{1/2}$$

(Eq. 7.22)

where:

v_i =   initial fragment velocity

E_k =   kinetic energy

M_C =   total mass of the empty vessel

Kinetic energy (E_k) is calculated from internal energy E. Internal energy can be calculated from Brode's equation (Eq .7.7):

Application of Eq. (7.22) and Brode's equation produces a large overestimation of the initial velocity v_i. As a result, refinements were developed in the methods for determining energy E. For a sudden rupture of a vessel filled with an ideal gas, decompression will occur so rapidly that heat exchange with surroundings will be negligible. Assuming adiabatic expansion, the highest fraction of energy available for translation to kinetic energy of fragments can be calculated with:

$$E_k = k \, \frac{p_1 V_1}{\gamma_1 - 1}$$

(Eq. 7.23)

where:

$$k = 1 - \left[\frac{p_0}{p_1}\right]^{(\gamma-1)/\gamma}$$

(Eq. 7.24)

Baum (1984) has refined this equation by incorporating the work of air pushed away by expanding gas:

$$k = 1 - \left[\frac{p_0}{p_1} \right]^{(\gamma-1)/\gamma} + (\gamma - 1) \frac{p_0}{p_1} \left[1 - \left(\frac{p_0}{p_1} \right)^{-1/\gamma} \right]$$

(Eq. 7.25)

For pressure ratios p_1/p_0 from 10 to 100 and γ ranges from 1.4 to 1.6, the factor k varies between 0.3 and 0.6, according to the refined equation proposed by Baum (1984). These refinements can reduce the calculated value of v_i by about 45%. According to Baum (1984) and Baker et al. (1978b), the kinetic energy calculated with the above equations is still an upper limit.

In Baum (1984), the fraction of total energy translated into kinetic energy is derived from data on fragment velocity measured in a large variety of experiments. (The experiments applied for this purpose include those described by Boyer et al. 1958; Boyer 1959; Glass 1960; Esparza and Baker 1977a; Moore 1967; Collins 1960; Moskowitz 1965; and Pittman 1972.) From these experiments, the fraction translated to kinetic energy was found to be between 0.2 to 0.5 of the total energy derived through Baum's refinement.

Based on these figures, it is appropriate to use k = 0.2 in Eq. (7.23) for rough initial calculations.

7.5.2.2.   Initial velocity based on theoretical considerations

A great deal of theoretical work has been performed to improve ability to predict initial fragment velocity. In the course of these efforts, a model introduced by Grodzovskii and Kukanov (1965) has been improved by various investigators. In this model, the acceleration force on fragments is determined by taking into account gas flow through ever-increasing gaps between fragments. This approach recognizes that not all available energy is translated to kinetic energy. Hence, calculated initial velocities are reduced.

Velocities of fragments from spherical pressure vessels bursting into two equal portions have been analytically determined for ideal gases by the work of Taylor and Price (1971). The theory was expanded to include a large number of fragments by Bessey (1974) and to cylindrical geometries by Bessey and Kulesz (1976). Baker et al. (1978b) modified the theory for unequal fragments. In calculations of initial velocity, the energy necessary to break vessel walls is neglected.

Baker et al. (1975) compared computer-code predictions of fragment velocity from spheres bursting into a large number of pieces and with some experimental data. Boyer et al. (1958) and Pittman (1972) measured fragment velocities from bursting glass spheres and bursting titanium alloy spheres, respectively. The calculated and measured velocities agree rather well after reported difficulties in velocity measurement are taken into account.

The results of a parameter study were used to compose a diagram (Figure 7.13) which can be used to determine initial fragment velocity (Baker et al. 1978a and 1983).

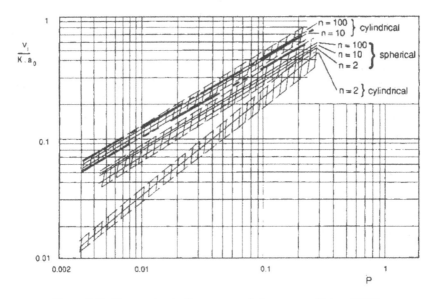

Figure 7.13. Fragment velocity versus scaled pressure. (Baker, 1983)

The scaled terms in the above figure are:

Scaled pressure:
$$\bar{P} = \frac{(p_1 - p_0) V_1}{M_c \, a_1^2}$$
(Eq. 7.26)

Scaled initial velocity:
$$\bar{v}_i = \frac{v_i}{K a_i}$$
(Eq. 7.27)

where k equals 1.0 for equal fragments. Figure 7.15 can be used to calculate the initial velocity v_i for bursting pressurized vessels filled with ideal gas. The quantities to be substituted, in addition to those already defined ($a_1, p_1, p_0,$ and V), are:

M_c =   mass of vessel

K =   factor for unequal fragments

$\bar{P}$ =   scaled pressure

$\bar{v}_i$ =   scaled initial velocity

Three separate regions in Figure 7.13 have been bounded to account for scatter of velocities of cylinders and spheres separating into 2, 10, or 100 fragments. The assumptions used in deriving the figure are from Baker et al. (1983), namely:

- The vessel under gas pressure bursts into equal fragments. If there are only two fragments and the vessel is cylindrical with hemispherical end-caps, the vessel bursts perpendicular to the axis of symmetry. If there are more than two fragments and the vessel is cylindrical, strip fragments are formed and expand radially about the axis of symmetry. (The end caps are ignored in this case.)
- Vessel thickness is uniform.
- Cylindrical vessels have a length-to-diameter ratio of 10.
- Contained gases used were hydrogen (H_2), air, argon (Ar), helium (He), or carbon dioxide (CO_2).

The sound speed a_1 of the contained gas has to be calculated for the temperature at failure:

$$a_1^2 = T\gamma R / m$$

(Eq. 7.28)

where:

R =   ideal gas constant

T =   absolute temperature inside vessel at failure

m =   molecular weight

Appendix B gives some specific characteristics for common gases.

When using Figure 7.13 for equal fragments, K has to be taken as 1 (unity). For the case of a cylinder breaking into two unequal parts perpendicular to the cylindrical axis, the value of K depends on the ratio of the fragment mass to the total mass of the cylinder, which was calculated by Baker et al. (1983). Factor K can be determined for a fragment with mass M_f with the aid of Figure 7.14. The dotted lines in the figure bound the scatter region.

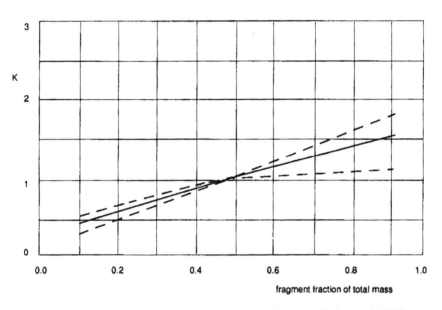

Figure 7.14. Adjustment factor for unequal mass fragments (Baker et al. 1983)

7.5.2.3. Initial velocity based on empirical relations

In addition to the theoretically derived Figure 7.15, an empirical formula developed by Moore (1967) can also be used for the calculation of the initial velocity:

$$v_i = 1.092 \left[\frac{E_k G}{M_C} \right]^{0.5}$$

(Eq. 7.29)

Where, for spherical vessels:

$$G = \frac{1}{1 + 3M_G / 5M_C}$$

(Eq. 7.30)

and for cylindrical vessels:

$$G = \frac{1}{1 + M_G / 2M_C}$$

(Eq. 7.31)

where:

M_G = total gas mass

E_k = kinetic energy

M_C = mass of casing or vessel

Moore's equation was derived for fragments accelerated from high explosives packed in a casing. The equation predicts velocities higher than actual, especially for low vessel pressures and few fragments. According to Baum (1984), the Moore equation predicts velocity values between the predictions of the equation based on total kinetic energy and the values derived from Figure 7.15.

Other empirical relations for ideal gases are given in Baum (1987); recommended velocities are upper limits. In each of these relations, a parameter F has been applied.

For a large number of fragments, F is given by:

$$F = \frac{(p_1 - p_0)r}{ma_1^2}$$

(Eq. 7.32)

where m is mass per unit area of vessel wall and r the radius of the vessel.

For a small number of fragments, F can be written as:

$$F = \frac{(p_1 - p_0)Ar}{M_f a_1^2}$$

(Eq. 7.33)

where:

r = radius of vessel

A = area of detached portion of vessel wall

M_f = mass of fragment

From these values of F, the following empirical relations for initial velocity have been derived:

- For an end-cap breaking from a cylindrical vessel:
$$v_i = 2a_1 F^{0.5}$$
(Eq.7.34)

- For a cylindrical vessel breaking into two parts in a plane perpendicular to its axis:
$$v_i = 2.18a_1[F(L/R)^{1/2}]^{2/3}$$
(Eq. 7.35)

where in F

A = πr^2

L = length of cylinder

- For a single small fragment ejected from a cylindrical vessel:

$$v_i = 2a_1 \left[\frac{Fs}{r}\right]^{0.38}$$
(Eq. 7.36)

Equation (7.34) is only valid under the following conditions:

$$20 < P_v/P_0 < 300; \quad \gamma = 1.4; \quad s < 0.3r$$

- For the disintegration of both cylindrical and spherical vessels into multiple fragments:
$$v_i = 0.88a_1 F^{0.55}$$
(Eq. 7.37)

7.5.2.4.   Discussions

In Baum (1984), a comparison is made between the models described in Section 7.5.2. This comparison is depicted in Figure 7.15. The energy E_k was calculated with k, according to Baum's refinement.

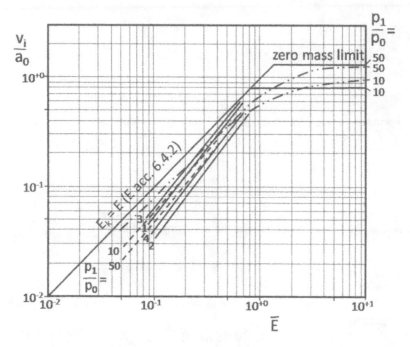

Figure 7.15. Calculated fragment velocities for a gas-filled sphere with $\gamma = 1.4$ (taken from Baum 1984; results of Baker et al. 1978a were added).

(– – –) Baum for $p_1 / p_0 = 10$ and 50;

(– · · –) Moore for $p_1 / p_0 = 10$ and 50;

Baker: curve 1 for $p_1/p_0 = 10$ and 2 fragments; 2 for $p_1/p_0 = 50$ and 2 fragments; 3 for $p_1/p_0 = 10$ and 100 fragments; 4 for $p_1/p_0 = 50$ and 100 fragments.

In Figure 7.17, lines have been added for a sphere bursting into 2 or 100 pieces for $p_1/p_0 = 50$ and 10, in accordance with Figure 7.15. Obviously, the simple relations proposed by Brode (1959) and Baum (1984) predict the highest velocity.

Differences between models become significant for small values of scaled energy $\bar{E}$, in the following equation:

$$\bar{E}_k = \left[\frac{2E_k}{M_c a_1^2} \right]^{0.5}$$

(Eq. 7.38)

In most industrial applications, scaled energy will be between 0.1 and 0.4 (Baum 1984), so under normal conditions with a relatively slow rate of pressure rise, few fragments are expected, and Figure 7.15 can be applied. However, if pressure rises rapidly, such as in a runaway reaction with an exponential increase, higher scaled-energy values can be reached.

In the relationships proposed by Brode (1959) and in Figure 7.15, velocity has no upper limit, although Figure 7.15 is approximately bounded by scaled pressures of 0.005 and 0.2 (scaled energies of approximately 0.1 and 0.7). Baum (1984) states, however, that there is an upper limit to velocity, as follows: the maximum velocity of massless fragments equals the maximum velocity of the expanding gas (the peak contact-surface velocity). In Figure 7.17, this maximum velocity is depicted by the horizontal lines for $p_1/p_0 = 10$ and 50. If values in Figure 7.15 are extrapolated to higher scaled pressures, velocity will be overestimated.

The equation proposed by Moore (1967) tends to follow the upper-limit velocity. This is not surprising, because the equation was based upon high levels of energy. Despite its simplicity, its results compare fairly well with other models for both low and high energy levels.

For lower scaled pressures, velocity can be calculated with the equation proposed by Baum (1987), which produces disintegration of both cylindrical and spherical vessels into multiple fragments ($v_i = 0.88 a_1 F^{0.55}$). Such a result can also be obtained by use of Figure 7.15. However, actual experience is that ruptures rarely produce a large number of fragments. The appearance of a large number of fragments in the low scaled-pressure regions of these equations or curves probably results from the nature of the laboratory tests from which the equations were derived. In those tests, small vessels made of special alloys were used; such alloys and sizes are not used in practice.

Baum's equation ($v_i = 0.88 a_1 F^{0.55}$) can be compared with curves in Figure 7.15 as F equals n times the scaled pressure, in which $n = 3$ for spheres and $n = 2$ for cylinders (end caps neglected). For spheres, Baum's equation gives higher velocities than the Baker et al. model (1983), but for cylinders, this equation gives lower velocities.

7.5.3.   Ranges for Free Flying Fragments

After a fragment has attained a certain initial velocity, that is, when the fragment is no longer accelerated by the explosion, the forces acting upon it during flight are those of gravity and fluid dynamics. Fluid-dynamic forces are subdivided into drag and lift components. The effects of these forces depend on the fragment's shape and direction of motion relative to the wind.

7.5.3.1.   Neglecting dynamic fluid forces

The simplest relationship for calculating fragment range neglects drag and lift forces. Vertical and horizontal range, z_v and z_h, then depend upon initial velocity and initial trajectory angle α_i:

$$H = \frac{v_i^2 \sin(\alpha_i)^2}{2g}$$

(Eq. 7.39)

$$R = \frac{v_i^2 \sin(2\alpha_i)}{g}$$

(Eq. 7.40)

where:

R = horizontal range

H = height fragment reaches

g = gravitational acceleration

α_i = initial angle between trajectory and a horizontal surface

v_i = initial fragment velocity

A fragment will travel the greatest horizontal distance when $\alpha_i = 45°$.

$$R_{max} = \frac{v_i^2}{g}$$

(Eq. 7.41)

7.5.3.2.   Incorporating dynamic fluid forces

Incorporating the effects of fluid-dynamic forces requires the composition of a set of differential equations. Using the Runge-Kutta method to solve these differential equations, Baker et al. (1975) determined fragment range for a number of conditions and plotted the results in a convenient form for practical use, shown in Figure 7.16. They assumed that the position of a fragment during its flight remains the same with respect to its trajectory; that is, that the angle of attack remained constant. In fact, fragments probably tumble during flight.

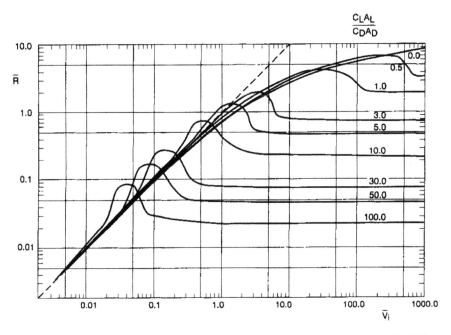

Figure 7.16. Scaled curves for fragment range predictions (taken from Baker et al. 1983)
(– – –): neglecting fluid dynamic forces.

Figure 7.16 plots scaled maximal range $\overline{R}$ and scaled initial velocity $\overline{v}_i$ given by:

$$\overline{R} = \frac{\rho_0 C_D A_D R}{M_f}$$

(Eq. 7.42)

$$\overline{v}_i = \frac{\rho_0 C_D A_D v_i^2}{M_f g}$$

(Eq. 7.43)

where:

$\overline{v}_i$	=	scaled initial velocity	(–)
$\overline{R}$	=	scaled maximal range	(–)
R	=	maximal range	(m)
ρ_0	=	density of ambient atmosphere	(kg/m^3)
C_D	=	drag coefficient	(–)
A_D	=	exposed area in plane perpendicular to trajectory	(m^2)
g	=	gravitational acceleration	(m/s^2)
M_f	=	mass of fragment	(kg)

In Figure 7.16, two more parameters are used, namely:

C_L	=	lift coefficient	
A_L	=	exposed area in plane parallel to trajectory	(m^2)

These curves were generated by maximization of range through variation of the initial trajectory angle. Each curve is for a specified value of lift-to-drag ratio, $C_L A_L/(C_D A_D)$. The dotted line denotes the case for which fluid dynamic forces are neglected, for which Eq 7.41 applies:

$$R_{max} = \frac{v_i^2}{g}$$

In most cases, "chunky" fragments are expected. The lift coefficient will be zero for these fragments, so only drag and gravity will act on them; the curve with $C_L A_L/(C_D A_D) = 0$ is then valid.

The procedure for determining fragment range by using Figure 7.16 is:

Step 1.

Calculate the lift-to-drag ratio, $C_L A_L/(C_D A_D)$ for the fragment.

Step 2.

Calculate the velocity term (Eq 7.43) $\bar{v}_i = \dfrac{p_0 C_D A_D v_i^2}{M_f g}$ for the fragment.

Step 3.

Select the curve on the graph for the appropriate lift/drag ratio; locate the velocity term on the horizontal axis; find the corresponding range term (Eq 7.42):

$$\bar{R} = \frac{p_0 C_D A_D R}{M_f}, \text{ and determine the range R.}$$

For lift-to-drag ratios, $C_L A_L / (C_D A_D)$, that are not on the curve, a linear interpolation procedure can be used to determine the range from the curve. However, interpolation in the steep areas of the curve can cause considerable error and it is recommended to calculate drag-to-lift ratio using a method such as the one presented in Baker et al. (1983).

Drag coefficients for various shapes can be found in Table 7.1. More information about lift and drag can be found in Hoerner (1958).

For fragments having plate-like shapes, the lift forces can be large, so predicted ranges can be much larger than the range calculated with $R_{max} = v_i^2/g$. This is called the "Frisbee" effect, especially when the angle of attack α_i is small (α_i = approximately 10°).

Table 7.1. Drag coefficients (Baker et al. 1983)

Shape	Sketch	C_D
Right Circular Cylinder (long rod), side-on		1.20
Sphere		0.47
Rod, end-on		0.82
Disc, face-on		1.17
Cube, face-on		1.05
Cube, edge-on		0.80
Long Rectangular Member, face-on		2.05
Long Rectangular Member, edge-on		1.55
Narrow Strip, face-on		1.98

7.5.4.   Ranges for Rocketing Fragments

Some accidents involving materials like propane and butane resulted in the propulsion of large fragments for unexpectedly long distances. Baker et al. (1978b) argued that these fragments developed a "rocketing" effect. In their model, a fragment retains a portion of the vessel's liquid contents. Liquid vaporizes during the initial stage of flight, thereby accelerating the fragment as vapor escapes through the opening. Baker et al. (1978b) provided equations for a simplified rocketing. Baker et al. (1983) applied this method to two cases, and compared predicted and actual ranges of assumed rocketed fragments.

Ranges for rocketing fragments can also be calculated from guidelines given by Baum (1987). As stated in Section 7.5.1for cases in which liquid flashes off, the initial-velocity calculation must take into account total energy. If this is done, rocketing fragments and fragments from a bursting vessel in which liquid flashes are assumed to be the same.

Ranges were calculated for a simulated accident with the methods of Baker et al. (1978a,b) and Baum (1987). It appears that the difference between these approaches is small. Initial trajectory angle has a great effect on results. In many cases (e.g., for horizontal cylinders) a small initial trajectory angle may be expected. If, however, the optimal angle is used, very long ranges are predicted.

7.5.5.   Statistical Analysis of Fragments from Accidental Explosions

Theoretical models presented in previous sections give no information on distributions of mass, velocity, or range of fragments, and little information on the number of fragments to be expected. These models are not developed to account for these parameters. More information can be found in the analysis of results of accidental explosions.

Baker (1978b) analyzed 25 accidental vessel explosions for mass and range distribution and fragment shape. Because data were limited, it was necessary to group like events into six groups in order to yield an adequate base for useful statistical analysis.

Information on each group is tabulated in Table 7.2. The values for energy range given in Table 7.2 require some discussion. In the reference, all energy values were calculated by use of Brode's equation (Eq. 7.7). Users should do the same in order to select the right event group.

Table 7.2. Groups of like PVB events used in fragmentation statistical analysis

Event Group Number	Number of Events	Explosion Material	Source Energy Range (J)	Vessel Shape	Vessel Mass (kg)	Number of Fragments
1	4	Propane, Anhydrous ammonia	1.487×10^5 to 5.95×10^5	Railroad tank car	25,542 to 83,900	14
2	9	LPG	3814 to 3921.3	Railroad tank car	25,464	28
3	1	Air	5.198×10^{11}	Cylinder pipe and spheres	145,842	35
4	2	LPG, Propylene	549.6	Semi-trailer (cylinder)	6343 to 7840	31
5	3	Argon	2438×10^9 to 1133×10^{10}	Sphere	48.26 to 187.33	14
6	1	Propane	24.78	Cylinder	511.7	11

Statistical analyses were performed on each of the groups to yield, as data availability permitted, estimates of fragment-range distributions and fragment-mass distributions. The next sections are dedicated to the statistical analysis according to the Baker et al. (1978b) method.

7.5.5.1.   Fragment range distribution

It was shown in the reference that the fragment range distribution for each of the six groups of events follows a normal, or Gaussian, distribution. It was then shown that the chosen distributions were statistically acceptable. The range distributions for each group are given in Figure 7.17 and Figure 7.18. With this information, it is possible to determine the percentage of all the fragments which would have a range smaller than, or equal to, a certain value.

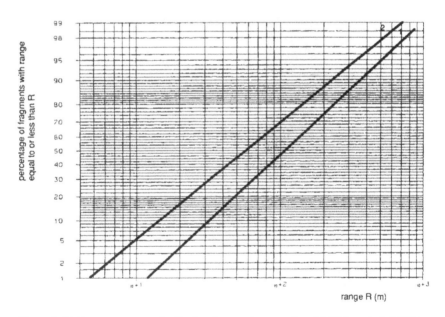

Figure 7.17. Fragment range distribution for event groups 1 and 2 (Baker et al. 1978b).

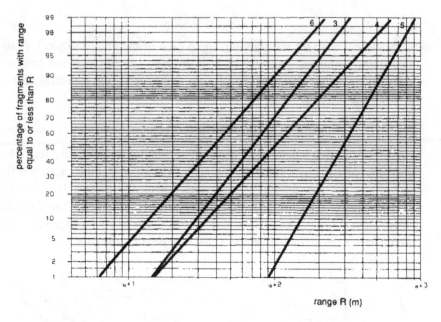

Figure 7.18. Fragment range distribution for event groups 3, 4, 5, and 6
(Baker et al. 1978b).

7.5.5.2.   Fragment-mass distribution

Pertinent fragment-mass distributions were available on three event groups
(2, 3, and 6). According to the reference, they follow a normal, or Gaussian,
distribution. These distributions are presented in Figure 7.19 and Figure 7.20.
The percentage of fragments having a mass smaller than or equal to a certain
value can be calculated using these figures.

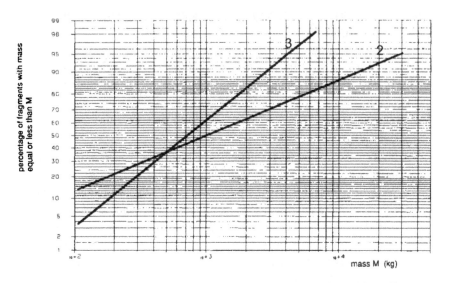

Figure 7.19. Fragment-mass distribution for event groups 2 and 3 (Baker et al. 1978b).

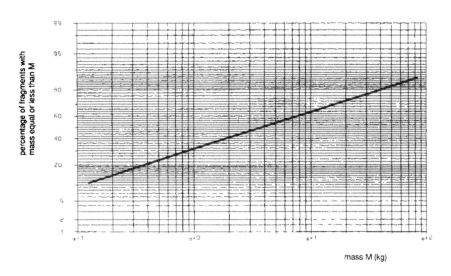

Figure 7.20. Fragment-mass distribution for event group 6 (Baker et al. 1978b).

7.6.   PREDICTING FRAGMENT EFFECTS FROM VESSEL BURSTS

The quantification of hazards associated with the fragments propelled from an exploding, pressurized vessel should involve the determination of their masses, velocities and ranges. Models have been developed for calculating fragments' initial velocities and ranges. These models were discussed earlier in this volume.

Neither statistical nor theoretical methods for determining fragment characteristics are fully adequate, and it is sometimes difficult to decide which one to use. For calculating maximum fragment range, the theoretical approach in Section 7.6.1 is more appropriate. For estimating mass and range distributions, the statistical approach in Section 7.5.5 should be applied. Another factor favoring choice of the statistical analysis approach is that it is faster, because the theoretical approach requires application of a number of equations and figures.

7.6.1.   Analytical Analysis

7.6.1.1.   Data collection

Before calculations can begin, a great deal of data must be collected describing the vessel itself, its contents, and the condition at failure. These data include:

For the vessel:

   shape (cylindrical or spherical)

   diameter D

   length L

   mass M

   wall thickness t

For vessel contents:

   molecular weight

   volume V

   chemical and physical properties

thermodynamic qualities

liquid/gas ratio

For the condition at failure:

internal pressure p_1 (Pa)

internal temperature T (K)

The pressure at failure is not always known. However, depending on the assumed cause of the failure, an estimate of pressure can be made:

- If failure is initiated by an increase in internal pressure in combination with a malfunctioning of the pressure relief, the pressure at failure will equal the failure pressure of the vessel. This failure pressure is usually the maximum working pressure multiplied by a safety factor as discussed in Section 7.1.1. More precise calculations are possible if the vessel's dimensions and material parameters are known.
- If failure is due to external heat applied to the vessel (e.g., from fire), the vessel's internal pressure rises, and at the same time its material strength drops. For initial calculations, it can be assumed that failure pressure equals PRV relief pressure at rated flow (see Section 7.1.1).
- If failure is initiated by corrosion or impact of a missile or fragment, it can be assumed that failure pressure will be the normal operating pressure.

Once a vessel's internal pressure at failure is determined or assumed, the temperature of its contents can be calculated.

7.6.1.2. Calculation of total energy

The first equation for a vessel filled with an ideal gas was derived by Brode (1959) [Eq. 7.7]:

$$E = \frac{(p_1 - p_0) V_1}{\gamma_1 - 1}$$

This equation is well known and often used to calculate initial fragment velocity, but its application can result in gross overestimation. Assuming adiabatic expansion of the ideal gas, energy can be calculated using Eq. (7.23):

$$E_k = k \frac{p_1 V_1}{\gamma_1 - 1}$$

Where factor k is calculated using Eq. (7.24):

$$k = 1 - (p_0 / p_1)^{(\gamma-1)/\gamma}$$

Baum (1984) uses a refined equation for k from Eq. (7.25):

$$k = 1 - (p_0 / p_1)^{(\gamma-1)/\gamma} + (\gamma - 1)(p_0 / p_1)[1 - (p_0 / p_1)^{-1/\gamma}]$$

If an energy value is found in literature, it is important to know which equation was used as a basis for calculation.

7.6.1.3.   Ranges for rocketing fragments

In some accidents, large fragments, usually consisting of either the vessel's end caps or half of the vessel itself, were reported to have traveled unexpectedly long distances. It was argued that these rocketing fragments were accelerated during their flight by expelling the liquid entrapped in the vessel section.

In Baker et al. (1983), a computer program to calculate the release of energy as a function of time was developed based on the rocketing problem. However, if one assumes that available energy is released instantaneously, as occurs in the case of flashing liquids, an upper limit of the initial velocity of the fragment is obtained. Apparently, rocketing fragments are equal to fragments of an exploding vessel where liquid flashing occurs. The unexpected long fragment ranges result from the extra available energy of the liquid. Therefore, no special method for calculating ranges of rocketing fragments is required.

7.6.1.4.   Determination of number of fragments

The number of fragments will usually be high for a high rate of pressure rise, such as a runaway reaction. When a runaway reaction involves condensed phase material in contact with a vessel wall, fragments can resemble high explosive detonation in which the casing disintegrates completely. The vessel wall not in contact with condensed phase material does not fragment as severely. Situations involving fragmentation of vessel walls in contact with condensed phase runaway materials should be addressed using high explosive fragment prediction methods, which are beyond the scope of this book.

For cases other than runaway reactions, the rate of pressurization is relatively low. The number of fragments has historically been in the range of about 30 to 100 when pressure reaches ultimate failure pressure as shown in Table 7.2 and the case histories in Chapter 3. For situations in which failure occurs near working pressure, such as loss of mechanical integrity due to fire exposure and other external agencies, the number of fragments from such events tend to be low, usually from 2 to 10 pieces.

7.6.1.5.   Calculation of initial velocity

As a vessel ruptures, its fragments accelerate rapidly to a maximum velocity. This value is the initial fragment velocity v_i. It is used to calculate either the range of fragment travel or, if collision with an obstacle occurs before maximum range is attained, impact velocity.

A number of methods and equations are available to determine the initial velocity. These are described elsewhere in this volume. To avoid confusion, only three methods are given here. Method 1 calculates the initial velocity, both for vessels filled with ideal gases and for vessels filled with liquid and vapor. In most cases, this method will give an upper velocity limit. Method 2 is only valid for gas-filled vessels, but there, velocity depends upon the shape of the vessel and expected number of fragments. For scaled energies larger than 0.8, Method 1 results in overestimates of velocity, and Method 2 is not valid in this region. Therefore, Method 3 is provided. Method 3 can also be applied for lower scaled energies, but Methods 1 and 2 are recommended.

Method 1

The simplest method is based on the total kinetic energy E_k of the fragments using Eq. (7.22):

$$v_i = \left[\frac{2E_k}{M_c} \right]^{\frac{1}{2}}$$

Converting the energy calculated with Eq. (7.23) and k [according to Eq. (7.25)] into fragment kinetic energy still results in an overestimate of the velocity when compared with experimental results. This is logical because a portion of the energy will be diverted into the creation of a blast wave. Negligible portions of the energy will produce rupturing of the vessel, causing noise and raising atmospheric temperature.

It appeared from experiments that the actual total kinetic energy generated is 0.2 to 0.5 times the energy calculated by Eqs. (7.23) and (7.25). Therefore, it is appropriate to adjust earlier calculations based on ideal gases as follows:

$$E_k = 0.2 \frac{p_1 V_1}{\gamma_1 - 1}$$

(Eq. 7.44)

Thus, E_k is 20% of the energy calculated for nonideal gases or for flash-vaporization situations. For scaled energies ($\bar{E}$) larger than about 0.8 as calculated by Eq. (7.36), the calculated velocity is too high, so Method 3 should be applied.

Method 2

Figure 7.13 can be used to determine initial velocity for vessels filled with an ideal gas (Baker et al. 1978a and 1983). The scaled pressure $\bar{P}$ on the horizontal axis of Figure 7.13 is determined using Equation 7.26:

$$\bar{P} = \frac{(p_1 - p_0) V_1}{M_c a_0^2}$$

where:

$\bar{P}$	=	scaled pressure	(-)
p_1	=	internal pressure at failure	(Pa)
p_0	=	ambient pressure	(Pa)
V	=	volume	(m³)
M_c	=	mass of the vessel	(kg)
a_0	=	speed of sound in gas at failure	(m/s)

The number of fragments must first be chosen, usually on the basis of scaled energy.

The restrictions under which Figure 7.13 was derived are as follows:
- Fragments are equal in size and shape. For two fragments only, the cylindrical vessel bursts perpendicularly to the axis of symmetry. For more than two fragments, the cylindrical vessel bursts into strip fragments which expand radially about the axis of symmetry, and end caps are neglected.
- Wall thickness is uniform.

- Cylindrical vessels have a length-to-diameter ratio of 10.
- Contained gases used were either hydrogen (H_2), air, argon (Ar), helium (He), or carbon dioxide (CO_2).

Figure 7.13 should be applied only with great caution to any situation where these restrictions are not valid.

The speed of sound a_1 of the contained gas at failure temperature must be calculated using Eq. (7.28):

$$a_1{}^2 = T\gamma R/m$$

where:

R	=	ideal gas constant	(J/kmol)
T	=	absolute temperature	(K)
m	=	molecular weight	(kg/kmol)

The vertical axis in Figure 7.13 is labeled the scaled velocity $\overline{v}_i$, provided in Eq. (7.27) as:

$$\overline{v}_t = \frac{v_i}{Ka_i}$$

where K = factor for unequal fragments from which v_i can be calculated.

The factor K takes unequal fragments into account. This factor is, however, open to discussion. One should usually assume equal fragments, that is, $K = 1$.

It is inadvisable to extrapolate outside the regions given in Figure 7.13. For high scaled-pressure values (i.e., scaled energy larger than 0.8), Method 3 should be used.

Method 3

This method employs Eq. (7.29), an empirical equation derived by Moore (1967):

$$v_i = 1.092 \left[\frac{E_k G}{M_C} \right]^{0.5}$$

where:

$$G = \frac{1}{1 + 3M_G / 5M_C}$$ for spherical vessels [from Eq. (7.30)]

$$G = \frac{1}{1 + M_G / 2M_C}$$ for cylindrical vessels [from Eq. (7.31)]

and

E_k	=	kinetic energy	(J)
M_G	=	total gas mass	(kg)
M_C	=	mass of casing or vessel	(kg)

Moore's equation was derived from fragments accelerated from high explosives packed in a casing. Baum (1984) showed, in comparing different models, that the Moore equation tends to follow the theoretical upper-velocity limit for high scaled energies.

7.6.1.6.   Ranges for free flying fragments

The simplest relationship for calculating the range of a free-flying obstacle with a given initial velocity is derived when fluid-dynamic frictional forces (lift and drag forces) are neglected. Then, the only force acting on the fragment is that of gravity, and the vertical and horizontal range, H and R, are dependent on the initial velocity v_i and the initial trajectory angle a_i, calculated as follows using Eq. (7.39)

$$H = \frac{v_i^2 \sin(a_i)^2}{2g}$$

and Eq. (7.40)

$$R = \frac{v_i^2 \sin(2a_i)}{g}$$

where:

H	=	vertical range	(m)
R	=	horizontal range	(m)

| a_i | = | initial trajectory angle | (deg) |
| g | = | gravitational acceleration | (m/s^2) |

The trajectory angle has a great influence on the range. The maximum range is found for an angle of 45° using Eq. (7.41):

$$R_{max} = \frac{v_i^2}{g}$$

It can be assumed that, should the vessel burst into two halves, fragments will travel parallel to their axes. If the vessel is initially positioned horizontally, the trajectory angle will be 5° to 10°.

The results of a computer analysis of parameters for fragment ranges, including drag and lift forces, are plotted in Figure 7.16. The developers assumed that fragment positions remain constant with respect to trajectory. Figure 7.16 plots the scaled maximum range $\bar{R}$ and the scaled initial velocity $\bar{v}_i$ given by Eq. (7.42)

$$\bar{R} = \frac{\rho_0 C_D A_D R}{M_f}$$

and Eq. (7.43)

$$\bar{v}_i = \frac{\rho_0 C_D A_D V_i^2}{M_f g}$$

where:

v_i	=	initial fragment velocity	(m/s)
$\bar{v}_i$	=	scaled initial velocity	(–)
$\bar{R}$	=	scaled range	(–)
R	=	actual range	(m)
ρ_0	=	density of ambient air	kg/m^3)
C_D	=	drag coefficient	(–)
A_D	=	exposed area in the plane perpendicular to the trajectory	(m^2)
M_f	=	mass of the fragment	(kg)

The curves in Figure 7.16 were generated by maximizing range through variation of initial trajectory angle, so the angle for maximum range does not necessarily equal 45°. In most cases, "chunky" fragments are expected. The lift coefficient is then zero, and the curve with $C_L A_L / C_D A_D = 0$ is valid. It can be seen from Figure 7.16 that for scaled velocities larger than 1, drag force becomes important and ranges will be shorter than those calculated with Eq. (7.51). Values for drag coefficients C_D can be found in Table 7.1.

Should a fragment be a thin plate, lift force becomes important, and the range will be greater than that calculated with Eq. (7.51). It is clear from Figure 7.16, however, that the range will only be greater for those regions of scaled velocity where this "Frisbee" effect occurs.

7.6.2.   Example Problem - Failure during Testing

This example problem is similar to the one in Section 7.4.5, blast effect prediction. A cylindrical vessel with a volume of 25 m³ and design pressure of 19.2 bar is used for the storage of propane. The wall thickness of the vessel is 3 mm, its material is carbon steel, and its length-to-diameter ratio is 10. The vessel is pressurized after fabrication with nitrogen to 24 bar. After testing, the safety valve will be set at 15 bar for normal operation.

Two different situations will be examined for maximum fragment range: failure during testing, and failure due to an external fire.

Because the maximum fragment range is required, the theoretical approach will be applied. Only the initial velocity and the maximum range of the fragments can be calculated with the theoretical approach.

First, energy must be calculated according to Brode's formula (Eq. 7.7):

$$E = \frac{(p_1 - p_0) V_1}{\gamma_1 - 1}$$

$$E = \frac{(25 \times 10^5\, Pa - 10^5\, Pa)\, 25\, m^3}{1.4 - 1} = 1.50 \times 10^8\, J$$

Equation 7.23 gives:

$$E_k = k\, \frac{p_1 V_1}{\gamma_1 - 1}$$

$$E = k\, \frac{25 \times 10^5\, Pa \times 25\, m^3}{1.4 - 1} = k \times 1.56 \times 10^8\, J$$

Using Eq. 7.24 to calculate factor k gives:

$$k = 1 - (p_0 / p_1)^{(\gamma-1)/\gamma}$$

$$k = 1 - (1\,bar / 25bar)^{0.4/1.4} = 0.601$$

so:

$$E = 9.38 \times 10^7 \, J$$

Alternatively, using Eq. 7.25 from Baum's refined method gives:

$$k = 1 - \left[\frac{p_0}{p_1} \right]^{(\gamma-1)/\gamma} + (\gamma-1)\frac{p_0}{p_1}\left[1 - \left(\frac{p_0}{p_1} \right)^{-1/\gamma} \right]$$

$$k = 0.601 + (1.4 - 1)\frac{1\,bar}{25\,bar}\left[1 - \left(\frac{1\,bar}{25\,bar} \right)^{-1/1.4} \right] = 0.458$$

so:

$$E = 0.458 \times 1.56 \times 10^8 \, J = 7.14 \times 10^7 \, J.$$

As was expected, the Brode equation gives higher values for energy than Eq. 7.23 with factor k calculated using either Eq. 7.24 or Eq. 7.25. The use of Eq. 7.23 is recommended as discussed in Section 7.5.2.1. Twenty to fifty percent of this energy will be translated into the kinetic energy of the fragments, so the maximum kinetic energy will be:

$$E_k = 0.5 \times 7.14 \times 10^7 \, J = 3.57 \times 10^7 \, J.$$

A quick estimate can be made with Eq.(7.44):

$$E_k = 0.2\frac{p_1 V}{\gamma-1}$$

Substituting,

$$E_k = 0.2\frac{25 \times 10^5 \, Pa \times 25 \, m^3}{1.4 - 1.0} = 3.125 \times 10^7 \, J$$

This value appears to be in good agreement with the one calculated with the theoretical approach.

In order to determine which method should be applied for the calculation of initial velocity, the scaled energy should first be determined using Eq. (7.38):

$$\overline{E} = \left[\frac{2E_k}{M_C a_1^2} \right]^{0.5}$$

The mass M_C of the vessel can be calculated, assuming hemispherical end caps of 5 mm thickness, to be 2723 kg. The speed of sound a_1 in nitrogen can be calculated with the Eq. (7.28):

$$a_1^2 = T\gamma R / m$$

where:

R	=	ideal gas constant	(J/(kmoleK))
T	=	absolute temperature	(K)
m	=	molecular weight	(kg/kmol)

$a_1 = (1.4 \times 8314.41 \text{ J/kmoleK} \times 293\text{K}/28\text{kg/kmol})^{1/2} = 349 \text{ m/s}$

Then

$$\overline{E} = \left[\frac{2(3.57 \times 10^7 J)}{(2723\,kg)(349\,m/s)^2} \right]^{0.5} = 0.46$$

Since the scaled energy is lower than 0.8 and nitrogen can be considered to be an ideal gas, both Methods 1 and 2 can be applied.

Method 1

According to Eq. (7.22):

$$v_i = \left[\frac{2E_k}{M_C} \right]^{0.5}$$

Then, the mean initial fragment velocity will be:

$$v_i = \left[\frac{2(3.57 \times 10^7 J)}{2723\,kg} \right]^{0.5} = 162 \text{m/s}$$

Method 2

The initial velocity can also be calculated from Figure 7.13. Calculation of scaled pressure using Eq. (7.26) yields:

$$\bar{P} = \frac{(p_1 - p_0)V_1}{M_C a_1^2}$$

$$\bar{P} = \frac{(25 \times 10^5 \, Pa - 10^5 \, Pa) \times 25m^3}{2723 kg (349 m/s)^2} = 0.181$$

Since the vessel is under a test in which pressure is increased slowly, it can be expected that the number of fragments generated will be low.

Assume that the vessel breaks into two equal parts at right angles to its axis. Use the graph in the Figure 7.13 to determine $\bar{v}_i$ For a vessel breaking into two parts, $\bar{v}_i = 0.3$, so:

$$v_i = \bar{v}_i \times 349 = 0.3 \times 349 = 105 \text{ m/s}$$

With the initial velocity determined, the horizontal range R can be calculated. If fluid-dynamic forces (lift and drag) are neglected, maximum range will be attained when the fragment is propelled at an angle of 45°. The range is then independent of the fragment's mass and shape and is simply the ratio of the velocity squared to gravitational acceleration, calculated using Eq. (7.41):

$$R_{max} = \frac{v_i^2}{g}$$

The initial trajectory angle is taken into account by Eq. (7.40):

$$R = \frac{v_i^2 \sin(2a_i)}{g}$$

For cylinders with horizontal axes, the initial trajectory will be low, typically 5° or 10°. Table 7.3 shows maximum ranges for initial velocities calculated by each method with various low trajectory angles assumed.

Table 7.3. Ranges for various initial trajectory angles

	Velocity v_i (m/s)	Range (m)		
		α= 5°	α= 70°	α= 45°
Method 1	162	465	916	2678
Method 2	105	195	385	1125

It is obvious that very long maximum ranges are attained if lift and drag forces are neglected. Taking these forces into account can reduce maximum ranges significantly. Fragments in this case are expected to be rather blunt, so the lift coefficient is taken as zero.

The scaled velocity can be calculated with Eq. (7.41). By applying the curve in Figure 7.16, a value for scaled range is found, from which the actual range can be calculated. This is performed for the initial velocity determined by Method 1 for 2 equal fragments. The density of the ambient atmosphere is assumed to be 1.3 kg/m³. The fragment area is A_d = 1.86 m² (diameter of the vessel = 1.53 m), and the drag coefficient is C_d = 0.47 (Table 7.1). Fragment mass is M_f = 1362 kg.

Apply Eq. (7.42):

$$\overline{R} = \frac{\rho_0 C_D A_D R}{M_f}$$

and Eq. (7.43):

$$\overline{v}_i = \frac{\rho_0 C_D A_D v_i^2}{M_f g}$$

For the end cap:

$$\overline{v}_i = \frac{(1.3kg/m^3)(0.47)(1.86m^2)(162m/s)^2}{(1362kg)(9.8m/s^2)} = 2.2$$

Figure 7.16 gives $\overline{R}$ = 1.2 for a lift-to-drag ration $C_L A_L / C_D A_D$ = 0, so

$$R = \frac{1.2(1362kg)}{(1.3kg/m^3)(0.47)(1.86m^2)} = 1438 \text{ m}$$

8. BASIC PRINCIPLES OF BLEVES

8.1. INTRODUCTION

The boiling liquid expanding vapor explosion or BLEVE was introduced in Chapter 2. In this chapter, we will discuss the BLEVE phenomenon in detail and present practical methods to estimate the hazards from BLEVEs.

8.2. DEFINITION OF A BLEVE

The term "BLEVE" (pronounced BLĔ-vĒ) was first introduced by J. B. Smith, W. S. Marsh, and W. L. Walls of Factory Mutual Research Corporation (now FM Global) in 1957. Various researchers have proposed alternative definitions to include or exclude observed phenomena. In the present context, BLEVE is defined as a sudden loss of containment of a pressure-liquefied gas existing above its normal atmospheric boiling point at the moment of its failure, which results in rapidly expanding vapor and flashing liquid. The release of energy from these processes (expanding vapor and flashing liquid) creates a pressure wave. A BLEVE requires three key elements:

- A liquid that exists above its normal atmospheric pressure boiling point
- Containment that causes the pressure on the liquid to be sufficiently high to suppress boiling
- A sudden loss of containment to rapidly drop the pressure on the liquid

Liquids normally stored under pressure have atmospheric pressure boiling points well below ambient temperature. Common examples of these types of materials include light hydrocarbons (e.g., propane, ethane, butane), ammonia, and refrigerants. Other liquids whose atmospheric pressure boiling points are above ambient temperature may be heated intentionally or unintentionally during use to above their boiling points and stored at pressure. These include many liquids, but common examples are water (e.g., in steam generation) and heavy hydrocarbons. If any of these commodities suddenly lose containment, the pressure will drop and the liquids will become superheated and flash to vapor.

The reason that these liquids are able to remain liquids despite being above their boiling point is due to some type of confinement. This confinement can take the form of process or storage vessels, or pipes.

While any loss of containment of a pressure-liquefied gas will lead to a reduction in pressure and boiling of the vessel contents, a BLEVE requires that this loss of containment be 'sudden' and 'significant' in size. A partial failure leading to a massive two-phase jet release would not typically be called a BLEVE since it does not represent a sudden loss of containment.

8.3.   THEORY

The theory behind gas expansion, blast and shock wave formation and propagation, and combustion has been addressed in other chapters. Due to the fact that BLEVEs involve the boiling of liquids and the interaction of this process with the failure mechanisms of the containment, the basics of the thermodynamics of boiling and of fracture mechanics are discussed in the following sections.

8.3.1.   Thermodynamics of Boiling

When heat is transferred to a liquid, the temperature of the liquid rises. When the boiling point is reached, the liquid starts to form vapor bubbles at active bubble nucleation sites. These active sites usually occur at interfaces with solids, including vessel walls. Nucleation at these sites is called heterogeneous nucleation. All boiling requires some level of superheat (defined as heating beyond the boiling point). If bubble nuclei are large (approaching millimeter sizes), then a very small degree of superheat is needed to make the bubble grow. If the bubble nuclei are very small (i.e., the vessel is very clean and smooth) then they will need a larger superheat for them to grow.

As mentioned above, boiling usually starts at the vessel wall because this is where bubble nuclei are able to be trapped in the surface crevices. When there is a shortage of nucleation sites, as in the case of a vessel with very clean and smooth walls (e.g., glass beakers or mirror-polished metal), the bubble nucleation must take place in the bulk of the liquid. Boiling in the bulk of the fluid generally takes place at submicron nucleation sites at suspended particles, impurities, crystals, or ions. This requires a large degree of superheat to make these very small bubble nuclei grow. This means the liquid normal boiling point can be exceeded without boiling; and the liquid can be significantly superheated while remaining a liquid. There is, however, a limit at a given pressure above which a liquid cannot be superheated, called the "superheat limit temperature", and when this limit is reached, microscopic vapor bubbles develop spontaneously at the molecular level in the pure liquid. This can produce a very violent phase change where the liquid changes phase instantaneously. This is called homogeneous nucleation.

Where a liquid is stored above its atmospheric boiling point under pressure, the pressure in the vessel suppresses boiling and prevents the formation of bubbles. If the pressure drops below the saturation pressure, boiling usually begins at the nucleation sites on the vessel wall (e.g., scale, machining marks, surface micro-fractures, scratches, impurities, etc.) and at the liquid vapor interface. This means the boiling starts at the walls and thereby avoids homogeneous nucleation. The phase change process is spread over a longer period of time and this slows the rate of release of expansion energy contained in the liquid. There is no experimental evidence with a real pressure vessel where the phase change was by homogeneous nucleation. Homogeneous nucleation is extremely unlikely in industrial use due to temperature gradients, impurities, surface roughness and other factors that cause localized nucleation.

8.3.2. Mechanics of Vessel Failure

For a BLEVE to take place, the pressure vessel must fail suddenly and catastrophically (i.e., complete loss of containment). This mechanism is similar or identical to the mechanisms involved in pressure vessel bursts (see Section 7.1.1) and the details of these mechanics are not repeated here. Small cracks formed by any of these mechanisms may grow due to a combination of factors including internal pressure, vessel material weakness, lack of material ductility, etc. The vessel material can also be softened as a result of heating and this can cause the material to lose strength and fail due to overpressure even if the pressure is below the normal capacity of the material.

A typical crack begins as a very small, local failure in the material where the containment is weakest (e.g., near welds, stress concentrations, heated zones, or flaws). Once the material fails locally, it no longer supports the material around it and can cause this surrounding material to be overloaded and fail. In this manner, the crack propagates along the vessel wall. If the failure is due to local effects (e.g., heating, corrosion, defects, inclusions, flaws, or local stress risers), the crack may arrest when it reaches a section of material that is strong enough to withstand the pressure without the support of the failed portion. This may occur at thickness changes (e.g., fittings, end caps, etc.) or where the material state changes (e.g., unheated, stronger material, etc.). In this case, the crack may become stable and the result will be a limited vessel failure and a leak or jet of the fluid from the vessel. On the other hand, if the crack continues to propagate, the vessel will fail completely.

While a full discussion of fracture mechanics is outside the scope of a handbook such as this one, it is useful to consider the concept of critical crack length to better understand catastrophic vessel failures. In brief, once a crack reaches a certain critical length (a function of the material and the stress condition), it will propagate into the remaining material in a brittle fashion, and will run at the speed of sound in the vessel wall material. This propagation mechanism contrasts with the local overload mechanism described above in that it does not require yielding and overload (i.e., ductile failure) to occur, the crack simply runs through the material along grain boundaries until it runs out of material or the loading condition suddenly changes. In most cases, the speed of sound in the vessel wall is much higher than the speed of sound in the vessel contents, and the failure of the vessel occurs much faster than the expansion of the liquid and faster still than the boiling of the liquid. For example, the speed of sound in steel is approximately four times the speed of sound in water and more than seventeen times the speed of sound in air (all at atmospheric temperature and pressure).

Failures of these kinds typically result in the vessel separating into two or more large pieces. In some cases, the vessel is left fully opened and flattened on the ground with the end caps attached (see Figure 8.1). In some cases, the vessel can fail into two or more sections with significant propulsion of each section by the ejection of the flashing liquid and vapor, called "rockets." In the case where a crack forms and then arrests, the result is a localized rupture with release of the vessel contents in the form of a powerful jet; this is not normally called a BLEVE type release. An example of this type of failure is shown in Figure 8.2.

Figure 8.1. 500-gallon (1.9 m³) pressure vessel opened and flattened on the ground after a fire-induced BLEVE. (Birk et al., 2003)

Figure 8.2. Fire test of 500-gallon (1.9 m³) propane pressure vessel resulting in massive jet release (not a BLEVE). (Birk et al., 2003)

The most common reported cause of industrial BLEVEs is overheating of a vessel due to fire. When a pressure vessel is engulfed in a fire, its metal is heated and it loses mechanical strength. The process of failure is by high temperature stress rupture as discussed by Birk and Yoon, 2006a.

At liquid wetted surfaces, the fire heat is rapidly conducted and convected to the liquid contents, thus raising the liquid temperature but keeping the liquid wetted portion of the vessel wall relatively cool. The liquid wetted wall (for the case of propane) will usually have a temperature within 100°C of the bulk liquid temperature unless the boiling critical heat flux of the liquid has been exceeded and the boiling has transitioned to a film boiling regime. However, for the wall in contact with the vapor, the process is very different. The convection of heat from the wall to the vapor is much slower due to the lower thermal conductivity, specific heat and density of the vapor. As a result the heat stays in the wall and much higher wall temperatures are achieved. The vapor space wall temperature can rise to very high levels where thermal radiation becomes the dominant mode of heat transfer to cool the wall. As a result, the wall in contact with vapor can reach temperatures above 600°C in a severe fire. At these temperatures, most ordinary pressure vessel steels have lost 60-80% of their strength. Local heating due to jet fire impingement can cause even higher wall temperatures in a much shorter period of time. Jet fires may even affect the mechanical strength of the metal below the liquid level if the critical heat flux of the liquid is exceeded. Due to these local heating effects, a pressure relief valve, even if properly sized and in working order, will not prevent a BLEVE. In fact, the pressure relief valve will generally act to reduce the amount of material in the vessel and would, over time, reduce the liquid wetted surface of the vessel and increases the area that is vulnerable to thermal degradation.

For a properly designed pressure vessel in good condition, fire induced rupture is possible within minutes of severe local fire contact. Predicting time to rupture is possible if the exact fire conditions, vessel pressure, and material properties are known. However, obtaining this information is effectively impossible outside of controlled testing programs. Estimates of time to rupture require knowledge of high temperature tensile and stress rupture properties. Figure 8.3 shows some sample high temperature stress rupture data for some pressure vessel steels from Birk and Yoon, 2006.

As can be seen, elevated temperatures can result in rapid vessel rupture. Where fire is possible, pressure vessels should be protected by means such as water spray, thermal insulation, earth mounding, etc. to limit the temperature rise in the vessel and preserve the strength of the material.

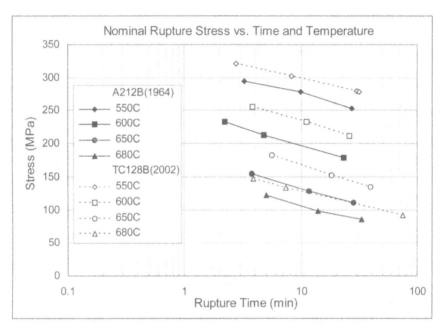

Figure 8.3. Sample of high temperature stress rupture data for two
pressure vessel steels. (Birk and Yoon, 2006)

8.3.3.   Description of a "Typical" BLEVE

While every accident has some unique aspects, when BLEVEs are discussed
in an industrial context, there is an implied sequence of events that is considered
"typical" of this type of explosion. It is important to note that this is not meant
to imply that all, or even any, BLEVE accidents proceed exactly as described
here, but that this sequence is being provided as a narrative of one type of
BLEVE.   For this example, consider the case of a metal storage vessel
containing a flammable pressure-liquefied gas, such as propane, initially stored
at atmospheric temperature.

Initiating Event

If a fire were to occur near the tank due to some other event such as a fuel
spill, structure fire, transport accident, or other source, the vessel wall would be
heated by the fire.

Liquid Response

The liquid near the vessel wall would increase in temperature and this would increase the vapor pressure of the liquid and raise the pressure in the vessel. The liquid near the wall would become less dense, rise, and produce a convection loop in the liquid that would transport heat from the wall.

If heating is applied to the surface of the vessel above the liquid level, heat transfer to the liquid is predominantly by radiation. This is not as effective a heat transfer mechanism as liquid convection and the wall temperature in the vapor space would increase more rapidly than the wall temperature in the liquid space.

Material Weakening and Crack Initiation

As the wall heats, it loses strength. As the fire continues, a combination of pressure increase and material softening would eventually lead to local yielding of the wall. As the material yields, it will elongate. This will create a bulge in the vessel wall. It will also thin the material where it is yielding. The thinning material has less strength and it will continue to yield. Eventually, this process will lead to a local failure and a crack will form. This is often seen as a 'knife edge' failure on the vessel fragments.

Crack Propagation

Where the crack forms, the fluid will flow through the hole thus created. The flow through the hole adds hydrodynamic forces to the bulged material. These forces, in addition to the pressure forces previously applied, cause the crack to progress into the thicker material. If the crack reaches the critical crack length, the vessel will fail catastrophically and a BLEVE will result. This is referred to as a "one step" BLEVE.

Re-pressurization

As the fluid is released, the pressure in the vessel may be decreased. As the pressure decreases, the liquid becomes superheated and boils. Due to the local heat transfer required to form bubbles, there is a time delay between the decrease in pressure and the boiling of the liquid. When the liquid boils, it can produce enough vapor to increase the pressure to or beyond the initial failure pressure and this can cause the crack to propagate. If this leads to a catastrophic loss of containment, a BLEVE will result. This type of re-pressurization event is referred to as a "two-step" BLEVE. If the re-pressurization does not cause a catastrophic failure of the containment, the result will be a jet of fluid.

Liquid Flashing

With the container removed, the superheated liquid flashes into two phases—saturated liquid at its atmospheric boiling point and vapor. The expansion of the liquid into these two phases and the expansion of the fluid from its initial pressure to atmospheric pressure creates a pressure wave and propels fragments into the surroundings. If flammable, the vapor and liquid will often ignite quickly when it contacts the surrounding fire and produce a fireball. The remaining liquid will burn as a pool or, if the fire is self-extinguished, will evaporate due to heat transfer to the liquid from the ground.

Consequences

The consequences of the BLEVE in this case would include a blast wave due to expansion of vapor and flashing liquid, a fireball due to ignition of the hydrocarbon by the nearby fire, and fragment throw or rocketing of vessel pieces.

It should be noted that this sequence does not require the pressure to exceed the design pressure, or even the rated working pressure, of the vessel. As a result, a BLEVE can occur even if the pressure relief device(s) installed on the vessel are working. In fact, a number of experiments have demonstrated this [Birk (1985); Birk, et al. (1997); Birk, et al. (2003); Holden, et al. (1985)]. It should also be noted that although a combustible material was used for this example, the fluid does not need to be flammable for a BLEVE to occur. If the material is flammable and does not ignite immediately, the result may be a flash fire or VCE rather than a fireball. Information on these consequences is available in chapters 5 and 6, respectively.

Other events can follow the BLEVE event, including:
- A vapor cloud explosion
- Pool fire
- Secondary equipment failures due to blast, thermal radiation or fragment loading
- Additional BLEVEs resulting from fires, blast, or impact damage created by an earlier BLEVE

In these cases, the secondary events would be separated in time from the BLEVE and would not be considered to be a concurrent event.

8.4.   BLEVE CONSEQUENCES

BLEVEs can directly lead to a number of consequences. These include:
- Airblast
- Fireball thermal hazards
- Fragment and debris throw

Since some of these consequences have been discussed in other chapters, this chapter will only address the differences between BLEVEs and other types of explosions. If the liquid is flammable, it will be dispersed by the BLEVE and its delayed ignition may lead to a flash fire (discussed in Chapter 5) or vapor cloud explosion (discussed in Chapter 6).

8.4.1.   Airblast

This section addresses the effects of BLEVE blasts and pressure vessel bursts. Actually, the blast effect of a BLEVE results not only from rapid evaporation (flashing) of liquid, but also from the expansion of vapor in the vessel's vapor (head) space. In many accidents, head-space vapor expansion produces most of the blast effects. Rapid expansion of vapor produces a blast wave identical to that of other pressure vessel ruptures containing only compressed gases and vapors. The flashing liquid also produces damaging effects through contribution to vapor and through jetting and rocketing effects.

8.4.1.1.   Energy Calculation

This section presents an overview of BLEVE blast and presents a literature review on pressure vessel bursts and BLEVEs. Evaluation of available energy from BLEVE explosions and pressure vessel bursts is emphasized because this is the most important parameter in determining the blast strength. Next, practical methods for estimating blast strength and duration are presented, followed by a discussion of the accuracy of each method. Example calculations are given in Section 7.3.1.3.

In the discussion of bursting pressure vessels (Chapter 7), the expansion energy of a bursting vessel, $E_{ex,Br}$ is calculated using Brode's equation (Eq. 7.7) (or others as described in that chapter):

$$E_{ex,Br} = \frac{(p_1 - p_0)V}{\gamma_1 - 1}$$

where:

$E_{ex,Br} =$ expansion energy   (J)

p_1 = the pressure inside the vessel   (Pa)

p_0 = the pressure outside the vessel (i.e., atmospheric pressure)   (Pa)

V = volume of the vessel   (m^3)

α_1 = ratio of specific heats

Some researchers use the available thermo mechanical energy or the energy able to do work on the surroundings, stored in the compressed gas. This is the work that could be extracted by expanding the gas without losses.

Determination of the energy released by flashing liquid was addressed by many investigators, including Baker et al. (1978b) and Giesbrecht et al. (1980). They all define explosion energy as the work done by the fluid on surrounding air as it expands isentropically. In this case, the change in internal energy must be calculated from thermodynamic data for the fluid. During this process, the system expands isentropically (no heat transfer, no irreversibilities) from state 1 (the initial state) to state 2, with p_2 equal to the ambient pressure p_0. After expansion, it has a residual internal energy U_2. The work which the system can perform is the difference between its initial and residual internal energies.

$$E_{ex,wo} = U_1 - U_2 \qquad \text{(Eq. 8.1)}$$

where:

$E_{ex,wo}$ =the work performed in expansion from state 1 to state 2.   (J)

Thus, for an ideal gas with a constant ratio of specific heat capacities (ã), the work is:

$$E_{ex,wo} = \frac{p_1 V_1 - p_0 V_2}{(\gamma - 1)} \qquad \text{(Eq. 8.2)}$$

For an ideal gas with a constant ratio of specific heat capacities, pV^{γ} is constant for isentropic expansion.

$$p_1 V_1^{\gamma_1} = p_2 V_2^{\gamma_2} = pV^{\gamma} \qquad \text{(Eq. 8.3)}$$

Therefore,

$$V_2 = V_1 \left[\frac{p_1}{p_0} \right]^{1/\gamma} \qquad \text{(Eq. 8.4)}$$

This gives, for the work:

$$E_{ex,wo} = \frac{p_1 V_1}{(\gamma_1 - 1)} \left[1 - \left(\frac{p_0}{p_1} \right)^{(\gamma-1)/\gamma} \right] \qquad \text{(Eq. 8.5)}$$

It is illustrative to compare work $E_{ex,wo}$ with the Brode expansion energy $E_{ex,Br}$. The ratio $E_{ex,wo}/ E_{ex,Br}$ can be written as

$$\frac{E_{ex,wo}}{E_{ex,Br}} = \frac{(\bar{p}_1 + 1) - (\bar{p}_1 + 1)^{1/\gamma}}{\bar{p}_1} \qquad \text{(Eq. 8.6)}$$

where:

$\bar{p}_1$ is the non-dimensional burst pressure in the initial state $(p_1/p_0) - 1$.

This function is depicted in Figure 8.4. It is instructive to consider that the liquid flashing fraction is most commonly calculated using an isentropic model to account for the fact that the system is not transferring heat or losing potential work to the surroundings over very small time scales. This corresponds to an assumption of an instantaneous change from one state to the other. To be consistent, many researchers use this same isentropic assumption when calculating the energy for flashing liquid systems. This assumption has the added advantage of being equally applicable to single and multi-phase systems as encountered in BLEVEs and can be applied using data for non-ideal fluids as readily as it can for ideal gases.

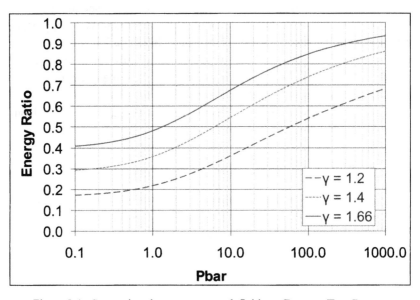

Figure 8.4. Comparison between energy definitions: Eex, wo/Eex, Br.

Analysis of Figure 8.4 makes it clear that the two definitions give widely varying results. The isentropic energy approaches the Brode energy for high initial overpressures, but for initial pressures of practical interest, the results may vary by 50% or more. Thus, there is no consensus on the definition of the most important variable of an explosion, its energy. All experimental and most numerical results given in the literature use Brode's definition. However, when the fluid is a non-ideal gas, a liquid, or a two-phase liquid/gas system, almost everyone uses the isentropic expansion energy as the explosion energy. The available data and prediction methods for blast parameters are based on these two, conflicting, definitions. It is important to understand which energy was used for the non-dimensionalization of results when using blast or other lookups to ensure that the correct energy definition is used.

8.4.1.2.   Selection of Failure Pressure

In the previous section and in Chapter 7, several definitions for burst energy were provided. All of these definitions fundamentally depend on three parameters: pressure, volume, and the ratio of specific heats. The volume is usually easy to determine based on the volume of the vessel and its fill level. The ratio of specific heats is also usually easy to determine as it is a property of the material in the vessel. The only parameter left to define is the pressure. In an accident investigation, the burst pressure may be known and, in this case, this would be the best pressure to use. In predicting the consequences of an accident that has yet to occur, it is necessary to predict the pressure that the vessel will contain at the time of failure. Several options are available based on the initiating scenario under consideration.

Maximum Allowable Working Pressure (MAWP)

If the initiating scenario is a corrosion, fatigue, or impact scenario, it is likely that the vessel will be at or below its maximum allowable working pressure at the time of failure, possibly at normal operating pressures.

Pressure Relief Device Set Pressure

It may be possible to use the pressure relief device (PRD) set pressure to determine the maximum pressure experienced by the vessel if a PRD is installed on the vessel and if the following conditions can be verified:
- that the PRDs are installed correctly,
- that the PRDs will be able to perform as designed,
- that the PRD will be passing the correct phase (i.e., will not be venting liquid or two-phase material if sized for vapor-only relief), and
- that the PRD is sized to handle all foreseeable causes.

Depending on the code that the vessel was designed to and to which section it was designed, the maximum pressure expected for a design condition fire can exceed the set pressure by a significant margin. Based on ASME codes, the pressure at which PRDs are required to achieve their rated flow rates can be up to 1.21 times the MAWP of the vessel. In some cases, the set pressure of the PRD may differ from the MAWP, by design or by achieved performance, and, in all cases, actual PRD performance (if known) should be considered when determining the possible failure pressure of a vessel.

Ultimate pressure

If there is a scenario in which the vessel may reach its ultimate tensile strength, the ultimate pressure may be used. These scenarios may include one or more of the following:

- overfill,
- runaway reaction,
- blocked PRD, or
- fire exceeds code-based design condition.

This can be estimated by assuming that the ultimate strength is some multiplier above the MAWP (typically a factor of ~3.5 to 4.0) or by calculating the pressure in the vessel when the ultimate stress is reached in the wall material. For example, Droste and Schoen (1988) describe an experiment in which an LPG tank failed at 39 bar, or 2.5 times the opening pressure of its safety valve.

8.4.1.3.   Selection of the Liquid Level at Failure

Since BLEVEs involve two-phase fluids, it is necessary to know the liquid level of the vessel at the time of failure. The best way to select this value is to use the statistics to assess the liquid level. In this approach, a probability distribution would be used to select a fill level (or levels) to meet a particular confidence interval. If the fill level probabilities are not known, the fill level should be conservatively assumed, with high and low values used, if appropriate, to provide high and low predictions of the consequences.

8.4.1.4.   Prediction of Blast Damage

A vessel filled with a pressure-liquefied gas can produce blast effects upon bursting in three ways. First, the vapor that is usually present above the liquid can generate a blast, as from a gas-filled vessel. Second, the liquid could boil violently upon sudden depressurization, and, if the boiling is very rapid and coherent it too could generate a shock. Third, if the released fluid is combustible and the BLEVE is not fire induced, a vapor cloud explosion or flash fire may occur (see Chapter 6). The vapor expansion and liquid flashing will be discussed in this section.

Experimental Work

Although a great many investigators studied the release of superheated liquids (that is, liquids that would boil if they were at ambient pressure), only a few have measured the blast effects that may result from release. Baker et al. (1978a) reports on a study done by Esparza and Baker (1977b) in which liquid CFC-12 was released from frangible glass spheres in the same manner as in their study of blast from gas-filled spheres. The CFC was below its superheat limit temperature. No significant blast was produced.

Investigators at BASF (Maurer et al. 1977; Giesbrecht et al. 1980) conducted many small-scale experiments on bursting cylindrical vessels filled with propylene. The vessels were completely filled with liquid propylene at a temperature of around 340 K (which is higher than the superheat limit temperature) and a pressure of around 60 bar. Vessel volumes ranged from 226 ml to 1.00 m^3. The vessels were ruptured with small explosive charges, and after each release, the resulting cloud was ignited. While the experiments focused on explosively dispersed vapor clouds and their subsequent deflagration, the pressure wave developed from the flashing liquid was measured.

The investigators found that overpressures from the flashing liquid compared well with those resulting from gaseous detonations of the same energy. Energy here means the work which can be done by the fluid in expansion, $E_{ex,wo}$. This means that energy release during flashing must have been very rapid. This would be expected because for a 100% liquid filled vessel the sudden rupture would lead to a near instantaneous drop to ambient pressure. This is not commonly the scenario for a pressure vessel rupture.

British Gas performed full-scale tests with LPG BLEVEs similar to those conducted by BASF. The experimenters measured very low overpressures from the flashing liquid, followed by the so-called "second shock," and by the pressure wave from the vapor cloud explosion. The pressure wave from the vapor cloud explosion probably resulted from experimental procedures involving ignition of the release. The liquid was below the superheat limit temperature at time of burst.

Analysis of an incident (Van Wees, 1989) involving a carbon dioxide storage vessel suggests that carbon dioxide can evaporate explosively even when its temperature is below the superheat limit temperature.

In Section 7.3.1.3., a method is given for calculating overpressure and impulse, given energy and distance. This method produces results which are in reasonable agreement with experimental results from BASF studies. The procedure is presented in more detail by Baker et al. (1978b).

Wiedermann (1986b) presents an alternative method for calculating work done by a fluid. The method uses the "lambda model" to describe isentropic expansion, and permits work to be expressed as a function of initial conditions and only one fluid parameter, lambda. Unfortunately, this parameter is known for very few fluids.

Other test series have been run by Barbone (1994), Ogiso (2004), and more recently by Chen (2006).

Theoretical Work

Relatively little theoretical work has been conducted to study the detailed mechanisms in a pressure vessel BLEVE. Reid's superheat theory (Reid 1979) postulates why a BLEVE takes place, but there is almost no analysis of the process in detail. In the literature it is often assumed that if the liquid temperature is at or above the atmospheric superheat limit, then the liquid will flash to vapor explosively (i.e. it will produce a shock) upon sudden loss of containment. If it is below the superheat limit, it will not produce a shock. Experimental evidence has shown this not be the case. There is no clear dividing line between a BLEVE and a non-BLEVE event based on the superheat limit temperature. If the fluid is above its atmospheric boiling point (i.e., it is superheated), there will be boiling. If there is large degree of superheat, there will be powerful boiling. As discussed in section 8.3.1, this boiling will almost certainly occur as heterogeneous boiling at nucleation sites and not as homogeneous boiling. As a result, the superheat limit temperature has no special significance to the event.

There have been attempts to predict the maximum possible pressure generated in a partially failed vessel during a two step BLEVE. In this analysis, it was assumed that the vessel begins to rupture and the pressure inside the vessel drops to atmospheric pressure before the liquid starts to flash. At this instant, it is assumed that the isentropic flash fraction of the liquid converts to vapor and this vapor is forced to fit within the available vessel volume. This type of analysis will yield a very high theoretical pressure. Birk et al. (2006) conducted such an analysis and suggested that very high pressures can be generated in a vessel that started with propane. This analysis, of course, assumes that the vessel holds together during this very rapid boiling process. If this idealized process was to occur in practice, the very high internal overpressure would produce a very high shock pressure when the vessel opens up. Field experiments have not produced shocks that would indicate that such internal pressures are being achieved.

Birk et al. (2006) attempted to measure any spike in pressure caused by explosive flashing inside the vessel during his fire testing of 400-liter ASME code propane pressure vessels. His results were non-conclusive. They did measure pressure spikes at the instant of failure but they were not certain that it was a true pressure – it might have been the instrument cables being destroyed at the time of failure. What they did measure was the shock overpressure at the side and ends of the vessels and these did not indicate any unusually high overpressures.

Van den Berg et al. (1994) studied overpressures generated by explosive flash evaporation by numerically solving the governing momentum equations. It was assumed that the blast produced by the flashing liquid was expansion controlled. This means the process is dominated by the expansion of the vapor, not the phase change of the liquid. In other words, it was assumed that the phase change was instantaneous. It was assumed that the liquid generates a shock. To do anything else requires very detailed models of the flashing process. Figure 8.5 from Venart (2000) shows the predicted blast decay curve for propane from two different calculations. One is the method of van den Berg (2006). The graph also shows the decay curve for a high explosive charge. The challenge is to accurately define the scaled distance and this depends on the energy available in the pressure vessel to start the shock.

Theoretical work to determine how the vessel opens and on how the liquid changes phase is very difficult because of the various processes involved, including:

- the vessel depressurization process which depends on exactly how it fails
- nucleation of bubbles at the vessel wall
- nucleation of bubbles on the liquid surface
- nucleation of bubbles on impurities or preexisting bubbles in the liquid
- homogeneous nucleation
- the complete vessel failure process

Practical theoretical models of these processes and their interactions do not currently exist. Since these detailed models do not exist, analysis should be conducted using simpler models with reasonable but conservative assumption.

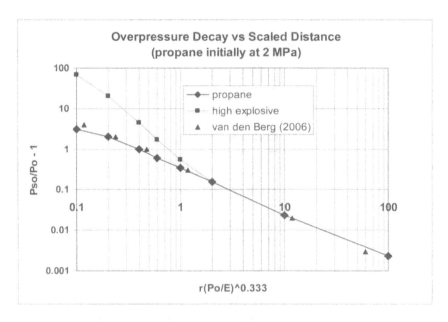

Figure 8.5. Overpressure Decay Curve for Propane Tank BLEVE.
(Birk et al., 2007)

8.4.1.5. Method for Predicting Blast Produced by Flashing Liquids

The work performed on surrounding air by the expansion can be calculated from thermodynamic data for the fluid (vapor initially in the vessel and vapor from flashing liquid) based on an assumption of an isentropic process.

In many cases, both liquid and vapor are present in a vessel. Experiments indicate that the blast wave from the expanding vapor is separate from that generated by flashing liquid. It has been suggested that the liquid may change phase too slowly to produce a shock of its own in a one-step BLEVE and that the primary blast loading in these events is often due to just the vapor phase. For safety predictive purposes, it is not known whether the event will be a one-step or two-step and event and the conservative approach is to assume that the blast waves are produced by the combined energy of the liquid and vapor. This method is given in Figure 8.6.

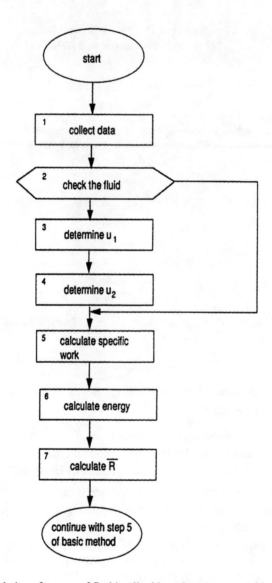

Figure 8.6. Calculation of energy of flashing liquids and pressure vessel bursts filled with vapor or nonideal gas.

Step 1: Collect the following data:

- Internal pressure p_1 (absolute) at failure. (see 8.4.1.2)
- Ambient pressure p_0.
- Quantity of the fluid (volume V_1 or mass).
- Distance from center of the vessel to the "target", r.
- Shape of the vessel: spherical or cylindrical.

As shown previously, it is possible for the internal pressure to rise until the bursting overpressure is reached, which may be much higher than the vessel's design pressure. Note also that this method assumes that the fluid is in thermodynamic equilibrium; yet, in practice, stratification of liquid and vapor will occur (Moodie et al. 1988; Birk et al. 1996a).

Expansion energy can be calculated using the methods outlined in 8.4.1.1. Thermodynamic data on various fluids can be found in Perry and Green (1984) or Edmister and Lee (1984), and other publications. Software is also commercially available which includes thermodynamic data for a wide range of commodities. The method for determining the thermodynamic data will be explained in detail in Step 3.

Step 2: Determine internal energy in initial state, u_1.

The work done by an expanding fluid is defined as the difference in internal energy between the fluid's initial and final states. Most thermodynamic tables and graphs do not present u_1, but only h, p, v, T (the absolute temperature), and s (the specific entropy). Therefore, u must be calculated with the following equation:

$$h = u + pv \qquad \qquad (Eq.\ 8.7)$$

where:

h	=	specific enthalpy (enthalpy per unit mass)	(J/kg)
u	=	specific internal energy	(J/kg)
p	=	absolute pressure	(Pa)
v	=	specific volume	(m^3/kg)

To use a thermodynamic graph, locate the fluid's initial state on the graph. (For a saturated fluid, this point lies either on the saturated liquid or on the saturated vapor curve, at a pressure p_1.) Read the enthalpy h_1, volume v_1, and entropy s_1 from the graph. If thermodynamic tables are used, interpolate these values from the tables. Calculate the specific internal energy in the initial state u_1 with Eq. (8.7). Note that it is assumed that the liquid will be at saturation, i.e., that the liquid will be at its boiling point for the vessel pressure at the time of failure.

Step 3: Determine internal energy in expanded state, u_2.

The specific internal energy of the fluid in the expanded state u_2 can be determined as follows: If a thermodynamic graph is used, assume an isentropic expansion (entropy s is constant) to atmospheric pressure p_0. Therefore, follow the constant-entropy line from the initial state to p_0. Read h_2 and v_2 at this point, and calculate the specific internal energy u_2.

When thermodynamic tables are used, read the enthalpy h_f, volume v_f, and entropy s_f of the saturated liquid at ambient pressure, p_0, interpolating if necessary. In the same way, read these values (h_g, v_g, s_g) for the saturated vapor state at ambient pressure. Then use the following equation to calculate the specific internal energy u_2:

$$u_2 = (1 - X) h_f + X h_g - (1 - X)p_0v_f - Xp_0v_g \qquad \text{(Eq. 8.8)}$$

where:

$$X \quad = \quad \text{quality or vapor mass fraction} \qquad (-)$$

$$= \quad \frac{m_g}{m_g + m_f}$$

$$= \quad \frac{s_1 - s_g}{s_g - s_f}$$

$$= \quad \frac{h_1 - h_g}{h_g - h_f}$$

$$= \quad \frac{v_1 - v_g}{v_g - v_f}, \text{ etc.}$$

$$S \quad = \quad \text{specific entropy}$$

Subscript 1 refers to initial state.

Subscript f refers to state of saturated liquid at ambient pressure.

Subscript g refers to state of saturated vapor at ambient pressure.
Equation (8.8) is only valid when X is between 0 and 1.

Step 4: Calculate the specific work.

The specific work done by an expanding fluid is defined as.

$$e_{ex} = u_1 - u_2 \qquad\qquad \text{(Eq. 8.9)}$$

where:

e_{ex} = specific work. (See Section 8.4.1.1)   (J/kg)

Step 5: Calculate the expansion energy.

To calculate expansion energy, multiply the specific expansion work by the mass of the fluid released. If energy per unit volume is used, multiply by the volume of fluid released. Multiply the energy by 2 to account for reflection of the shock wave on the ground, if applicable as follows:

$$E_{ex} = (1\text{-frag})(\text{gnd})\, e_{ex}\, m_1 \qquad\qquad \text{(Eq. 8.10)}$$

where:

m_1 = mass of the fluid   (kg)

gnd = factor for ground reflection (see Chapter 4)   (-)

frag = fragment reduction factor   (-)

As discussed in Chapter 6, the fragment reduction factor is a means of accounting for energy consumed in the production and throw of fragments of the vessel. In the case of BLEVEs, this factor may be increased since some of the energy may be consumed by propelling the liquid.

Repeat Steps 3 to 5 for each material present in the vessel, and add the energies to find the total energy E_{ex} of the explosion.

Step 6: Calculate the nondimensional range to the receptor

The nondimensional range $\bar{R}$ of the receptor can be calculated using Eq. (7.21), as follows:

$$\bar{R} = R\left[\frac{p_0}{E_{ex}}\right]^{1/3}$$

where:

R =   the distance at which blast parameters are to be determined.   (m)

Step 7: Calculate the pressure and impulse at the receptor

Use the dimensionless stand-off along with the selected burst pressure, temperature and ratio of specific heats, as necessary to calculate the blast using the methodologies presented in chapter 7.

8.4.1.6.   Accuracy

The method presented above usually give upper limit estimates of blast parameters from BLEVEs. However, the detailed mechanism of the vessel failure may lead to directional effects that can sometimes enhance local overpressures, and this will not be captured by this method (see Chapter 7 for more details). The failure process of a pressure vessel can be slow and this also affects the production of the shock.

Figure 8.7 and Figure 8.8 from Birk et al. (2007) show predicted and observed blast overpressures from BLEVEs of 2 m^3 ASME code propane pressure vessels. These vessels were horizontal cylinders and no adjustment has been made for shape factor enhancement in the predictions shown. The explosion energy was multiplied by 2 to account for the ground reflection.

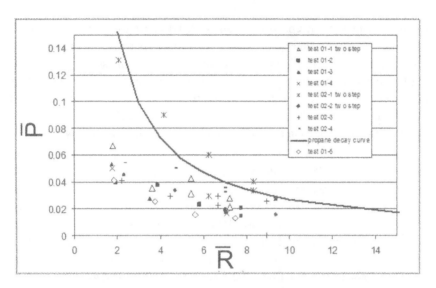

Figure 8.7. Measured first peak overpressures vs scaled distance (based on vapor energy) from 2000-liter propane tank BLEVEs. (Birk et al., 2007)

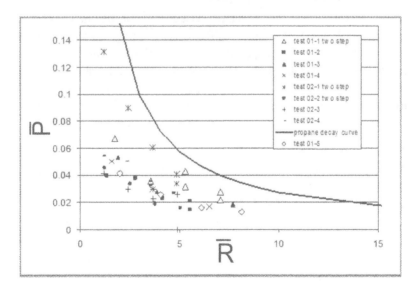

Figure 8.8. Measured first peak overpressures vs scaled distance (based on liquid energy) from 2000-liter propane tank BLEVEs. (Birk et al., 2003)

Figure 8.7 shows the overpressure versus scaled distance for the tests conducted. The solid line is the predicted decay in overpressure assuming the vapor was the source of the shock. In Figure 8.8, the same plot is given assuming the liquid was the source of the shock. These figures show that the prediction methodology shown here is generally conservative with just one test falling above the predictions when just the vapor energy is used in the predictions. When the liquid energy is included in the predictions, the results are conservative.

The main sources of deviation between the predicted and observed pressures lie in the release-process details (i.e. directional effects). The isentropic expansion energy method provides a good estimate of the energy for the tests where data is available. Note that there has been one test of a 22% propane filled full scale rail tank-car in Germany (Balke et al., 1999) and the methods used here seem to agree well with the observed blast from that test as well. In these predictions, the energy translated into kinetic energy of fragments and ejected liquid was not subtracted from blast energy (i.e., zero energy consumed by fragments). This may artificially increase the expansion energy available for blast production by as much as 50%. For these reasons it is concluded that these calculation methods are generally conservative.

As noted earlier, for safety planning calculations, it may be conservatively assumed that all of the isentropic liquid and vapor energy goes into generating blast. This will produce a conservative estimate of blast damage. For accident reconstructions, calculations should be calculated for the liquid and liquid + vapor energies separately. The predictions based on both phases will be conservative and can be considered an upper bound while the prediction using just the vapor phase will be a reasonable estimate for a one-step BLEVE.

8.4.2.   Thermal Hazards

In general, structural damage due to blast waves produced by BLEVEs is limited the near-field of the vessel. If the vessel has a significant fill level with a combustible fluid when the BLEVE occurs, the resulting fireball can often impose damaging thermal loads at greater distances than the blast waves. For this reason, it is important to predict the size and duration of the fireball from a BLEVE where the vessel contents are combustible or flammable.

The radiation hazard from a BLEVE fireball can be estimated once the following fireball properties are known:
- the maximum diameter of the fireball, that is, fuel mass contributing to fireball generation;
- the surface-emissive power of the fireball;

- the total duration of the combustion.

For each of these properties, data and calculation techniques are available. A summary of each is presented in the sections below.

8.4.2.1.   Fuel Contribution to Fireball

Hasegawa and Sato (1977) showed that when the isentropic flash fraction of the liquid at the temperature at time of vessel failure equals 36% or more, all of the contained fuel contributes to the fireball. For lower flash fractions, part of the fuel is consumed in the fireball and the remainder forms a pool. In this model, it is assumed that the amount of fuel that will contribute to the fireball is equal to three times the isentropic flash fraction up to a maximum of 100% of available liquid fuel. To simplify the calculations, a conservative approach can be taken in which it is assumed that all available liquid fuel will contribute to the BLEVE fireball.

8.4.2.2.   Fireball Size and Duration

Many small-scale experiments have been carried out to measure the durations and maximum diameters of fireballs. These experiments have resulted in the development of empirical relations among the total mass of fuel in the fireball and its duration and diameter. Fireball diameter estimates, as published by several investigators and modelers, are presented in Table 8.1 and Table 8.2.

Average values for calculation of fireball diameter and duration are available from Roberts (1982) and Pape et al. (1988), which produced the following equations:

$$D_c = 5.8 m_f^{1/3} \qquad\qquad \text{(Eq. 8.11)}$$

And

$$t_c = 0.45 m_f^{1/3} \text{ for } m_f < 30{,}000 \text{ kg}$$
$$t_c = 2.6 m_f^{1/6} \text{ for } m_f > 30{,}000 \text{ kg} \qquad \text{(Eq. 8.12)}$$

where:

D_c	=	final fireball diameter	(m)
t_c	=	fireball duration	(s)
m_f	=	mass of fuel in fireball	(kg)

Because this relationship also reflects the average of all relations from Table 8.1 and Table 8.2, its use is recommended for calculating the final diameter and duration of a spherical fireball.

Table 8.1. Empirical relationships for fireball durations and diameters
(adapted from Abbasi and Abbasi, 2007) [see nomenclature in Table 8.2]

Source Empirical Correlations	Material	Diameter, D_{max} (m)	Duration, t_B (s)
Hardee and Lee 1973	Propane	$5.55\,M^{0.333}$	—
Fay and Lewis 1977	Propane	$6.28\,M^{0.333}$	$2.53\,M^{0.167}$
Hasegawa and Sato 1977	Pentane	$5.28\,M^{0.277}$	$1.10\,M^{0.097}$
Hasegawa and Sato 1978	n-Pentane	$5.25\,M^{0.314}$	$1.07\,M^{0.181}$
Williamson and Mann 1981	Not provided	$5.88\,M^{0.333}$	$1.09\,M^{0.167}$
Lihou and Maund 1982	Butane	$5.72\,M^{0.333}$	$0.45\,M^{0.333}$
Lihou and Maund 1982	Rocket Fuel	$6.20\,M^{0.320}$	$0.49\,M^{0.320}$
Lihou and Maund 1982	Propylene	$3.51\,M^{0.333}$	$0.32\,M^{0.333}$
Lihou and Maund 1982	Methane	$6.36\,M^{0.325}$	$2.57\,M^{0.167}$
Moorhouse and Pritchard 1982	Flammable Liquid	$5.33\,M^{0.327}$	$1.09\,M^{0.327}$
Lihou and Maund 1982	Propane	$3.46\,M^{0.333}$	$0.31\,M^{0.333}$
Duiser 1985	Flammable Liquid	$5.45\,M^{1.3}$	$1.34\,M^{0.167}$
Marshall 1987	Hydrocarbon	$5.50\,M^{0.333}$	$0.38\,M^{0.333}$
Gayle and Bransford 1965 and Bagster and Pitblado 1989	Flammable Liquid	$6.14\,M^{0.325}$	$0.41\,M^{0.340}$
Pietersen 1985, CCPS 1989, Prugh 1994 and TNO 1997	Flammable Liquid	$6.48\,M^{0.325}$	$0.852\,M^{0.260}$
Roberts 1982 and CCPS 1999	Flammable Liquid	$5.80\,M^{0.333}$	$0.45\,M^{0.333}\;(M<3\times10^4)$ $2.60\,M^{0.167}\;(M>3\times10^4)$
Martinsen and Marx 1999	Flammable Liquid	$8.66\,M^{0.25}\,t^{0.333}$ $0\le t\le t_b/3$	$0.9\,M^{0.25}$

Table 8.2. Analytical relationships for fireball durations and diameters
(adapted from Abbasi and Abbasi, 2007)

Source	Material	Diameter, D_{max} (m)	Duration, t_B (s)
Bader et al., 1971	Propellant	$0.61\left(\dfrac{3}{4\pi\rho}\right)^{1/3}(W_b)^{1/3}$	$0.572(W_b)^{1/6}$
Hardee and Lee 1973	LNG	$6.24M^{0.333}$	$1.11M^{0.167}$
Fay and Lewis 1977	Flammable Liquid	$\dfrac{g\beta t^2(\rho_a - \rho_p)}{7\rho_p}$	$\left(\dfrac{14\rho_p}{g\beta(\rho_a - \rho_p)}\right)^{0.5}\left(\dfrac{3V}{4\pi}\right)^{0.167}$

M: mass of fuel in fireball (kg);
t: time elapsed after BLEVE (s);
ρ: density of fireball gas (lb/ft^3);
W_b: mass of propellant (lb);
g: acceleration due to gravity (m/s^2);
$â$: entrainment coefficient;
ρ_a: density of air (kg/m^3);
ρ_p: density of products of combustion (kg/m^3)

8.4.2.3.   Radiation

For a receptor not normal to the fireball, radiation received can be calculated based on the solid flame model as follows:

$$q = EF\tau_a \qquad\qquad \text{(Eq. 8.13)}$$

where:

q	=	radiation received by receptor	(W/m^2)
E	=	surface emissive power	(W/m^2)
F	=	view factor	$(-)$
τ_a	=	atmospheric attenuation factor (transmissivity)	$(-)$

The surface-emissive power E, the radiation per unit time emitted per unit area of fireball surface, can be assumed to be equal to the emissive powers measured in full-scale BLEVE experiments by British Gas (Johnson et al. 1990). These entailed the release of 1000 and 2000 kg of butane and propane at 7.5 and 15 bar. Test results revealed average surface-emissive powers of 320 to 370

kW/m^2. A value of 350 kW/m^2 seems to be a reasonable value to assume for BLEVEs for most hydrocarbons involving a vapor mass of 1000 kg or more.

For a point on a plane surface located at a distance L from the center of a sphere (fireball) that can "see" all of the fireball, the view factor (F) is given by Figure A-1 in Appendix A:

$$F = \frac{r^2}{L^2} \cos \Theta \qquad \text{(Eq. 8.14)}$$

where:

r	$=$	the radius of the fireball ($r = D_c/2$)	(m)
L	$=$	the distance from the center of the fireball	(m)
Θ	$=$	the angle between the normal to the surface and the connection of the point to the center of the sphere.	(deg)

In the general situation, the fireball center has a height (z_c) above the ground ($z_c \geq D_c/2$), and the distance (X) is measured from a point at the ground directly beneath the center of the fireball to the receptor at ground level. When this distance is greater than the radius of the fireball, the view factor can be calculated as follows:

For a vertical surface:

$$F = \frac{X(D_c/2)^2}{(X^2 + z_c^2)^{3/2}} \qquad \text{(Eq. 8.15)}$$

For a horizontal surface:

$$F = \frac{z_c(D_c/2)^2}{(X^2 + z_c^2)^{3/2}} \qquad \text{(Eq. 8.16)}$$

In most cases, the BLEVE fireball is assumed to touch the ground ($z_c = D_c/2$). For large scale BLEVEs, the assumption that the fireball is at its maximum diameter and "rests" on the ground will predict thermal hazard quite accurately.

Atmospheric transmissivity τ_a can be estimated by using Eq. 5.7:

$$\tau_a = \log(14.1 \, h_{rel}^{-0.108} X^{-0.13})$$

where

| τ_a | = | transmissivity | (–) |
| h_{rel} | = | relative humidity | (–) |

The point-source model can also be used to calculate the radiation received by a receptor at some distance from the fireball center. Hymes (1983) presents a fireball-specific formulation of the point-source model developed from the generalized formulation (presented in Chapter 5) and Roberts' (1982) correlation of the duration of the combustion phase of a fireball. According to this approach, the peak thermal input at distance L is given by

$$q = \frac{2.2\tau_a RH_c m_f^{0.67}}{4\pi L^2} \qquad \text{(Eq. 8.17)}$$

where:

m_f	=	mass of fuel in the fireball	(kg)
τ_a	=	atmospheric transmissivity	(–)
H_c	=	net heat of combustion per unit mass	(J/kg)
R	=	radiative fraction of heat of combustion	(–)
L	=	distance from fireball center to receptor	(m)
q	=	radiation received by the receptor	(W/m^2)

Hyme suggests the following values of R

| R | = | 0.3, fireballs for vessels bursting below relief valve pressure; |
| R | = | 0.4, fireballs for vessels bursting at or above relief valve pressure. |

8.4.2.4.   Calculation Procedure

The following procedure can be followed for estimating radiation hazards:

Estimate the mass of liquid in the vessel at the time of failure

Estimate the isentropic flash fraction of the liquid and use three times the flash fraction as the mass that contributes to the fireball orconservatively assume that all liquid contributes to the fireball.

Estimate the fireball diameter (D_c) and duration (t_c) using one of the models from Table 8.1 or another equivalent model.

Assume a surface-emissive power of 350 kW/m^2.

Estimate the geometric view factor on the basis of the fireball diameter and the position of the receptor using the relationships presented in Section 8.4.2.3 or Chapter 5.

Estimate the atmospheric transmissivity τ_a using the relationships presented in Section 8.4.2.3 or Chapter 5.

Estimate the received thermal flux q using the relationships presented in Section 8.4.2.3 or Chapter 5.

8.4.3.   Fragment and Debris Throw

A BLEVE can produce fragments that fly far from the explosion source. Primary fragments, which are part of the original vessel, can be highly hazardous and may result in damage to structures and injuries to people. Where the vessel breaks into large pieces, they may be thrown while carrying some of the flashing liquid with them. As the liquid flashes, it will expand and propel the pieces in a manner similar to that of a rocket. For this reason, these self-propelled fragments are called 'rockets'. Primary fragment and rocket effects are determined by the number, shape, velocity, and trajectory of fragments; these fragments and rockets generally represent the farthest ranging hazard due to BLEVEs.

When a high explosive detonates, a large number of small fragments with high velocity and chunky shape result. In contrast, a BLEVE produces only a few fragments, varying in size (small, large), shape (chunky, disk-shaped), and initial velocities. Fragments can travel long distances, because large, half-vessel fragments can "rocket" and disk-shaped fragments can "frisbee." The results of an experimental investigation described by Schulz-Forberg et al. (1984) illustrate BLEVE-induced vessel fragmentation.

All parameters of interest with respect to fragmentation will be discussed. The primary discussion regarding the calculation of the range and damage potential of vessel fragments is provided in Chapter 7 and is not repeated here. The following discussion is limited to areas where BLEVEs differ from failures of gas-filled pressure vessels.

8.4.3.1.   Experimental Results

Figure 8.9 illustrates results of three fragmentation tests of 4.85-m^3 vessels

50% full of liquid propane. The vessels were constructed of steel (StE 36; unalloyed fine-grained steel with a minimum yield strength of 360 N/mm²), and had wall thicknesses of 5.9 mm (Test 1) and 6.4 mm (Tests 2 and 3). Vessel overpressure at moment of rupture was 24.5 bar in the first test, 39 bar in the second test, and 30.5 bar in the third.

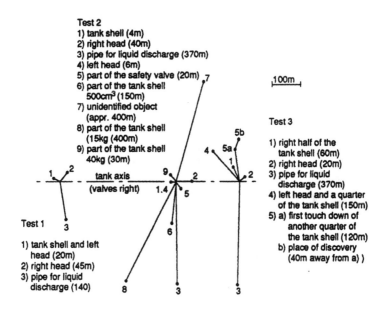

Figure 8.9. Schematic view of vessel fragments' flight after vessel bursts in three BLEVE tests. (Schulz-Forberg et al. 1984)

8.4.3.2.   Theoretical Models

Research on predictions of fragment velocity and range has heretofore been concentrated on the idealized situation of the gas-filled, pressurized vessel. Other cases, including those of nonideal gas-filled vessels and vessels containing combinations of gas and liquid, were investigated (Johnson et al. 1990). Fragment velocity and range can be assumed to depend on the total available energy of a vessel's contents. If this energy is known, the vessel's contents are not significant. Since the fragment production for ideal and non-ideal gases have already been discussed in Chapter 7, it is not necessary to replicate such a discussion here. As discussed above, the isentropic energy can be calculated for the vessel and this can be used to predict the range and size distribution of fragments using the models presented in Chapter 7.

A unique aspect of BLEVEs is that, where the containment breaks into a small number of large pieces, some of these pieces may be propelled by flashing liquid to great distances. This 'rocketing' phenomenon is discussed below.

8.4.4.   Ranges for Rocketing Fragments

Ranges for rocketing fragments can be calculated from guidelines given by Baum (1987) for cases in which liquid flashes off. The initial-velocity calculation must take into account total energy. If this is done, rocketing fragments and fragments from a bursting vessel in which liquid flashes are assumed to be the same.

Ranges were calculated for a simulated accident with the methods of Baker et al. (1978a,b) and Baum (1987). It appears that the difference between these approaches is small. Initial trajectory angle has a great effect on results. In many cases (e.g., for horizontal cylinders) a small initial trajectory angle may be expected. If the optimal angle is used, very long ranges are predicted.

Birk et al. (1996b) presented a simplified approach to estimate the distances of rockets. He presented the following simple formulas for estimates of rocket ranges.

for vessels smaller than 5 m³

the maximum likely range is $R = 90\ m^{0.333}$

for vessels larger than 5 m³

the maximum likely range is $R = 465\ m^{0.10}$ (Eq. 8.18)

where:

R   =   rocket range   (m)

m   =   mass of the liquid and vapor lading in the vessel at the time
          of failure   (kg)

To illustrate the full range and distribution of data available on the issue of fragment throw from BLEVEs, a series of plots of observations is provided in Figure 8.10 through Figure 8.17. It is particularly noteworthy that, while most fragments do fall along the long axis of horizontal vessels, the range of the fragments is similar in both the axial and radial directions. This means that the probability density may be skewed to provide a higher density of fragments along the long axis but there are still many long-ranging fragments launched towards the 'side' or radial direction of these vessels.

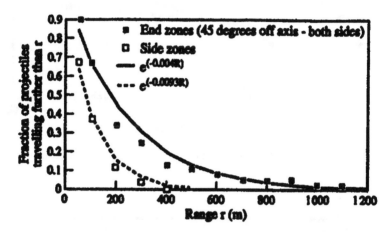

Figure 8.10. Fragment range distribution for Mexico City accident.
(Pietersen, 1988)

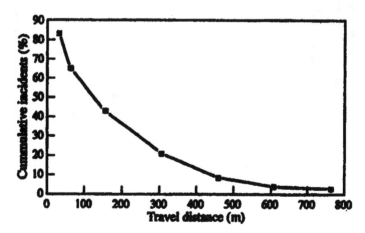

Figure 8.11. Fragment throw from LPG rail car BLEVEs (52 incidents).
(Campbell, 1981)

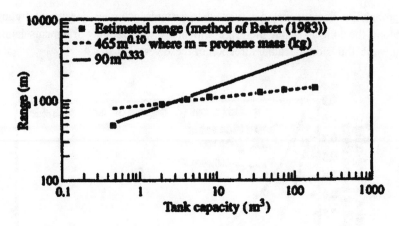

Figure 8.12. Predicted propane BLEVE rocket range versus tank capacity.

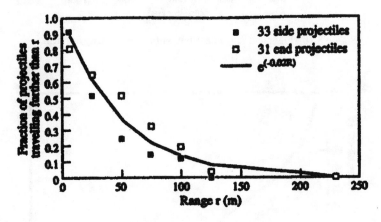

Figure 8.13. Fragment range distribution from BLEVEs of 400-liter propane tanks
(64 primary and secondary projectiles).

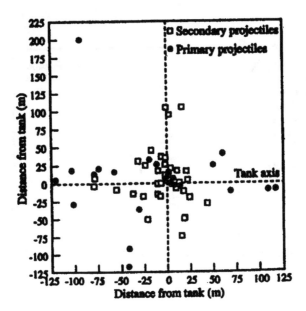

Figure 8.14. Observed projectiles from 13 BLEVEs of 400-liter propane tanks.
(Birk et al. 1995)

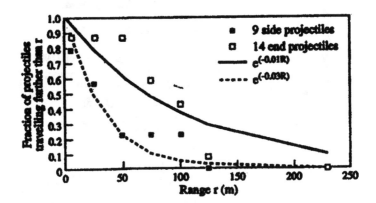

Figure 8.15. Fragment range distribution from BLEVEs of 400-liter propane tanks
(23 primary projectiles).

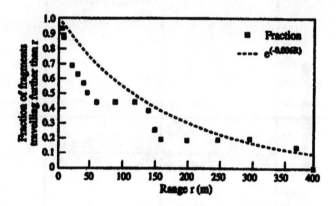

Figure 8.16. Fragment range distribution for BLEVEs of 4850-liter tanks with propane.
(Schulz-Forberg et al., 1984)

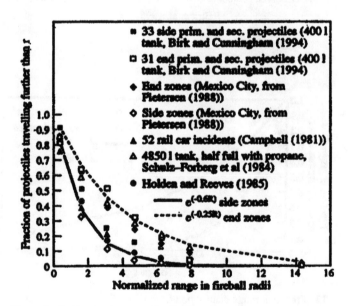

Figure 8.17. Fragment range distribution normalized by fireball radii.

8.5.   ANALYTICAL MODELS

As with other areas of engineering, numerical models are being developed for the prediction and analysis of BLEVEs. Commercial models have been produced that use a combination of empirical correlations and thermodynamics to relax some of the assumptions made by simpler models. While these models provide more information than the simple approaches, they are still developmental.

8.6.   SAMPLE PROBLEMS

This section illustrates the application of the calculations presented in this chapter. Each problem is intended to illustrate one part of the computation. In a full safety analysis, all hazards should be assessed.

8.6.1.   Sample Problem #1: Calculation of Air Blast from BLEVEs

Problem Definition

After a successful pressure test, a 25 m^3 spherical pressure vessel is put back into service. The safety valve is set at 1.5 MPa (15 bar). What would the blast load be at a control room 15 meters away if the vessel failed when filled with propane at a pressure of 1.21 times MAWP?

Consider two cases, one in which the vessel is almost completely (80%) filled and one in which the vessel is almost empty (10% filled). Assume that all the energy contributes to blast generation (i.e., no loss due to fragment generation and throw).

Solution for 80% Filled Vessel

The boiling point of propane is T_b = 231 K, so it is clear that the liquid's temperature can easily rise above the boiling point when the vessel is exposed to a fire and is likely to be above the boiling point at ambient temperatures.. Therefore, the BLEVE prediction method must be selected.

Step 1: Collect the data.

- The failure overpressure is assumed to be 1.21 times the opening pressure of the safety valve. Thus:
  $p_1 = 1.21 \times (1.5 \text{ MPa}) + (0.1 \text{ MPa}) = 1.9 \text{ MPa} (19 \text{ bar})$
- Thermodynamic data are read from thermodynamics tables. In this case, the NIST data is being used from the website http://webbook.nist.gov/chemistry/. Subscript "f" denotes the saturated liquid (fluid) state, and subscript "g" the saturated vapor (gaseous) state.

Table 8.3. Thermodynamic data for propane

State	P_1 (MPa)	u_f (kJ/kg)	u_g (kJ/kg)	v_f (m³/kg)	v_g (m³/kg)	s_f (kJ/kg.K)	s_g (kJ/kg.K)
1	1.916	348.94	581.79	2.2811×10^{-3}	0.022754	1.4996	2.3282
2	0.1013	99.816	483.94	1.7214×10^{-3}	0.41422	0.60559	2.4491

Step 2: Determine internal energy in initial state, u_1.

In this case, the NIST database provided the internal energy directly. If this was not the case, the specific internal energy of each phase of the fluid at the failure state would be calculated with Eq. (8.7):

$$h = u + pv$$

Step 3: Determine internal energy in expanded state, u_2.

As in step 3, the internal energy was obtained directly from the NIST database. It is still necessary to calculate the flash fraction of the liquid as it expands isentropically from its initial pressure to atmospheric pressure.

As the liquid is depressurized, it partially vaporizes; as the vapor is depressurized, it partially condenses. The vapor ratio X can in both cases be calculated from:

$$X = \frac{s_1 - s_f}{s_g - s_f}$$

For the saturated liquid:

$$X = (1.4996 \text{ kJ/kg.K} - 0.60559 \text{ kJ/kg.K})$$
$$/ (2.4491 \text{ kJ/kg.K} - 0.60559 \text{ kJ/kg.K})$$
$$= 0.485$$
$$u_{2f} = (1 - 0.485) \times 99.816 \text{ kJ/kg} + 0.485 \times 483.94 \text{ kJ/kg}$$
$$= 286.1 \text{ kJ/kg}$$

For the saturated vapor:

$$X = (2.4491 \text{ kJ/kg.K} - 2.3282 \text{ kJ/kg.K})$$
$$/ (2.4491 \text{ kJ/kg.K} - 0.60559 \text{ kJ/kg.K})$$
$$= 0.066$$
$$u_{2g} = (0.066) \times 99.816 \text{ kJ/kg} + (1-0.066) \times 483.94 \text{ kJ.kg}$$
$$= 458.75 \text{ kJ/kg}$$

Step 4: Calculate the specific work.

The specific work done by a fluid in expansion is calculated with Eq. (8.9) as follows:

$$e_{ex} = u_1 - u_2$$

Substitution of values for the saturated liquid gives:

$$e_{ex} = 348.94 \text{ kJ/kg} - 286.1 \text{ kJ/kg}$$
$$= 62.84 \text{ kJ/kg}$$

and for the saturated vapor:

$$e_{ex} = 581.79 \text{ kJ/kg} - 458.75 \text{ kJ/kg}$$
$$= 123.04 \text{ kJ/kg}$$

Step 5: Calculate the explosion energy.

The explosion energy is calculated with Eq. (8.10), assuming no fragment reduction factor (frag=0) and a ground reflection factor of 2 (gnd=2):

$$E_{ex} = 2e_{ex}m_1$$

The mass of the released fluid is:

$$m_1 = V_1/v_1$$

When the vessel is full, 80% of the volume is occupied by liquid. (This fraction changes only marginally when the vessel is heated by fire.) The mass of the liquid is:

$$
\begin{aligned}
m_{1f} \; &= \; \text{(fill) x (volume) / } v_f \\
&= \; 0.80 \times 25 \text{ m}^3 / (2.2811 \times 10^{-3} \text{ m}^3/\text{kg}) \\
&= \; 8768 \text{ kg}
\end{aligned}
$$

and the mass of the vapor is:

$$m_{1g} = 0.20 \times 25 \text{ m}^3 / 0.022754 \text{ m}^3/\text{kg} = 219.7 \text{ kg}$$

This gives, for explosion energy of the saturated liquid,

$$E_{ex} = 2 \times 62.84 \text{ kJ/kg} \times 8768 \text{ kg} = 1102.0 \text{ MJ}$$

and, of the saturated vapor:

$$E_{ex} = 2 \times 123.04 \text{ kJ/kg} \times 219.7 \text{ kg} = 54.1 \text{ MJ}$$

Assuming that the blasts from vapor expansion and liquid flashing are simultaneous, the total energy of the surface explosion is:

$$E_{ex} = 1102.0 \text{ MJ} + 54.1 \text{ MJ} = 1156.1 \text{ MJ}$$

At this point, fragment energy can be deducted. Since the fragment energy is unknown, it is neglected. This is a conservative assumption.

Step 6: Calculate the nondimensional range of the receptor

The non-dimensional range of the receptor is calculated with Eq. (7.21):

$$\bar{R} = R\left[\frac{p_0}{E_{ex}}\right]^{1/3}$$

This gives, for the BLEVE of the full vessel at the control building:

$$\bar{R} = 100\,m\left[\frac{101325\ Pa}{1156.1\times10^6\ J}\right]^{1/3} = 4.44$$

Computations are continued with Step 5 of the PVB method.

Step 7: Determine the pressure and impulse at the receptor

Figure 7.6 gives a nondimensional overpressure $\bar{P}_s$ of 0.058 for $\bar{R} = 4.44$

The nondimensional side-on impulse $\bar{I}$ at the control building is read from Figure 7.7: for $\bar{R} = 4.4$, $\bar{I} = 0.014$.

The side on overpressure (p_s) and impulse, i_s, can then be calculated from the non-dimensional parameters as follows:

$$p_s - p_0 = 0.058 \times 101.325\ \text{kPa} = 5.9\ \text{kPa}$$
$$i_s = 0.014 \times (101325\ \text{Pa})^{2/3} \times (1156.1 \times 10^6\ \text{J})^{1/3} / 340\ \text{m/s} = 94\ \text{Pa-s}$$

No corrections are necessary for aspect ratio. Hence, the calculated blast parameters at the control building are as follows: a side-on peak overpressure of 5.9 kPa and a side-on impulse of 94 Pa-s.

Solution for 10% Filled Vessel

When the vessel is only 10% filled, the energy of the explosion is different. Therefore, calculations must be repeated starting from Step 5 above.

Step 5: Calculate the expansion energy.

Explosion energy is calculated with Eq. (8.10), assuming no fragment reduction factor (frag=0) and a ground reflection factor of 2 (gnd=2):

$$E_{ex} = 2e_{ex}m_1$$

The mass of released fluid is:

$$m_1 = V_1/v_1$$

When the vessel is empty, 10% of the volume is occupied by liquid. The liquid mass is:

$$m_1 = \frac{0.10 \times 25m^3}{2.2811 \times 10^{-3}m^3/kg} = 1096 \text{ kg}$$

While the vapor mass is:

$$m_1 = \frac{0.90 \times 25m^3}{0.02275m^3/kg} = 988.8 \text{ kg}$$

This gives, for explosion energy of the saturated liquid:

$$E_{ex} = 2 \times (62.84 \text{ kJ/kg}) \times (1096 \text{ kg}) = 137.7 \text{ MJ}$$

And, for the saturated vapor:

$$E_{ex} = 2 \times (123.04 \text{ kJ/kg}) \times (988.8 \text{ kg}) = 243.3 \text{ MJ}.$$

Assuming that the blasts from vapor expansion and flashing liquid are simultaneous, the total energy of the surface explosion is:

$$E_{ex} = 137.7 \text{ MJ} + 243.3 \text{ MJ} = 381.1 \text{ MJ}$$

The other calculations are performed as described above. The results of these computations are summarized in Table 8.4. Note the following points:

- The explosion of a vessel full of liquid above its normal atmospheric boiling temperature has much more energy, and therefore, causes a much more severe blast than a dry gas- or vapor-filled vessel.
- This calculation takes into account only the blast from the expansion of vessel contents. In fact, this blast may be followed by one from a vapor cloud explosion. This possibility must be considered separately with the methods presented in earlier chapters.

Table 8.4. Results of sample problem #1

	At control building		
Vessel Status	$\bar{R}$	$p_s - p_0$ (kPa)	i_s (Pa.s)
80% filled	4.4	5.0	79
10% filled	6.4	3.6	46

8.6.2.   Sample Problem #2: Calculation of Fragments from BLEVEs

Problem Definition

Assume the conditions are the same as Problem #1. The cylindrical vessel with a volume of 25 m³ and design pressure of 19.2 bar is used for the storage of propane. The empty vessel mass M is 2723 kg, and its length-to-diameter ratio is 10. Assume that 40% of the energy contributes to fragment generation.

Assume that he safety valve of the vessel is sized in such a way that the maximum internal pressure will be 1.21 times the MAWP.

Solution

Calculate the Energy: Using the same procedure as in Sample Problem #1, Steps 1-5 the almost filled vessel was found to have an energy of 1156.1 MJ, and, for the almost empty vessel, it had an energy of 381.1 MJ.

Since these values were calculated in order to determine blast for a vessel placed at grade level; a factor of 2 was applied to account for surface reflection. This factor should not be applied in determining available internal energy for fragment throw calculations. Therefore, the available internal energy for the 80% filled vessel is

$$E = 1156.1 \text{ MJ} / 2 = 578.1 \text{ MJ}$$

For the 10% filled vessel, the internal energy is

$$E = 381.1 \text{ MJ} / 2 = 190.6 \text{ MJ}$$

As specified in the problem definition, 40% of the energy contributes to fragment generation. As a result, the fragment energy is

$$E = 578.1 \text{ MJ} \times 0.4 = 231.2 \text{ MJ} \quad \text{(for the 80\% filled vessel)}$$

$$E = 190.6 \text{ MJ} \times 0.4 = 76.2 \text{ MJ} \quad \text{(for the 10\% filled vessel)}$$

Select a fragment throw method: In order to determine which method should be applied in the calculation of the initial velocity, the scaled energy should be determined (see Section 7.5.2.4). Applying Eq. (7.38):

$$\overline{E}_k = \left[\frac{2E_k}{Ma_1^2} \right]^{0.5}$$

where:

$\overline{E}_K$	=	scaled energy	(–)
E_K	=	energy	(J)
M	=	vessel mass	(kg)
a_1	=	speed of sound in gas	(m/s)

The speed of sound a_1 in propane can be calculated [from Eq. (7.28)] as follows:

$$a_1^2 = \gamma RT/m$$

where:

R	=	ideal gas constant	[J/(kmol.K)]
T	=	absolute temperature	(K)
m	=	molecular mass	(kg/kmol)

Because the temperature is not known, it is assumed to be 500 K. Then,
$$a_0^2 = 1.13 \times (8314.41 \text{ J/kmol.K}) \times (500 \text{ K}) / (44 \text{ kg/kmol})$$
$$= 1.07 \times 10^5 \text{ (m/s)}^2$$

For the 10% filled case,
$$\overline{E} = \left[\frac{2 \times \left(76.2 \times 10^6 J\right)}{\left(2723 kg\right) \times \left(1.07 \times 10^5 m^2 / s^2\right)} \right]^{0.5} = 0.72$$

And for the 80% filled case,
$$\overline{E} = \left[\frac{2 \times \left(231.2 \times 10^6 J\right)}{\left(2723 kg\right) \times \left(1.07 \times 10^5 m^2 / s^2\right)} \right]^{0.5} = 1.26$$

Because the scaled energy is higher than 0.7, Method 3 has to be applied for both cases.

Method 3 [Eq. (7.29)] gives
$$v_i = 1.092 \left[\frac{E_k G}{M_C} \right]^{0.5}$$

where, from Eq. (7.31)
$$G = \frac{1}{1 + C / 2M} \qquad \text{for cylindrical vessels}$$

and C is total mass of gas and M is the mass of the vessel.

The mass C is assumed to be the entire liquid inventory converted to gas. Liquid propane has a specific weight of 585.3 kg/m³, and the volume of the vessel was 25 m³. Therefore, for the 80% filled case with a large number of fragments:
$$C = 0.8 \times (585.3 \text{ kg/m}^3) \times (25 \text{ m}^3) = 11,706 \text{ kg}$$
$$G = 1/[1 + (11,706 \text{ kg}) / (2 \times 2723 \text{ kg})] = 0.32$$

and
$$v_i = 1.092[(231.2 \times 10^6 \text{ J}) \times 0.32 / (2723 \text{ kg})]^{0.5} = 180 \text{ m/s}$$

For the 10% filled case:
$$C = 0.1 \times (585.3 \text{ kg/m}^3) \times (25 \text{ m}^3) = 1463 \text{ kg}$$
$$G = 1 / [1 + (1463 \text{ kg})/(2 \times 2723 \text{ kg})] = 0.79$$

and
$$v_i = 1.092[(76.2 \times 10^6 \text{ J}) \times 0.79 / (2723 \text{ kg})]^{0.5} = 162 \text{ m/s}$$

Predict Fragment Ranges: Fragment ranges will be calculated by neglecting lift and drag forces for different initial trajectory angles with Eq. (7.40):

$$R = \frac{v_i^2 \sin(2a_i)}{g}$$

where:

R	=	horizontal range	(m)
a_i	=	initial trajectory angle	(rad)
g	=	gravitational acceleration	(m/s^2)

To assume an initial trajectory angle of 45° is probably too conservative. Directional effects can be expected. Large fragments like an end cap will travel in a direction parallel to the axis of the vessel. The trajectory angle will therefore be 5 to 10°.

Table 8.5. Results of sample problem #2

Fill Level (%)	v_i (m/s)	$R(m)$		
		$a = 5°$	$a = 10°$	$a = 45°$
10	162	467	920	2689
80	180	574	1130	3305

8.6.3.   Sample Problem #3: Thermal Radiation from a BLEVE

Problem Definition

A liquefied propane tank truck with a volume of 6000 U.S. gallons (22.7 m^3) is involved in a traffic accident, and the tank truck is engulfed by fire from burning gasoline. The tank is 90% filled with propane. Assume that all of the propane will contribute to the fireball and that the atmospheric transmissivity is 1.

Solution

Estimate fireball diameter and duration. Liquid propane has a specific weight of 585.3 kg/m^3, so the total mass of propane in the tank is:
$$m = 0.9 \times (22.7 \text{ m}^3) \times (585.3 \text{ kg/m}^3) = 11{,}958 \text{ kg}$$

With the relations given above, the diameter (D_c) and the duration (t_c) of the fireball can be calculated:
$$D_c = 5.8 \times m_f^{1/3} = 5.8 \times (11958 \text{ kg})^{1/3} = 133 \text{ m}$$
$$t_c = 0.45 \times m_f^{1/3} = 0.45 \times (11958 \text{ kg})^{1/3} = 10.3 \text{ s}$$

- Assume a surface-emissive power of 350 kW/m^2.
- Estimate the geometric view factor. The center of the fireball has a height of 66.5 m, and thus the view factor (for a vertical object) follows from the relation given in Appendix A:
$$F_v = (X \times (66.5 \text{ m})^2)/(X^2 + (66.5 \text{ m})^2)^{3/2}$$

where X = the distance measured along the ground from the object to a point directly below the center of the fireball, that is, the position of the tank. This distance must be greater than the radius of the fireball, because actual development of the fireball often involves, first, an initial hemispherical shape, which would engulf near-field receptors, and second, ascent of the fireball over time, which would significantly affect radiation distances to near-field receptors. Therefore, near-field radiation estimates are of questionable accuracy.

- Estimate the radiation received at a receptor. With an attenuation factor of 1, the radiation received by a vertical receptor at a distance X from the tank can be calculated from Eq. (5.6):
$$q = EF_v \, \tau_a = (350 \text{ kW/m}^3) \times (X \times (66.5 \text{ m})^2)/(X^2 + (66.5 \text{ m})^2)^{3/2}$$

The results of this calculation for various distances X are tabulated in Table 8.6.

- Alternative approach: point-source model. Another method of calculating the radiation received by an object relatively distant from the fireball is to use the point-source model. From this approach, the peak thermal input at distance L from the center of the fireball is given by Eq. 8.17:

$$q = \frac{2.2\tau_a RH_c m_f^{0.67}}{4\pi L^2}$$

Substituting the appropriate values for the variables yields the following:

m_f = 11,958 kg

τ_a = 1.0

H_c = 4.636×10^7 J/kg

R = 0.4 (It is assumed that the relief valve operated prior to vessel rupture.)

Therefore,

$$q = \frac{2.2(1.0)(0.4)(4.636x10^7 \, J/kg)(11958 \, kg)^{0.67}}{4\pi L^2}$$

$$q = 1.7 \times 10^9/L^2 \text{ W/m}^2 = 1.7 \times 10^9/L^2 \text{ kW/m}^2$$

The radiation received by an object normal to the fireball at distances of X = 100, 200, 500, and 1000 m from the tank is presented in Table 8.6.

Table 8.6. Results of sample problem #3

Ground Distance (m)	View Factor	Solid Flame Radiation (kW/m²)	Point Source Radiation (Hymes) (kW/m²)
100	0.255	89	122
200	0.0945	33	39
500	0.0172	6.0	6.8
1000	0.00439	1.5	1.7

9. REFERENCES

Abbasi, T., Abbasi, S.A., "The boiling liquid expanding vapour explosion (BLEVE): Mechanism, consequence assessment, management," Journal of Hazardous Materials, 141 (2007) 489-519.

Adamczyk, A. A. 1976. An investigation of blast waves generated from non-ideal energy sources. *UILU-ENG 76–0506.* Urbana: University of Illinois.

AIChE/CCPS. 1996. *Guidelines for Evaluating Process Plant buildings for External Explosions and Fires.* American Institute of Chemical Engineers, New York.

American Petroleum Institute and Chemical Manufacturers Association. 1995. *Management of Hazards Associated with Location of Process Plant Buildings.* API RP 752 and CMA Manager's Guide, May, 1995.

American Petroleum Institute. 1982. Recommended Practice 521.

American Society of Civil Engineers. 1997. *Design of Blast Resistant Buildings in Petrochemical Facilities.* New York.

Anderson, C, W. Townsend, R. Markland, and J. Zook. 1975. Comparison of various thermal systems for the protection of railroad tank cars tested at the FRA/BRL torching facility. Interim Memorandum Report No. 459, Ballistic Research Laboratories.

Andrews, G.E., 1997, "Fundamentals of combustion, explosion and mitigation", Seminar notes, Leeds University

API, "Pressure-Relieving and Depressuring Systems," American Petroleum Institute, STD 521, Washington, D.C., 2008.

Army, Navy, and Air Force Manual. 1990. "Structures to resist the effects of accidental explosions." TM 5–1300, NAVFAC P-397, AFR 88–22. Revision 1.

Aslanov, S. K., and O. S. Golinskii. 1989. Energy of an asymptotically equivalent point detonation for the detonation of a charge of finite volume in an ideal gas. *Combustion, Explosion, and Shock Waves,* pp. 801–808.

Auton, T. R., and J. H. Pickles. 1978, "The calculation of blast waves from the explosion of pancake-shaped vapor clouds." Central Electricity Research Laboratories note No. RD/L/N 210/78.

Auton, T. R., and J. H. Pickles. 1980. Deflagration in heavy flammable vapors. *Inst. Math. Appl.Bull.* 16:126–133.

Bader, B. E., A. B. Donaldson, and H. C. Hardee. 1971. Liquid-propellent rocket abort fire model. *J. Spacecraft and Rockets* 8:1216–1219.

Bagster, D.F., R.M. Pitblado, Thermal hazards in the process industry, Chemical Engineering Progress 85 (1989) 69–77.

Baker, Q.A., Baker, W.E., "Pros and Cons of TNT Equivalence for Industrial Explosion Accidents," International Conference and Workshop on Modeling and Mitigating the Consequences of Accidental Releases of Hazardous Materials, Center for Chemical Process Safety of AICHE, 1991.

Baker, Q.A., Doolittle, D.M., Fitzgerald, G.A., and Tang, M.J.., "Recent Developments in the Baker-Strehlow VCE Analysis Methodology," 31st Loss Prevention Symposium, American Institute of Chemical Engineers, Paper 42f, 1997.

Baker, Q.A., Tang, M.J., "Effect of Vessel Temperature on Blast Loads," 2004 ASME/JSME Pressure Vessel and Piping Conference, July 2004.

Baker, Q.A., Tang, M.J., Scheier, E., Silva, G.J., "Vapor Cloud Explosion Analysis," American Institute of Chemical Engineers, 28th Annual Loss Prevention Symposium, 1994.

Baker, W. E. 1973. *Explosions in Air.* Austin: University of Texas Press.

Baker, W. E., J. J. Kulesz, R. E. Richer, R. L. Bessey, P. S. Westine, V. B. Parr, and G. A. Oldham. 1975 and 1977. *Workbook for Predicting Pressure Wave and Fragment Effects of Exploding Propellant Tanks and Gas Storage Vessels.* NASA CR-134906. Washington: NASA Scientific and Technical Information Office.

Baker, W. E., J. J. Kulesz, R. E. Ricker, P. S. Westine, V. B. Parr, L. M. Vargas, and P. K. Moseley. 1978b. *Workbook for Estimating the Effects of Accidental Explosions in Propellant Handling Systems.* NASA Contractor report no. 3023.

Baker, W. E., P. A. Cox, P. S. Westine, J. J. Kulesz, and R. A. Strehlow. 1983. *Explosion Hazards and Evaluation.* New York: Elsevier Scientific.

Bakke, J. R. 1986. "Numerical simulations of gas explosions." Ph.D. Thesis, University of Bergen, Norway.

Bakke, J. R., and B. H. Hjertager. 1986a. Quasi-laminar/turbulent combustion modeling, real cloud generation and boundary conditions in the FLACS-ICE code. *CMI No. 865402–2.* Chr. Michelsen Institute, 1986. Also in Bakke's Ph.D. thesis "Numerical simulation of gas explosions in two-dimensional geometries." University of Bergen, Bergen, 1986.

Bakke, J. R., and B. H. Hjertager. 1986b. The effect of explosion venting in obstructed channels. In *Modeling and Simulation in Engineering.* New York: Elsevier, pp. 237–241.

Bakke, J. R., and B. J. Hjertager. 1987. The effect of explosion venting in empty vessels. *Int. J. Num. Meth. Eng.* 24:129–140.

Balcerzak, M. H., M. R. Johnson, and F. R. Kurz. 1966. "Nuclear blast simulation. Part I—Detonable gas explosion." Final eport DASA 1972–1. Niles, Ill.: General American Research Division.

Balke, C., Heller, W., Konersmann, R., Ludwig J., 1999, *Study of the Failure Limits of a Tank Car Filled with Liquefied Petroleum Gas Subjected to an Open Pool Fire Test, BAM Project 3215,* Federal Institute for Materials Research and Testing (BAM)

Barbone R, Frost DL, Makis A, Nerenberg J, Explosive Boiling of a Depressurized Volatile Liquid, Proceedings of IUTAM Symposium on Waves in Liquid/Gas and Liquid Vapour Two-Phase Systems , 1994, Kyoto, Japan

Baum, M. R. 1984. The velocity of missiles generated by the disintegration of gas pressurized vessels and pipes. *Trans. ASME.* 106:362–368.

Baum, M. R. 1987. Disruptive failure of pressure vessels: preliminary design guide lines for fragment velocity and the extent of the hazard zone. In *Advances in Impact, Blast Ballistics, and Dynamic Analysis of Structures.* ASME PVP. 124. New York: ASME.

Baum, M.R., Disruptive Failure of Pressure Vessels: Preliminary Design Guidelines for Fragment Velocity and the extent of the Hazard Zone, Journal of Pressure Vessel Technology 110, 168-176, 1988, ASME

Benedick, W.B., "High Explosive Initiation of Methane-air Detonations," Combustion and Flame, 35 (1979) pp. 89-94

Benedick, W. B., J. D. Kennedy, and B. Morosin. 1970. Detonation limits of unconfined hydrocarbon-air mixtures. *Combust. and Flame.* 15:83–84.

Benedick, W. B., R. Knystautas, and J. H. S. Lee. 1984. "Large-scale experiments on the transmission of fuel-air detonations from two-dimensional channels." *Progress in Astronautics and Aeronautics.* 94:546–555, AIAA Inc., New York.

Berufsgenossenschaft der Chemischen Industrie. 1972. Richtlinien zur Vermeidug von Zünd-gefahren infolge elektrostatischer Aufladungen. Richtlinie Nr. 4.

Bessey, R. L. 1974. Fragment velocities from exploding liquid propellant tanks. *Shock Vibrat. Bull.* 44.

Bessey, R. L., and J. J. Kulesz. 1976. Fragment velocities from bursting cylindrical and spherical pressure vessels. *Shock Vibrat. Bull.* 46.

Birk, A.M., 1995, Scale Effects with Fire Exposure of Pressure Liquefied Gas Tanks, *Journal of Loss Prevention in the Process Industries* 8(5):275-90

Birk, A.M., 1996, Hazards from BLEVEs: An Update and Proposal for Emergency Responders, *Journal of Loss Prevention in the Process Industries* 9(2):173-81

Birk, A.M., Cunningham, M.H., 1994, *A Medium Scale Experimental Study of the Boiling Liquid Expanding Vapour Explosion, TP 11995E*, Transport Canada

Birk, A.M., Cunningham, M.H., 1994, The Boiling Liquid Expanding Vapour Explosion, *International Journal of Loss Prevention in the Process Industries* 7(6):474-80

Birk, A.M., Cunningham, M.H., 1996, Liquid Temperature Stratification and its Effect on BLEVEs and their Hazards, *Journal of Hazardous Materials* 48:219-37

Birk, A.M., Cunningham, M.H., Ostic, P., Hiscoke, B., 1997, *Fire Tests of Propane Tanks to Study BLEVEs and Other Thermal Ruptures: Detailed Analysis of Medium Scale Test Results, TP 12498E*, Transport Canada

Birk, A.M., Davison, C., Cunningham, M.H., Blast Overpressures from Medium Scale BLEVE Tests, Journal of Loss Prevention in the Process Industries 20, 194-206, 2007

Birk, A.M., Poirier, D., Davison, C., Wakelam, C., 2005, *Tank-Car Thermal Protection Defect Assessment: Fire Tests of 500 gal Tanks with Thermal Protection Defects, TP 14366E*, Transportation Development Centre, Transport Canada

Birk, A.M., VanderSteen, J.D.J., Davison, C., Cunningham, M.H., Mirzazadeh, I., 2003, *PRV Field Trials -- The Effects of Fire Conditions and PRV Blowdown on Propane Tank Survivability in a Fire, TP 14045E*, Transport Canada

Birk, A.M., VanderSteen, J.D.J., The Effect of Pressure Relief Valve Blowdown and Fire Conditions on the Thermo-Hydraulics within a Pressure Vessel, ASME Journal of Pressure Vessel Technology 128, 467-475, 2006,

Birk, A.M., Yoon, K.T., High Temperature Stress-Rupture Data for the Analysis of Dangerous Goods Tank-Cars Exposed to Fire, Journal of Loss Prevention in the Process Industries 19, 442-451, 2006

Bjerketvedt, D., and O. K. Sonju. 1984. "Detonation transmission across an inert region." *Progress in Astronautics and Aeronautics.* 95, AIAA Inc., New York.

Bjerketvedt, D., O. K. Sonju, and I. O. Moen. 1986. "The influence of experimental condition on the re-initiation of detonation across an inert region." *Progress in Astronautics and Aeronautics.* 106:109–130. AIAA Inc., New York.

Blackmore, D. R., J. A. Eyre, and G. G. Summers. 1982. Dispersion and combustion behavior of gas clouds resulting from large spillages of LNG and LPG onto the sea. *Trans. I. Mar. E. (TM).* 94: (29).

Blevins, R.D., 1984, *Applied Fluid Dynamics Handbook*, New York, Van Nostrand Reinhold Company Inc.

Board, S. J., R. W. Hall, and R. S. Hall. 1975. Detonation of fuel coolant explosions. *Nature* 254:319–320.

Boris, J. P. 1976. "Flux-Corrected Transport modules for solving generalized continuity equations." *NRL Memorandum report 3237.* Naval Research Laboratory, Washington, D.C.

Boris, J. P., and Book D. L. 1976. "Solution of continuity equations by the method of Flux-Corrected Transport." *Meth. Computat. Phys.* Vol. 16. New York: Academic Press.

Boris, J. P., and D. L. Book. 1973. Flux-corrected transport I: SHASTA-A fluid transport method that works. *J. Comp. Phys.* 11:38.

Bowen, J. G., E. R. Fletcher, and D. R. Richmond. 1968. Estimate of man's tolerance to the direct effects of air blast. Lovelace Foundation for Medical Education and Research. Albuquerque, NM.

Boyer, D. W., H. L. Brode, I. I. Glass, and J. G. Hall. 1958. "Blast from a pressurized sphere." UTIA Report No. 48. Toronto: Institute of Aerophysics, University of Toronto.

Bradley, D., Lau, A.K.C. and Lawes, M., 1992, Phil Trans R Soc, Lond, A 338-359

Brasie, W. C, and D. W. Simpson. 1968. "Guidelines for estimating damage explosion." *Proc. 63rd Nat. AIChE Meeting.* AIChE. New York.

Brasie, W. C. 1976. "The hazard potential of chemicals." AIChE Loss Prevention. 10:135–140.

Britton, L.G., Williams, T.J., "Some Characteristics of Liquid-to-Metal Discharges Involving a Charged 'Low Risk' Oil," J. Electrostatics, 13 (1982) 185-207.

Brode, H. L. 1955. "Numerical solutions of a spherical blast wave." *J. Appl. Phys.* 26:766–775.

Brode, H. L. 1959. "Blast wave from a spherical charge." *Physics of Fluids.* 2(2):217–229.

Brossard, J., D. Desbordes, N. Difabio, J. L. Garnier, A. Lannoy, J. C. Leyer, J. Perrot, and J. P. Saint-Cloud. 1985. "Truly unconfined deflagrations of ethylene-air mixtures." Paper presented at the 10th Int. Coll. on Dynamics of Explosions and Reactive Systems. Berkeley, California.

Brossard, J., S. Hendrickx, J.L. Garnier, A. Lannoy and J.L. Perrot, "Air Blast From Unconfined Gaseous Detonations", Proceedings of the 9th International Colloquium on Dynamics of Explosion and Reactive Systems", The American Institute of Aeronautics and Astronautics, Inc., 1984

Bull, D. C, J. E. Elsworth, and P. J. Shuff. 1982. Detonation cell structures in fuel-air mixtures. *Combustion and Flame* 45:7 - 22.

Bull, D. C, J. E. Elsworth, M. A. McCleod, and D. Hughes. 1981. "Initiation of unconfined gas detonations in hydrocarbon-air mixtures by a sympathetic mechanism." *Progress in Astronautics and Aeronautics.* 75:61–72. AIAA Inc., New York.

Bull, D.C., J.E. Elsworth and G. Hooper, "Initiation of Spherical Detonation in Hydrocarbon-Air Mixtures", Acta Astronautica, 5 (1978) pp. 997-1008

Burgess, D. S., and M. G. Zabetakis, 1973. "Detonation of a flammable cloud following a propane pipeline break, the December 9, 1970 explosion in Port Hudson (MO)." *Bureau of Mines Report of Investigations No. 7752.* United States Department of the Interior.

Burgess, D. S., and M. Hertzberg. 1974. *Advances in Thermal Engineering.* New York: John Wiley and Sons.

Burgoyne, J. H. 1963. The flammability of mists and sprays. *Second Symposium on Chemical Process Hazards.*

Buschman, Jr., A. J., and C. M. Pittman. 1961. Configuration factors for exchange of radiant energy between antisymmetrical sections of cylinders, cones and hemispheres and their bases. *NASA, Technical Note D-944.*

Cambray, P., and B. Deshaies. 1978, "Ecoulement engendre par un piston spherique: solution analytique approchee." *Acta Astronautica.*5:611–617.

Cambray, P., B. Deshaies, and P. Clavin. 1979. "Solution des equations d'Euler associees a l'expansion d'une sphere a vitesse constante." *Journal de Physique.* Coll. C8, 40(11):19–24.

Campbell, J.A., "Estimating the magnitude of macro-hazards," Society of Fire Protection Engineers, Report 81-2, Boston, Massachusetts, 1981.

Cates, A.T., Fuel gas explosion guidelines, Int. Conf. Fire and Explosion Hazards Inst Energy, 1991.

Catlin, C.A. and Johnson, D.M., 1992, Combust Flame, 88:15

CCPS/AIChE. 1991. *International conference and workshop on modeling and mitigating the consequences of accidental releases of hazardous material.* New York: CCPS/AIChE.

Center For Chemical Process Safety, *Guidelines for Evaluating The Characteristics Of Vapor Cloud Explosions, Flash Fires, And BLEVEs,* AIChE, 1994

Center For Chemical Process Safety, *Guidelines for Consequence Analysis of Chemical Releases,* American Institute of Chemical Engineers, New York, 1999.

Center for Chemical Process Safety. 1989. *Guidelines for Chemical Process Quantitative Risk Analysis.* New York: AIChE/CCPS.

Chan, C, J. H. S. Lee, I. O. Moen, and P. Thibault. 1980. "Turbulent flame acceleration and pressure development in tubes." *Proceedings of the First Specialists Meeting of the Combustion Institute,* Bordeaux, France, pp. 479–484.

Chan, C., I. O. Moen, and J. H. S. Lee, 1983. "Influence of confinement on flame acceleration due to repeated obstacles." *Combust. and Flame.* 49:27–39.

Chapman, W. R., and R. V. Wheeler. 1926. "The propagation of flame in mixtures of methane and air. Part IV: The effect of restrictions in the path of the flame." *J. Chem. Soc.* pp. 2139–2147.

Chapman, W. R., and R. V. Wheeler. 1927. "The propagation of flame in mixtures of methane and air. Part V: The movement of the medium in which the flame travels." *J. Chem. Soc.* pp. 38–47.

CHEETAH 1.39 User's Manual, UCRL-MA-117541 Rev.3, 1996

Chushkin, P. I., and L. V. Shurshalov. 1982. Numerical computations of explosions in gases. (Lecture Notes in Physics 170). *Proc. 8th Int. Conf. on Num. Meth. in Fluid Dynam.*, 21—42. Berlin: Springer Verlag.

Clayton, W.E., Griffin, M.L., "Catastrophic failure of a liquid carbon dioxide storage vessel," *Process Safety Progress*, 1994, 13, 202-209.

Cloutman, L. D., C. W. Hirt, and N. C. Romero. 1976. "SOLA-ICE: a numerical solution algorithm for transient compressible fluid flows." *Los Alamos Scientific Laboratory report LA-6236.*

Clutter, JK and Mathis, J. Computational modeling of vapor cloud explosions in off-shore rigs using a flame –speed based on combustion model. *J. of Loss Prevention in the Process Industries 15: 391-401*, 2002

Combustion, 1973, pp. 1201-1215

Coward, H. F., and G. W. Jones. 1952. Limits of flammability of bases and vapors. *Bureau of Mines Bulletin* 503.

Cowperthwaite, M., and W.H. Zwisler, "Tiger Computer Program Documentation", Report No. Z106, Stanford Res. Inst. Menlo Park, CA 1973

Cracknell, R. F. and Carsley, A. J., "Cloud Fires: A methodology for hazardous consequence modeling," IChemE Symposium Series 141, 139 pp, 1997.

Crowl, D.A., "Calculating the Energy of Explosion Using Thermodynamic Availability," J. Loss Prevention. Process Ind., 5(2): 109-118, 1992.

Crowl, D.A., Understanding Explosions, CCPS Concept Book, American Institute of Chemical Engineers, New York, NY, ISBN 0-8169-0779-X, 2003

Davenport, J. A. 1977. "A study of vapor cloud incidents." *AIChE Loss Prevention Symposium,* Houston, Texas.

Davenport, J. A. 1977. "A survey of vapor cloud incidents." *Chemical Engineering Progress.* Sept. 1977, 54–63.

Davenport, J. A. 1983. "A study of vapor cloud incidents—an update," 4th Int. Symp. Loss Prevention and Safety Promotion in the Process Industries. Harrogate (UK*), IChemE Symp. Series No. 80.*

Davenport, J. A. 1986. "Hazards and protection of pressure storage of liquefied petroleum gases." *Fifth International Symposium on Loss Prevention and Safety Promotion in the Process Industries*, European Federation of Chemical Engineering, Canner, France.

Denisov, Yu. N., K. I. Shchelkin, and Ya. K. Troshin. 1962. Some questions of analogy between combustion in a thrust chamber and a detonation wave. *8th Symposium (International) on Combustion*, pp. 1152–1159. Pittsburgh: PA: The Combustion Institute.

Department of Labor and Workforce Development - Division of Boiler and Elevator Inspection, Accident Report, Dana Corporation, Paris Extrusion Plant, Tennessee, June 18, 2007.

Desbordes, D., and N. Manson. 1978, "Explosion dans Fair de charges spheriques non confinees de melanges reactifs gazeux." *Acta Astronautica.* 5:1009–1026.

Deshaies, B., and J. D. Leyer. 1981. "Flow field induced by unconfined spherical accelerating flames." *Combust. and Flame.* 40:141–153.

Deshaies, B., and P. Clavin. 1979. "Effets dynamiques engendres par une flamme spherique a vitesse constante." *Journal de Mecanique.* 18(2):213–223.

Dorge, K. J., D. Pangritz and H. Gg. Wagner. 1976. "Experiments on velocity augmentation of spherical flames by grids." *Acta Astronautica.* 3:1067–1076.

Dorge, K. J., D. Pangritz, and H. Gg. Wagner. 1979. "Uber die Wirkung von Hindernissen auf die Ausbreitung von Flammen." *ICI Jahrestagung.* S.441–453.

Dorge, K. J., D. Pangritz, and H. Gg. Wanger. 1981. "Uber den Einfluss von mehreren Blenden auf die Ausbreitung von Flammen: Eine Fortsetzung der Wheelerschen Versuche." *Z. fur Phys. Chemie Neue Folge.* Bd. 127, S.61 -78.

Dorofeev, S.B., "Blast Effects Of Confined And Unconfined Explosions," Proceedings of the 20th International Symposium on Shock Waves, 1995

Droste, B., and W. Schoen. 1988. Full-scale fire tests with unprotected and thermal insulated LPG storage tanks. *J. Haz. Mat.* 20:41–53.

Duiser, J. A. 1989. Warmteuitstraling (Radiation of heat). Method for the calculation of the physical effects of the escape of dangerous materials (liquids and gases). Report of the Committee for the Prevention of Disasters, Ministry of Social Affairs, The Netherlands, 2nd Edition.

Edmister, W. C, and B. I. Lee. 1984. *Applied Hydrocarbon Thermodynamics,* 2nd ed. Houston: Gulf Publishing Company.

Eggen, J.B.M.M., "GAME: Development of Guidance for the Application of the Multi-Energy Method," TNO Prins Maurits Laboratory report prepared for the Health and Safety Executive, HSE Contract Research Report 202/1998, ISBN 0 7176 1651 7, 1998

Eichler, T. V., and H. S. Napadensky. 1977. "Accidental vapor phase explosions on transportation routes near nuclear power plants." *IIT Research Institute final report no. J6405*. Chicago, Illinois.

Eisenberg, N. A., C. J. Lynch, and R. J. Breeding. 1975. "Vulnerability model. A simulation system for assessing damage resulting from marine spills." *U.S. Department of Commerce Report No. AD/A015/245*. Washington: National Technical Information Service.

Elsworth, J., J. Eyre, and D. Wayne. 1983. Combustion of refrigerated liquefied propane in partially confined spaces, Int. Sym. "Loss Prevention and Safety Promotion in the Process Industries." Harrogate (UK), *IChemE Symp. Series No. 81*. pp. C35-C48.

Ermak, D.L. and R. P. Koopman, "Results of 40-m3 LNG Spills onto Water", S. Hartwig (ed.) Heavy Gas Assessment - II, Battelle-Institute e.V., Frankfurt am Main, Germany, 163-179, Jan 1983.

Ermak D.L. et al., Heavy gas dispersion test summary report', Lawrence Livermore National Laboratory Report No. UCRL-21210 October, 1988.

Esparza, E. D., and W. E. Baker. 1977a. *Measurement of Blast Waves from Bursting Pressurized Frangible Spheres.* NASA CR-2843. Washington: NASA Scientific and Technical Information Office.

Esparza, E. D., and W. E. Baker. 1977b. *Measurement of Blast Waves from Bursting Frangible Spheres Pressurized with Flash-evaporating Vapor or Liquid.* NASA CR-2811. Washington: NASA Scientific and Technical Information Office.

Exxon (unpublished). Damage estimates from BLEVEs, UVCEs and spill fires. Factory Mutual Research Corporation. 1990. Private Communication. Fishburn, B. 1976. "Some aspects of blast from fuel-air explosives." *Acta Astronautica.* 3:1049–1065.

Factory Mutual Research Corporation. 1990. "Guidelines for the estimation of property damage from outdoor vapor cloud explosions in chemical processing facilities." Technical Report, March 1990.

Fay, J. A., and D. H. Lewis, Jr. 1977. *Unsteady burning of unconfined fuel vapor clouds. 16th Symposium (International) on Combustion*, pp. 1397–1405. Pittsburgh, PA: The Combustion Institute.

Felbauer GF, Heigl JH, McQueen W, Whipp RH, May WG (1972). Spills of LNG on water- vaporization and downwind drift of combustible mixtures, Report No. EE61E-72, Florham Park, NJ: Esso Res. and Development Lab.

Ficket, W., and W. C. Davis. 1979. *Detonation.* Berkeley: University of California Press.

Fishburn, B., N. Slagg, and P. Lu. 1981. "Blast effect from a pancake-shaped fuel drop-air cloud detonation (theory and experiment)." *J. of Hazardous Materials.* 5:65–75.

Gayle J.B., J.W. Bransford, *Size and duration of fireballs from propellant explosions*, ReportNASATMX-53314. George C. Marshall Space Flight Center, Huntsville, USA, 1965.

Geng, J.H. and J.K. Thomas, "Simulation and Application of Blast Wave-Target Interaction." Presented at the 41st Annual Loss Prevention Symposium, Houston, Texas, April 2007a.

Geng, J.H. and J.K. Thomas, "Reflection of Blast Waves Off Cylindrical Pipes," ASME Pressure Vessel and Piping Conference (PVP '07), San Antonio, TX, July 22-26, 2007b

Gibbs, G. J., and H. F. Calcote. 1959. Effect on molecular structure on burning velocity. *Jr. Chem. Eng. Data.* 4(3):226–237.

Giesbrecht, H., K. Hess, W. Leuckel, and B. Maurer, 1981. "Analysis of explosion hazards on spontaneous release of inflammable gases into the atmosphere. Part 1: Propagation and deflagration of vapor clouds on the basis of bursting tests on model vessels." *Ger. Chem. Eng.* 4:305–314.

Giesbrecht, H., K. Hess, W. Leuckel, and B. Maurer. 1980. Analyse der potentiellen Explosionswirkung von kurzzeitig in de Atmosphaere freigesetzen Brenngasmengen. *Chem. Ing. Tech.* 52(2): 114–122.

Giesbrecht, H., K. Hess, W. Leuckel, and B. Maurer. 1981. "Analysis of explosion hazards on spontaneous release of inflammable gases into the atmosphere." Part 1: Propagation and deflagration of vapor clouds on the basis of bursting tests on model vessels. Part 2: Comparison of explosion model derived from experiments with damage effects of explosion accidents. *Ger. Chem. Eng.* 4:305–325.

Girard, P., M. Huneau, C. Rabasse, and J. C. Leyer. 1979. "Flame propagation through unconfined and confined hemispherical stratified gaseous mixtures." *17th Symp. (Int.) on Combustion.* pp. 1247–1255. The Combustion Institute, Pittsburgh, PA.

Giroux, E. D. 1971. HEMP users manual. *Lawrence Liver more Laboratory report no. UCRL-51079.* University of California, Livermore, California.

Glass, I. I. 1960. UTIA *Report No. 58.* Toronto: Institute of Aerophysics, University of Toronto.

Glasstone, S. 1957. The effects of nuclear weapons. USAEC.

Glasstone, S. 1966. *The Effects of Nuclear Weapons.* US Atomic Energy Commission, Revised edition 1966.

Glasstone, S., and P. J. Dolan. 1977. *The Effects of Nuclear Weapons.* US Dept. of Defense, Third edition.

Godunov, S. K., A. V. Zabrodin and G. P. Propokov. 1962. *J. of USSR Comp. Math., Math.Phys.* 1:1187.

Goldwire, H. C. Jr., H. C. Rodean, R. T. Cederwall, E. J. Kansa, R. P. Koopman, J. W. McClure, T. G. McRae, L. K. Morris, L. Kamppiner, R. D. Kiefer, P. A. Urtiew, and C. D. Lind. 1983. "Coyote series data report LLNL/NWC 1981 LNG spill tests, dispersion, vapor burn, and rapid phase transition." *Lawrence Livermore National Laboratory Report UCID—19953.* Vols. 1 and 2.

Gorev, V. A., and Bystrov S. A. 1985. "Explosion waves generated by deflagration combustion." *Comb., Explosion and Shock Waves.* 20:(6):614–620.

Gouldin, F.C., 1987, Combust Flame, 68:249

Green Book 1989. Methods for the determination of possible damage to people and objects resulting from releases of hazardous materials. Published by the Dutch Ministry of Housing, Physical Planning and Environment. Voorburg, The Netherlands. Code: CPR.6E

Grodzovskii, G. L., and F. A. Kukanov. 1965. Motions of fragments of a vessel bursting in a vacuum. *Inzhenemyi Zhumal* 5(2):352–355.

Grossel, S.S., Deflagration and Detonation Flame Arresters, Center for Chemical Process Safety, American Institute of Chemical Engineers, New York, NY, 2002.

Gugan, K. 1978. *Unconfined vapor cloud explosions.* IChemE, London.

Guirao, C. M., G. G. Bach, and J. H. Lee. 1976. "Pressure waves generated by spherical flames." *Combustion and Flame.* 27:341–351.

Guirao, C. M., G. G. Bach, and J. H. S. Lee. 1979. "On the scaling of blast waves from fuel-air explosives." *6th Symp. on Blast Simulation.* Cahors, France.

Guirao, C.M., Knystautas, R., Lee, J.H., Benedick, W., and Berman, M., "Hydrogen-air Detonations," 19th Symposium (International) on Combustion, pp 583-590, The Combustion Institute, Pittsburgh, PA, 1982.

Guirguis, R. H., M. M. Kamel, and A. K. Oppenheim. 1983. "Self-similar blast waves incorporating deflagrations of variable speed." *Progess in Astronautics and Aeronautics.* 87:121–156, AIAA Inc., New York.

Hanna, S. R., and P. J. Drivas. 1987. *Guidelines for Use of Vapor Cloud Dispersion Models.* New York: American Institute for Chemical Engineers, CCPS.

Hardee, H. C, and D. O. Lee. 1973. Thermal hazard from propane fireballs. *Trans. Plan. Tech.* 2:121–128.

Hardee, H. C, and D. O. Lee. 1978. A simple conduction model for skin burns resulting from exposure to chemical fireballs. *Fire Res.* 1:199–205.

Hardee, H. C, D. O. Lee, and W. B. Benedick. 1978. Thermal hazards from LNG fireball. *Combust. Sci. Tech.* 17:189–197.

Hargrave, G.K., Jarvis, S., Williams, T.C., "A study of transient flow turbulence generation during flame/wall interactions in explosions," *Measurement Science and Technology, Institute of Physics Publishing,* Measurement Science and Technology 13 (2002) 1036-1042.

Harlow, F. H., and A. A. Amsden. 1971. "A numerical fluid dynamics calculation method for all flow speeds." *J. of Computational Physics.* 8(2): 197–213.

Harris, R. J. 1983. *The investigation and control of gas explosions in buildings and heating plant.* New York: E & FN Spon in association with British Gas Corporation

Harris, R. J., and M. J. Wickens. 1989. "Understanding vapor cloud explosions—an experimental study." *55th Autumn Meeting of the Institution of Gas Engineers,* Kensington, UK.

Harrison, A. J., and J. A. Eyre. 1986. "Vapor cloud explosions—The effect of obstacles and jet ignition on the combustion of gas clouds, 5th Int. Symp." *Proc. Loss Prevention and Safety Promotion in the Process Industries.* Cannes, France. 38:1, 38:13.

Harrison, A. J., and J. A. Eyre. 1987. "The effect of obstacle arrays on the combustion of large premixed gas/air clouds." *Comb. Sci. Tech.* 52:121–137.

Harrison, A.J., Eyre, J.A., "The Effect of Obstacle Arrays on the Combustion of Large Premixed Gas/Air Clouds," Coubust. Sci. Technol. 52 (1987) 121-137.

Harrison, A.J., Eyre, J.A., "Vapour cloud explosions – the Effect of Obstacles and Jet Ignition on the Combustion of Gas Clouds," Proc. 5th Int. Symp. on Loss Prevention and Safety Promotion in the Process Industries, Cannes, France, 1986, pp 38-1, 38-13.

Hasegawa, K., and K. Sato 1987. Experimental investigation of unconfined vapor cloud explosions and hydrocarbons. Technical Memorandum No. 16, Fire Research Institute, Tokyo.

Hasegawa, K., and Sato, K. 1977. Study on the fireball following steam explosion of *n*-pentane. *Second International Symposium on Loss Prevention and Safety Promotion in the Process Industries*, pp. 297–304.

Hasegawa K., K. Sato, Fireballs, 12, Technical Memos of Fire Research Institute of Japan, Tokyo, 1978, pp. 3–9.

Health and Safety Executive, 1975. 'The Flixborough Disaster: Report of the Court of Inquiry,' HMSO, ISBN 0113610750.

Health and Safety Executive. 1979. *Second Report. Advisory Committee Major Hazards*. U.K. Health and Safety Commission, 1979.

Health and Safety Executive. 1986. "The effect of explosions in the process industries." *Loss Prevention Bulletin*. 1986. 68:37–47.

Health & Safety Executive. 1986. The effect of explosions in the process industries. *Loss Prevention Bulletin*. 68:37–47.

Health and Safety Laboratory, HSL, C.J. Butler and Royle M., "Experimental data acquisition for validation of a new vapor cloud fire (VCF) modeling approach," Report HSL/2001/15.

High, R. 1968. The Saturn fireball. *Ann. N.Y. Acad. Sci.* 152:441–451.

Hinman, E. E., and Hammonds, D. J. 1997. *Lessons from the Oklahoma City Bombing: Defensive Design Techniques*. American Society of Civil Engineers, New York, 1997.

Hirsch, F. G. 1968. Effects of overpressure on the ear, a review. *Ann. NY Acad. Sci.*

Hirst, W. J. S., and J. A. Eyre. 1983. "Maplin Sands experiments 1980: Combustion of large LNG and refrigerated liquid propane spills on the sea." *Heavy Gas and Risk Assessment II*. Ed. by S. Hartwig. pp. 211–224. Boston: D. Reidel.

Hjertager, B. H. 1982a. Simulation of transient compressible turbulent reactive flows. *Comb. Sci. Tech.* 41:159–170.

Hjertager, B. H. 1982b. Numerical simulation of flame and pressure development in gas explosions. *SM study No. 16*. Ontario, Canada: University of Waterloo Press. 407–426.

Hjertager, B. H. 1984. "Influence of turbulence on gas explosions." *J. Haz. Mat.* 9:315–346.

Hjertager, B. H. 1985. "Computer simulation of turbulent reactive gas dynamics." *Modeling, Identification and Control.* 5(4):211-236.

Hjertager, B. H. 1989. "Simulation of gas explosions." *Modeling, Identification and Control.* 1989. 10(4):227–247.

Hjertager, B. H. 1991. "Explosions in offshore modules." *IChemE Symposium Series No. 124,* pp. 19–35. Also in *Process Safety and Environmental Protection,* Vol. 69, Part B, May 1991.

Hjertager, B. H., K. Fuhre, and M. Bjorkhaug. 1988a. "Concentration effects on flame acceleration by obstacles in large-scale methane-air and propane-air explosions." *Comb. Sci. Tech.,* 62:239–256.

Hjertager, B. H., K. Fuhre, S. J. Parker, and J. R. Bakke. 1984. "Flame acceleration of propane-air in a large-scale obstructed tube." *Progress in Astronautics and Aeronautics.* 94:504–522. AIAA Inc., New York.

Hjertager, B. H., M. Bjorkhaug, and K. Fuhre. 1988b. "Explosion propagation of non-homogeneous methane-air clouds inside an obstructed $50m^3$ vented vessel." *J. Haz. Mat.* 19:139–153.

Hjertager, B. H., T. Solberg, and J. E. Forrisdahl. 1991b. Computer simulation of the 'Piper Alpha' gas explosion accident."

Hjertager, B. H., T. Solberg, and K. O. Nymoen. 1991a. Computer modeling of gas explosion propagation in offshore modules.

Hjertager, B.H., "Explosions in Obstructed Vessels," Course on Explosion Prediction and Mitigation, University of Leeds, UK, June 28-30, 1993.

Hoerner, S. F. 1958. *Fluid Dynamic Drag.* Midland Park, NJ: Author.

Hoff, A. B. M. 1983. "An experimental study of the ignition of natural gas in a simulated pipeline rupture." *Comb. and Flame.* 49:51–58.

Hogan, W. J. 1982. The liquefied gaseous fuels spill effects program: a status report. *Fuel-air explosions,* pp. 949–968. Waterloo, Canada: University of Waterloo Press, 1982.

Holden, P.L., Reeves, A.B., 1985, *Fragment Hazards from Failures of Pressurized Liquefied Gas Vessels,* IChemE Symposium Series, No. 93, Manchester, UK

Hopkinson, B. 1915. British Ordnance Board Minutes 13565.

HSE Flash Fire Model Specification, WS Atkins Report No. AM5222-R1, Issue No. 2, September, 2000.

Hymes, I. 1983. The physiological and pathological effects of thermal radiation. *United Kingdom Atomic Energy Authority,* SRD R 275.

IChemE. 1987. The Feyzin disaster, *Loss Prevention Bulletin No. 077:* 1–10.

Industrial Risk Insurers. Oil and Chemical Properties Loss Potential Estimation Guide. *IRI-Information February 1, 1990.*

Istratov, A. G., and V. B. Librovich. 1969. "On the stability of gas-dynamic discontinuities associated with chemical reactions. The case of a spherical flame." *Astronautica Acta* 14:453–467.

Jaggers, H. C, O. P. Franklin, D. R. Wad, and F. G. Roper. 1986. Factors controlling burning time for non-mixed clouds of fuel gas. *I. Chem. E. Symp. Ser.* No. 97.

Janet, D. E. 1968. Derivation of the British Explosives Safety Distances. *Ann. NY Acad. Sci.* 152.

Jarrett, D. E. 1968. "Derivation of the British explosives safety distances." *Ann. N. Y. Acad. Sci.* Vol. 152.

Johansson, O. 1986. BLEVES a San Juanico. *Face au Risque.* 222(4):35–37, 55–58.

Johnson, D. M., M. J. Pritchard, and M. J. Wickens. 1990. Large scale catastrophic releases of flammable liquids. *Commission of the European Communities report, Contract No.: EV4T. 0014. UK(H).*

Johnson, D.M., Pritchard, M.J., 1991, *Large Scale Catastrophic Releases of Flammable Liquids,* Commision of the European Communities Report EV4T.0014

Karlovitz, B. 1951. "Investigation of turbulent flames." *J. Chem. Phys.* 19:541–547.

Khitrin, L.N. and Goldenberg, S.A. 1962, "Gas Dynamics and Combustion", IPST, p.139

Kingery, C.N., Bulmash, G., "Air Blast Parameters versus Distance for TNT Spherical Air Burst and Hemispherical Surface Burst," Tech. Rep. ARBRL-TR 02555, US Army, Ballistics Research Laboratory, Aberdeen Proving Grounds, MD, 1984.

Kjäldman, L., and R. Huhtanen. 1985. "Simulation of flame acceleration in unconfined vapor cloud explosions." *Research Report No. 357.* Technical Research Centre of Finland.

Kjäldman, L., and R. Huhtanen. 1986. Numerical simulation of vapour cloud and dust explosions. *Numerical Simulation of Fluid Flow and Heat/Mass Transfer Processes.* Vol. 18, Lecture Notes in Engineering, 148–158.

Kletz, T. A. 1977. "Unconfined vapor cloud explosions—an attempt to quantify some of the factors involved." *AIChE Loss Prevention Symposium.* Houston, TX. 1977.

Knystautas, R., J. H. Lee, and C. M. Guirao. 1982. The critical tube diameter for detonation failure in hydrocarbon-air mixtures. *Combustion and Flame.* 48:63–83.

Knystautas, R., J. H. Lee, and I. O. Moen. 1979. "Direct initiation of spherical detonation by a hot turbulent gas jet." *17th Symp. (Int.) on Combustion.* pp. 1235–1245. The Combustion Institute, Pittsburgh, PA.

Knystautas, R., Guirao, C.M., Lee, J.H., and Sulmistras, A., "Measurements of cell size in hydrocarbon-air mixtures and predictions of critical tube diameter, critical initiation energy, and detonability limits," presented at the 9th International Colloquium on the Dynamics of Explosions and Reactive Systems, Poitiers, France, 1983.

Kogarko, S. M., V. V. Adushkin, and A. G. Lyamin. 1966. "An investigation of spherical detonations of gas mixtures." *Int. Chem. Eng.* 6(3):393–401.

Kuchta, J. M. 1985. Investigation of fire and explosion accidents in the chemical, mining, and fuel-related industries-A manual. *Bureau of Mines Bulletin 680.*

Kuhl, A. L. 1983. "On the use of general equations of state in similarity, analysis of flamedriven blast waves." *Progress in Astronautics and Aeronautics.* 87:175–195, AIAA Inc., New York.

Kuhl, A. L., M. M. Kamel, and A. K. Oppenheim. 1973. "Pressure waves generated by steady flames." *14th Symp. (Int.) on Combustion.* pp. 1201–1214, The Combustion Institute, Pittsburgh, PA.

Launder, B. D., and D. B. Spalding. 1974. The numerical computation of turbulent flows. *Comput. Meth. Appl. Mech. Eng.* 3:269–289.

Launder, B. E., and D. B. Spalding. 1972. *Mathematical models of turbulence,* London: Academic Press.

Lee, J. H. S. 1983. "Gas cloud explosion—Current status." *Fire Safety Journal.* 5:251–263.

Lee, J. H. S., and I. O. Moen. 1980. "The mechanism of transition from deflagration to detonation in vapor cloud explosions." *Prog. Energy Comb. Sci.* 6:359–389.

Lee, J. H. S., and K. Ramamurthi. 1976. "On the concept of the critical size of a detonation kernel." *Comb. and Flame.* 27:331–340.

Lee, J. H. S., R. Knystautas, and A. Freiman. 1984. "High speed turbulent deflagrations and transition to detonation in H2-air mixtures." *Combustion and Flame.* 56:227–239.

Lee, J. H. S., R. Knystautas, and CK. Chan. 1984. "Turbulent flame propagation in obstacle-filled tubes." *20th Symp. (Int.) on Combustion.* pp. 1663–1672. The Combustion Institute, Pittsburgh, PA.

Lee, J. H. S., R. Knystautas, and N. Yoshikawa. 1978. "Photochemical initiation of gaseous detonations." *Acta Astronautica.* 5:971–982.

Lees, F. P. 1980. Loss Prevention in the Process Industries. London: Butterworths.

Lees, F.P., Loss Prevention in the Process Industry, Butterworth Heinemann, Second Edition, ISBN 0 7506 1547 8, 1996.

Lee's Loss Prevention in the Process Industries, Third Ed, S. Mannan, Editor, Elsevier-Butterworth Heineman, New York, p 16/172, 2005.

Leiber, C. O. 1980. Explosionen von Flüsigkeitstanken. Empirische Ergebnisse—Typische Unfälle. J. *Occ. Acc.* 3:21–43.

Lenoir, E. M., and J. A. Davenport. 1993. "A Survey of Vapor Cloud Explosions: Second Update." *Process Safety Progress.* 12:12–33.

Lewis, D. 1985. New definition for BLEVEs. *Haz. Cargo Bull.* April, 1985: 28–31.

Lewis, D. J. 1980. "Unconfined vapor cloud explosions—Historical perspective and predictive method based on incident records." *Prog. Energy Comb. Sci.,* 1980. 6:151 -165.

Lewis, D. J. 1981. "Estimating damage from aerial explosion type incidents—Problems with a detailed assessment and an approximate method." *Euromech 139.* Aberystwyth (UK).

Lewis, D. J. 1989. Soviet blast—the worst yet? *Hazardous Cargo Bulletin.* August 1989. 59–60.

Leyer, J. C. 1981. "Effets de pression engendres par l'explosion dans l'atmosphere de melanges gazeux d'hydrocarbures et d'air." *Revue Generale de Thermique Fr.* 243:191–208.

Leyer, J. C. 1982. "An experimental study of pressure fields by exploding cylindrical clouds." *Combustion and Flame.* 48:251–263.

Liepmann, H. W, and A. Roshko. 1967. *Elements of Gas Dynamics.* New York: John Wiley and Sons.

Lighthill, J. 1978. *Waves in fluids*. Cambridge: Cambridge University Press.

Lihou, D. A., and J. K. Maund. 1982. Thermal radiation hazard from fireballs. *I. Chem. E. Symp. Ser.* No. 71, pp. 191–225.

Lind, C. D. 1975. "What causes unconfined vapor cloud explosions." *AIChE Loss Prevention Symp.* Houston, proceedings pp. 101–105.

Lind, C. D., and J. Whitson. 1977. "Explosion hazards associated with spills of large quantities of hazardous materials (Phase 3)." *Report Number CG-D-85–77.* United States Dept. of Transportation, U.S. Coast Guard, *Final Report ADA047585.*

Linney, R. E. 1990. Air Products and Chemicals, Inc. Personal communication.

Love, T. J. 1968. *Radiative heat transfer.* Cincinnati, OH: C. E. Merrill.

Luckritz, R. T. 1977. "An investigation of blast waves generated by constant velocity flames." Aeronautical and Astronautical Engineering Department. University of Illinois. Urbana, Illinois *Technical report no. AAE 77–2.*

Luckritz, R.T., "An investigation of blast waves generated by constant velocity flames," Dissertation for Doctor of Philosophy, University of Maryland, 1977

Lumley, J. L., and H. A. Panofsky. 1964. *The Structure of Atmospheric Turbulence.* New York: John Wiley and Sons.

Mackenzie, J., and D. Martin. 1982. "GASEXl—A general one-dimensional code for gas cloud explosions." UK Atomic Energy Authority, Safety and Reliability Directorate, *Report No. SRD R251.*

Magnussen, B. F., and B. H. Hjertager. 1976. "On the mathematical modelling of turbulent combustion with special emphasis on soot formation and combustion." *16th Symp. (Int.) on Combustion.* pp. 719–729. The Combustion Institute, Pittsburgh, PA.

Magnussen, B. F., and B. H. Hjertager. 1976. On the mathematical modeling of turbulent combustion with special emphasis on soot formation and combustion. *16th Symp. (Int) on Combustion.* Combustion Institute, PA, pp. 719–729.

Makhviladze, G.M., Yakush, S.E., "Large-Scale Unconfined Fires and Explosions," Proceedings of the Combustion Institute, Volume 29, 2002/pp.195-210

Mancini, R. A. 1991. Private communication.

Markstein, G. H. 1964. *Non-steady flame propagation.* New York: Pergamon.

Marsh Global Marine and Energy, *Practice Loss Control Newsletter*, Issue 1, 2007

Marshall, V. C 1976. "The siting and construction of control buildings—a strategic approach." *I.Chem.E. Symp. Series, No. 47.*

Marshall, V. C, 1986. "Ludwigshafen—Two case histories." *Loss Prevention Bulletin* 67:21–33.

Marshall, V.C., *Major Chemical Hazards,* Ellis Horwood, Chichester, 1987.

Martin, D. 1986. Some calculations using the two-dimensional turbulent combustion code Flare. *SRD Report R373.* UK Atomic Energy Authority.

Martinsen W.E., J.D. Marx, *An improved model for the prediction of radiant heat flux from fireball,* Proceedings of the International Conference and Workshop on Modeling the Consequences of Accidental Releases of Hazardous Materials, San Francisco, 1999.

Marx, K. D., J. H. S. Lee, and J. C. Cummings. 1985. Modeling of flame acceleration in tubes with obstacles. *Proc. of 11th IMACS World Congress on Simulation and Scientific Computation.* 5:13–16.

Matsui, H and Lee, J.H, "On the measure of the relative detonation hazards of gaseous fuel-oxygen and air mixtures", 17th symposium (Int.) on combustion, pp. 1269-1279, 1978

Matsui, H., and J. H. S. Lee. 1979. On the measure of relative detonation hazards of gaseous fuel-oxygen and air mixtures. *Seventeenth Symposium (International) on Combustion,* pp. 1269–1280. Pittsburgh, PA: The Combustion Institute.

Maurer, B., K. Hess, H. Giesbrecht, and W. Leuckel. 1977. Modeling vapor cloud dispersion and deflagration after bursting of tanks filled with liquefied gas. *Second Int. Symp. on Loss Prevention and Safety Promotion in the Process Ind.,* pp. 305–321. Heidelberg.

McBride B. J. and S. Gordon, "Computer Program For Calculation of Complex Chemical Equilibrium Compositions And Applications", NASA Reference Publication 1311, 1996

McDevitt, C. A., F. R. Steward, and J. E. S. Venart. 1987. What is a BLEVE? *Proc. 4th Tech. Seminar Chem. Spills,* pp. 137–147. Toronto.

McKay, D. J., S. B. Murray, I. O. Moen, and P. A. Thibault. 1989. "Flame-jet ignition of large fuel-air clouds." *Twenty-Second Symposium on Combustion,* pp. 1339–1353, The Combustion Institute, Pittsburgh.

Mercx, M., Modeling and experimental research into gas explosions; overall final report of the Merge project, Commission on the European Communities, contract STEP-CT-011, (SSMA), 1993.

Mercx, W.P.M., "Large scale experimental investigation into vapor cloud explosions, comparison with the small scale DISCOE trials", 7th Int. Symposium. On Loss Prevention and Safety Promotion in the Process Industries, Italy, May 1992

Mercx, W.P.M., "Modeling and Experimental Research into Gas Explosions: overall final report of the MERGE project", Commission of the European Communities, contract STEP-CT-011 (SSMA)m 1993

Mercx, W.P.M., "Modeling and Experimental Research into Gas Explosions," Proc. Symp. On Loss Prevention and Safety Promoting in the Process Industries, Antwerp, Belgium, June 1995.

Mercx, W.P.M., A.C. van den Berg, and D. van Leeuwen, Application of correlations to quantify the source strength of vapour cloud explosions in realistic situations. Final report for the project: 'GAMES',TNO Report PML 1998-C53, Rijswijk, The Netherlands (1998).

Mercx, W.P.M., et al., "Developments in Vapour Cloud Explosion Blast Modeling," Journal of Hazardous Materials, 71 (2000) 301-319.

Mercx, W.P.M., N.R. Popat and H. Linga, "Experiments to Investigate the Influence of an Initial Turbulence Field on the Explosion Process," Final Summary Report for EMERGE

Mercx, W.P.M., van den Berg, A.C., van Leeuwen, D., "Application of Corelations to Quantify the Source Strength of Vapour Cloud Explosions in Realistic Situations," Final repor to the GAMES project, TNO Prins Maurits Laboratory report PML 1998-C53, The Netherlands, 1998

Miller, T., Birk, A.M., 1997, A Re-examination of Propane Tank Tub Rockets Including Field Trial Results, *ASME Pressure Vessels and Piping* 119:356-64

Moen, I. O., D. B. Bjerketvedt, A. Jenssen, and P. A. Thibault. 1985. "Transition to detonation in a large fuel-air cloud." *Comb. and Flame.* 61:285–291.

Mitrofanov, V.V. and Soloukhin, R.I., "The Diffraction of Multifront Detonation Waves," *Soviet Physics-Doklady*, Vol. 9, No. 12, 1965, pp. 1055-1058.

Moen, I. O., D. Bjerketvedt, T. Engebretsen, A. Jenssen, B. H. Hjertager, and J. R. Bakke. 1989. "Transition to detonation in a flame jet." *Comb. and Flame.* 75:297–308.

Moen, I. O., J. H. S. Lee, B. H. Hjertager, K. Fuhre, and R. K. Eckhoff. 1982. "Pressure development due to turbulent flame propagation in large-scale methane-air explosions." *Comb. and Flame.* 47:31–52.

Moen, I. O., J. W. Funk, S. A. Ward, G. M. Rude, and P. A. Thibault. 1984. Detonation length scales for fuel-air explosives. *Prog. Astronaut. Aeronaut.* 94:55–79.

Moen, I. O., M. Donato, R. Knystautas, and J. H. Lee. 1980a. "Flame acceleration due to turbulence produced by obstacles." *Combust. Flame.* 39:21–32.

Moen, I. O., M. Donato, R. Knystautas, J. H. Lee, and H. Gg. Wagner. 1980b. "Turbulent flame propagation and acceleration in the presence of obstacles." *Progress in Astronautics and Aeronautics.* 75:33–47, AIAA Inc., New York.

Moen, I.O., Funk, J.W., Ward, S.A., Thibault, P.A., "Detonation length scales for fuel-air explosives", the 9th ICODERS, 1983, AIAA

Mogford, J., Fatal Accident Investigation Report, Isomerization Unit Explosion, Texas City Accident, BP Public Release Final Report, December 9, 2005.

Moodie, K., L. T. Cowley, R. B. Denny, L. M. Small, and I. Williams. 1988. Fire engulfment tests on a 5-ton tank. *J. Haz. Mat.* 20:55–71.

Moore, C. V. 1967. *Nuclear Eng. Des.* 5:81–97.

Moorhouse, J., and M. J. Pritchard. 1982. Thermal radiation from large pool fires and thermals—Literature review. *IChemE Symp. Series No. 71.* p. 123.

Moskowitz, H. 1965. AIAA paper no. 65–195.

MSHA (Mine Safety and Health Administration), United States Department of Labor, Metal and Nonmetal Mine Safety and Health Report of Investigation, Surface Metal Mine (Alumina), Nonfatal Exploding Vessels Accident July 5, 1999, Gramercy Works, Kaiser Aluminum and Chemical Corporation Gramercy, St. James Parish, Louisiana, ID No. 16-00352.

Mudan, K. S. 1984. Thermal radiation hazards from hydrocarbon pool fires. *Progr. Energy Combust. Sci.* 10(1):59–80.

Munday, G., and L. Cave. 1975. "Evaluation of blast wave damage from very large unconfined vapor cloud explosions." International Atomic Energy Agency, Vienna.

Nabert, K., and G. Schön. 1963. *Sicherheitstechnische Kennzahle brennbarer Gase und Dämpfe.* Berlin: Deutscher Eichverlag GmbH.

National Transportation Safety Board. 1971. "Highway Accident Report: Liquefied Oxygen tank truck explosion followed by fires in Brooklyn, New York, May 30, 1970." *NTSB-HAR-71-6.*

National Transportation Safety Board. 1972. "Pipeline Accident Report, Phillips Pipe Line Company propane gas explosion, Franklin County, MO, December 9, 1970." National Transportation Safety Board, Washington, DC, *Report No. NTSB-PAR-72–1.*

National Transportation Safety Board. 1972. "Railroad Accident Report—Derailment of Toledo, Peoria and Western Railroad Company's Train No. 20 with Resultant Fire and Tank Car Ruptures, Crescent City, Illinois, June 21, 1970. *NTSB-RAR-72–2.*

National Transportation Safety Board. 1973. "Hazardous materials railroad accident in the Alton and Southern Gateway Yard, East St. Louis, Illinois, January 22, 1972." Report No. NTSB-RAR-73–1. National Transportation Safety Board, Washington, DC.

National Transportation Safety Board. 1973. "Highway Accident Report—Propane Tractor-Semitrailer overturn and fire, U.S. Route 501, Lynchburg, Virginia, March 9, 1972." *NTSB-HAR-73–3.*

National Transportation Safety Board. 1975. "Hazardous material accidents at the Southern Pacific Transportation Company's Englewood Yard, Houston, Texas, September 21, 1974." *Report No. NTSB-RAR-75–7.* National Transportation Safety Board, Washington, DC.

National Transportation Safety Board. 1975. "Hazardous materials accident in the railroad yard of the Norfolk and Western Railway, Decatur, Illinois, July 19, 1974." *Report No. NTSB-RAR-75–4.* National Transportation Safety Board, Washington, DC.

National Transportation Safety Board. 1979. "Pipeline Accident report—Mid-America Pipeline System—Liquefied petroleum gas pipeline rupture and fire, Donnellson, Iowa, August 4, 1978." *NTSB-Report NTSB-PAR-79-I.*

Nedelka, D., J. Moorhouse, and R. F. Tucket, "The Montoir 35 m diameter LNG pool fire experiments", Proc. 9th Intl. Conf. On LNG, Nice, Fr, 17-20 Oct 1989, published by Instit. Gas Technology, Chicago, 2, pp III-3 1-23 (1990).

Nettleton, M. A. 1987. *Gaseous Detonations.* New York: Chapman and Hall.

Okasaki, S., J. C. Leyer, and T. Kageyama. 1981. "Effets de pression induits par l'explosion de charges combustibles cylindriques non confinees." First Specialists Meeting of the Combustion Institute. Bordeaux, France, proceedings, pp. 485–490.

Oppenheim, A. K. 1973. "Elementary blast wave theory and computations." *Proc. of the Conf. on Mechanisms of Explosions and Blast Waves.* Yorktown, Virginia.

Oppenheim, A. K., J. Kurylo, L. M. Cohen, and M. M. Kamel. 1977. "Blast waves generated by exploding clouds." *Proc. 11th Int. Symp. on Shock Tubes and Waves*. pp. 465–473. Seattle.

Opschoor, G. 1974. Onderzoek naar de explosieve verdamping van op water uitspreidend LNG. Report Centraal Technisch Instituut TNO, Ref. 74–03386.

OSHA (Occupational Safety and Health Administration) U.S. Department of Labor, "Phillips 66 Company Houston Chemical Complex Explosion and Fire," April 1999.

Pape, R. P. (Working Group Thermal Radiation). 1988. Calculation of the intensity of thermal radiation from large fires. *Loss Prevention Bulletin*. 82:1–11.

Parker, R. J. (Chairman), 1975. The Flixborough Disaster. Report of the Court of Inquiry. London: HM Stationery Office.

Patankar, S. V. 1980. *Numerical heat transfer and fluid flow*, Washington: Hemisphere.

Patankar, S. V., and D. B. Spalding. 1972. A calculation procedure for heat, mass and momentum transfer in three-dimensional parabolic flows. *Int. J. Heat and Mass Transfer*. 15:1787–1806.

Patankar, S. V., and D. B. Spalding. 1974. A calculation procedure for the transient and steady-state behavior of shell-and-tube heat exchangers. In N. H. Afgan and E. V. Schlünder (eds.), *Heat Exchangers: Design and Theory Sourcebook*. New York: McGraw-Hill, pp. 155–176.

Perry, R. H., and D. Green. 1984. *Perry's Chemical Engineers' Handbook*, 6th ed. New York: McGraw-Hill.

Petit, G.N., Harms, J.D., Woodward, J.L., "Post-Mortem Risk Modeling of the Mexico City Disaster," International Association of Probabilistic Safety Assessment and Management, International Conference on Probabilistic Safety Assessment and Management, PSAM-2 Conference, San Diego, CA, March 1994.

Pförtner, H. 1985. "The effects of gas explosions in free and partially confined fuel/air mixtures." *Propellants, Explosives, Pyrotechnics*. 10:151–155.

Phillips, H. 1980. "Decay of spherical detonations and shocks." Health and Safety Laboratories *Technical Paper No. 7*.

Phylaktou, H., *Gas explosions in long closed vessels with obstacles*, Ph.D. Thesis, University of Leeds (1993).

Phylaktou, H. and G.E Andrews, "Application of Turbulent Combustion Models to Explosion Scaling", Trans IChemE, Vol. 73, Part B, February 1995

Phylaktou, H. and G.E Andrews, "Prediction of the Maximum Turbulence Intensities Generated by Grid-Plate Obstacles in Explosion-Induced Flows", 25th Symposium on Combustion (International), pp 103-110, 1994

Phylaktou, H. and G.E. Andrews, "Application of turbulent combustion models to explosion scaling," *Trans IChemE*, v.73 part B, pp. 3-10 (1995).

Phylaktou, H. and G.E. Andrews, "Gas explosions in linked vessels," *Journal of Loss Prevention in the Process Industries*, v.6, pp.16 (1993).

Phylaktou, H. and G.E. Andrews, "Prediction of the maximum turbulence intensities generated by grid plate obstacles in explosion induced flows," *Proceedings of the 25th International Symposium on Combustion*, Irvine, CA USA (1994).

Phylaktou, H., and Andrews, G.E., 1994, 25th Symposium (International) on Combustion

Phylaktou, H.; Liu, Y.; Andrews, G.E., "Turbulent Explosions - A Study of the Influence of the Obstacle Scale," Hazards XII - European Advances in Process Safety Symposium, pp269-284, 1994

Phylaktou, H., Liu, Y. and Andrews, G.E., 1994, IChemE Symposium Series No. 134, 271

Pickles, J. H., and S. H. Bittleston. 1983. "Unconfined vapor cloud explosions— The asymmetrical blast from an elongated explosion." *Combustion and Flame.* 51:45–53.

Pierorazio, A.J., Thomas, J.K., Baker, Q.A, and Ketchum, D.E., "An Update to the Baker-Strehlow-Tang Vapor Cloud Explosion Prediction Methodology Flame Speed Table," 38th Annual Loss Prevention Symposium, AIChE Spring National Meeting, 2004.

Pietersen, C. M. 1985. Analysis of the LPG incident in San Juan Ixhuatepec, Mexico City, 19 November 1984. *Report—TNO Division of Technology for Society.*

Pietersen, C. M. 1988. Analysis of the LPG disaster in Mexico City. *J. Haz. Mat.* 20:85–108.

Pitblado, R. M. 1986. Consequence models for BLEVE incidents. Major Industrial Hazards Project, NSW 2006. University of Sydney.

Pittman, J. F. 1972. *Blast and Fragment Hazards from Bursting High Pressure Tanks.* NOLTR 72–102. Silver Spring, Maryland: U.S. Naval Ordnance Laboratory.

Pittman, J. F. 1976. *Blast and Fragments from Superpressure Vessel Rupture.* NSWC/WOL/TR 75–87. White Oak, Silver Spring, Maryland: Naval Surface Weapons Center.

Porteous, W., Blander, M., Limits of Superheat and Explosive Boiling of Light Hydrocarbons, Halocarbons, and Hydrocarbon Mixtures, AIChE Journal 21[3], 560-566, 1975

Pritchard, D. K. 1989. "A review of methods for predicting blast damage from vapor cloud explosions." *J. Loss Prev. Proc. Ind.* 2(4):187–193.

Prugh R.W., *Quantitative evaluation of fireball hazards*, Process Safety Progress 13 (1994) 83.

Prugh, R. W. 1987. "Evaluation of unconfined vapor cloud explosion hazards." *Int. Conf. on Vapor Cloud Modeling.* Cambridge, MA. pp. 713–755, AIChE, New York.

Puttock, J. S., "Fuel Gas Explosion Guidelines - the Congestion Assessment Method," 2nd European Conference on Major Hazards On and Off-shore, Manchester, UK, October 24, 1995.

Puttock, J., Fuel Gas Explosion Guidelines - the Congestion Assessment Method, 2nd European Conference on Major Hazards On- and Off- shore, Manchester, October 1995.

Puttock, J., Developments for the Congestion Assessment Method for the Prediction of Vapour-Cloud Explosions 10th International Symposium on Loss Prevention and safety Promotion in the Process Industries, Stockholm, June 2001.

Puttock JS., Blackmore DR, Colenbrander GW, Davis PT, Evans A, Homer JB, Redfern JJ, Van't Sant WC, Wilson RP. *Spill tests of LNG and Refrigerated Liquid Propane on the sea, Maplin sands, Experimental details of the dispersion tests,* Shell International Research Report TNER.84.046, May 1984.

Puttock, J., Yardley, M., Cresswell, T., Prediction of vapor cloud explosions using the SCOPE model, J. Loss Prev. Process Ind., 13 (2000) 419.

Puttock, J.S. "Fuel Gas Explosion Guidelines - The Congestion Assessment Method", ICHEME Symposium Series No. 139

Raj, P. K. 1977. Calculation of thermal radiation hazards from LNG fires. *A Review of the State of the Art, AGA Transmission Conference T135–148.*

Raj, P. K. 1982. MIT-GRI Safety & Res. Workshop, LNG-fires, Combustion and Radiation, Technology & Management Systems, Inc., Mass.

Raj, P. K., and H. W. Emmons. 1975. On the burning of a large flammable vapor cloud. Paper presented at the *Joint Technical Meeting of the Western and Central States Section of the Combustion Institute.* San Antonio, TX.

Raj, P. K., and K. Attalah. 1974. "Thermal radiation from LNG fires." *Adv. Cryogen. Eng.* 20:143.

Raj P.K., Moussa, N.A., Aravamudan, K.S. (1979). Experiments involving pool and vapor fires from spills of LNG on water, USCG Report CG-D-55-79, Washington DC 20590, NTIS AD-A077073.

Raju, M. S., and R. A. Strehlow. 1984. "Numerical investigations of non-ideal explosions." *J. Haz. Mat.* 9:265–290.

Ramier, S., Venart, J.E.S., Boiling Liquid Expanding Vapour Explosions:Dynamic Re-Pressurization and Two Phase Discharge, IChemE Symposium Series No. 147, 527-537, 2000

Reid, R. C. 1976. Superheated liquids. *Amer. Scientist.* 64:146–156.

Reid, R. C. 1979. Possible mechanism for pressurized-liquid tank explosions or BLEVE's. *Science.* 203(3).

Reid, R. C. 1980. Some theories on boiling liquid expanding vapor explosions. *Fire.* March 1980: 525–526.

Reid, R.C., Possible Mechanism for Pressurized Liquid Tank Explosions or BLEVEs, Science 203, 1263-1265, 1979

Reider, R., H. J. Otway, and H. T. Knight. 1965. "An unconfined large volume hydrogen/ air explosion." *Pyrodynamics.* 2:249–261.

Richtmyer, R. D. and K. W. Morton. 1967. *Difference methods for initial value problems.* New York: Interscience.

Ritter, K. 1984. Mechanisch erzeugte Funken als Zündquellen. *VDI-Berichte Nr.494.* pp. 129–144.

Roberts, A. F. 1982. Thermal radiation hazards from release of LPG fires from pressurized storage. *Fire Safety J.* 4:197–212.

Roberts, A. F., and D. K. Pritchard. 1982. Blast effects from unconfined vapor cloud explosions. *J. Occ. Acc.* 3:231–247.

Robinson, C. S. 1944. *Explosions, their anatomy and destructiveness.* New York: McGraw-Hill.

Rosenblatt, M., and P. J. Hassig. 1986. "Numerical simulation of the combustion of an unconfined LNG vapor cloud at a high constant burning velocity." *Combust. Science and Tech.* 45:245–259.

Sachs, R. G. 1944. The dependence of blast on ambient pressure and temperature. BRL Report no. 466, Aberdeen Proving Ground. Maryland.

Sadèe, C, D. E. Samuels, and T. P. O'Brien. 1976/1977. "The characteristics of the explosion of cyclohexane at the Nypro (U.K.) Flixborough plant on June 1st 1974." *J. Occ. Accid.* 1:203–235.

Schardin, H. 1954. *Ziviler Luftschutz.* 12:291–293.

Schildknecht, M. 1984. Versuche zur Freistrahlzöndung von Wasserstoff-Luft-Gemischen im Hinblick auf den Übergang Deflagration-Detonation, report BIeV-R-65.769–1, Battelle Institut e.V., Frankfurt, West Germany.

Schildknecht, M., and W. Geiger. 1982. Detonationsähnliche Explosionsformen-Mögliche Intiierung Detonationsähnlicher Explosionsformen durch partiellen Einschluss, Teilaufgabe 1 des Teilforschungsprogramm Gasexplosionen, report BIeV-R-64.176–2, Battelle Institut e.V., Frankfurt, West Germany.

Schildknecht, M., W. Geiger, and M. Stock. 1984. "Flame propagation and pressure buildup in a free gas-air mixture due to jet ignition." *Progress in Astronautics and Aeronautics.* 94:474–490.

Schmidli, J., S. Banerjee, and G. Yadigaroglu. 1990. Effects of vapor/aerosol and pool formation on rupture of vessel containing superheated liquid. *J. Loss Prev. Proc. Ind.* 3(1):104–111.

Schneider, H., and H. Pförtner. 1981. Flammen und Druckwellenausbreitung bei der Deflagration von Wasserstoff-Luft-Gemischen, Fraunhofer-Institute für Treib- und Explosiv-stoffe (ICT), Pfinztal-Berghaven, West Germany.

Schoen, W., U. Probst, and B. Droste. 1989. Experimental investigations of fire protection measures for LPG storage tanks. *Proc. 6th Int. Symp. on Loss Prevention and Safety Promotion in the Process Ind.* 51:1 – 17.

Schulz-Forberg, B., B. Droste, and H. Charlett. 1984. Failure mechanics of propane tanks under thermal stresses including fire engulfment. *Proc. Int. Symp. on Transport and Storage of LPG and LNG.* 1:295–305.

Seifert, H., and H. Giesbrecht. 1986. "Safer design of inflammable gas vents." 5th Int. Symp. *Loss Prevention and Safety Promotion in the Process Industries.* Cannes, France, proceedings, pp. 70–1, 70–21.

Sha, W. T., C. I. Yang, T. T. Kao, and S. M. Cho. 1982. Multi-dimensional numerical modeling of heat exchangers. *J. Heat Trans.* 104:417–425.

Sherman, M. P., S. R. Tiezsen, W. B. Bendick, W. Fisk, and M. Carcassi. 1985. "The effect of transverse venting on flame acceleration and transition to detonation in a large channel." Paper presented at the 10th Int. Coll. on Dynamics of Explosions and Reactive Systems. Berkeley, California.

Sherry, W.L., "LP-Gas Distribution Plant Fire," Fire Journal, 1974, pp 52-57.

Shurshalov, L. V. 1973. *J. of USSR Comp. Math., Math. Phys.* 13:186. Sichel, M. 1977. "A simple analysis of blast initiation of detonations." *Acta Astronautica.* 4:409–424.

Simpson, I.C. (1984). Atmospheric transmissivity—the effects of atmospheric attenuation on thermal radiation, SRD (Safety and Reliability Directorate) Report R304, UK Atomic Energy Authority, Culcheth, Warrington, UK.

Sivashinsky, G. I. 1979. "On self-turbulization of a laminar flame." *Acta Astronautica.* 6:569–591.

Smith, J.M., van Ness, H.C., *Introduction to Chemical Engineering Thermodynamics*, 4th ed. New York: McGraw-Hill, 1987.

Snowdon, P., Puttock, J.S., Provost, E.T., Cresswell, T.M., Rowson, JJ, Johnson, RA, Masters, AP, Bimson, SJ., Critical design of validation experiments for vapour cloud explosion assessment methods, Proc. Intl. Conf. "Modeling the Consequences of Accidental Releases of Hazardous Materials, San Francisco, Sept. 1999.

Sokolik, A. S. 1963. *Self-ignition, flame and detonation in gases.* Israel Program of Scientific Translations. Jerusalem.

Spalding, D. B. 1981. A general purpose computer program for multi-dimensional one- and two-phase flow. *Mathematics and Computers in Simulation, IMACS, XXII.* 267–276.

Stephens, M. M. 1970. *Minimizing Damage from Nuclear Attack, Natural and Other Disasters.* Washington: The Office of Oil and Gas, Department of the Interior.

Steunenberg, C. F., G. W. Hoftijzer, and J. B. R. van der Schaaf. 1981. Onderzoek naar aanleiding van een ongeval met een tankauto te Nijmegen. *Pt-Procestechniek.* 36(4): 175–182.

Stewart, F. R. 1964. Linear flame heights for various fuels. *Combustion and Flame* 8: 171–178.

Stinton, H. G. 1983. Spanish camp site disaster. *J, Haz. Mat.* 7:393–401.

Stock, M. 1987. "Fortschritte der Sicherheitstechnik II." *Dechema monographic.* Vol. 111.

Stock, M., and W. Geiger. 1984. "Assessment of vapor cloud explosion hazards based on recent research results." 9th Int. Symp. on the Prevention of Occupational Accidents and Diseases in the Chemical Industry, Luzern, Switzerland.

Stock, M., W. Geiger, and H. Giesbrecht. 1989. "Scaling of vapor cloud explosions after turbulent jet release." 12th Int. Symp. on the Dynamics of Explosions and Reactive Systems. Ann Arbor, MI.

Stokes, G. G. 1849. "On some points in the received theory of sound." *Phil. Mag.* XXXIV(3):52.

Stoll, A. M., and M. A. Chianta. 1971. *Trans. N.Y. Acad. Sci.*, 649–670.

Strehlow, R. A. 1970. Multi-dimensional detonation wave structure. *Astronautica Acta* 15:345–357.

Strehlow, R. A. 1975. "Blast waves generated by constant velocity flames: A simplified approach." *Combustion and Flame.* 24:257–261.

Strehlow, R. A. 1981. "Blast wave from deflagrative explosions: an acoustic approach." *AIChE Loss Prevention.* 14:145–152.

Strehlow, R. A., and W. E. Baker. 1976. The characterization and evaluation of accidental explosions. *Prog. Energy Combust. Sci.* 2:27–60.

Strehlow, R. A., R. T. Luckritz, A. A. Adamczyk, and S. A. Shimpi. 1979. "The blast wave generated by spherical flames." *Combustion and Flame.* 35:297–310.

Strehlow, R.A., and Ricker, R.E., "The Blast Wave from a Bursting Sphere," AIChE, 10, pp 115-121, 1976.

Tang, M.J and Baker, Q.A., "Predicting Blast Effects From Fast Flames", 32th Loss Prevention Symposium, AIChE March 1998

Tang, M.J. and Q.A. Baker," Blast Effects from Vapor Cloud Explosions", Internal Report, Wilfred Baker Engineering, Inc, 1997

Tang, M.J. and Baker, Q.A., "A New Set of Blast Curves for Vapor Cloud Explosions," Center for Chemical Process Safety/American Institute of Chemical Engineers, 33rd Loss Prevention Symposium,1999

Tang, M.J., Cao, C.Y., and Baker, Q.A., "Blast Effects From Vapor Cloud Explosions", International Loss Prevention Symposium, Bergen, Norway, June 1996

Taylor, D. B., and C. F. Price. 1971. Velocity of Fragments from Bursting Gas Reservoirs. *ASME Trans. J. Eng. Ind.* 93B:981–985.

Taylor, G. I. 1946. "The air wave surrounding an expanding sphere." *Proc. Roy. Soc. London.* Series A, 186:273–292.

Taylor, P. H. 1985. "Vapor cloud explosions—The directional blast wave from an elongated cloud with edge ignition." *Comb. Sci. Tech.* 44:207–219.

Taylor, P. H. 1987. "Fast flames in a vented duct." *21st Symp. (Int.) on Combustion.* The Combustion Institute, Pittsburgh, PA.

Taylor, P.H. and Hirst, W.J.S., 1989, 22nd Symposium (International) on Combustion

Taylor, P.H., Hirst, W.J.S., "The Scaling of Vapour Cloud Explosions: a Fractal Model for Size and Fuel Type," 22nd International Symposium on Combustion, 1988.

TM5-1300, "Structures to Resist the Effects of Accidental Explosions," U.S Department of the Army Technical Manual TM5-1300, November, 1990.

Thomas, J.K., A.J. Pierorazio, M. Goodrich, M. Kolbe, Q.A. Baker and D.E. Ketchum (2003) "Deflagration to Detonation Transition in Unconfined Vapor Cloud Explosions,"Center for Chemical Process Safety (CCPS) 18th Annual International Conference & Workshop, Scottsdale, AZ, 23-25 September 2003.

Tweeddale, M. 1989. Conference report on the 6th Int. Symp. on Loss Prevention and Safety Promotion in the Process Industries, *J. of Loss Prevention in the Process Industries.* 1989. 2(4):241.

U.S. Chemical Safety and Hazard Investigation Board, "Investigation Report of Refinery Explosion and Fire, BP Texas City, Texas" Report No. 2005-04-I-TX, March 2007.

Urtiew, P. A. 1981. "Flame propagation in gaseous fuel mixtures in semiconfined geometries." *report no.* UCID-19000. Lawrence Livermore Laboratory.

Urtiew, P. A. 1982. Recent flame propagation experiments at LLNL within the liquefied gaseous fuels spill safety program. *Fuel-air explosions,* pp. 929–948. Waterloo, Canada: University of Waterloo Press.

Urtiew, P. A., and A. K. Oppenheim. 1966. "Experimental observations of the transition to detonation in an explosive gas." *Proc. Roy. Soc. London.* A295:13–28.

Van den Berg, A. C, C. J. M. van Wingerden, and H. G. The. 1991. "Vapor cloud explosion blast modeling." International Conference and Workshop on Modeling and Mitigation the Consequences of Accidental Releases of Hazardous Materials, May 21–24, 1991. New Orleans, USA. proceedings, pp. 543–562.

Van den Berg, A. C, C. J. M. van Wingerden, J. P. Zeeuwen, and H. J. Pasman. 1987. "Current research at TNO on vapor cloud explosion modeling.*"* *Int. Conf. on Vapor Cloud Modeling.* Cambridge, MA. Proceedings, pp. 687–711, AIChE, New York.

Van den Berg, A. C. 1980. "BLAST—a 1-D variable flame speed blast simulation code using a 'Flux-Corrected Transport' algorithm." Prins Maurits Laboratory *TNO report no. PML 1980–162*.

Van den Berg, A. C. 1984. "Blast effects from vapor cloud explosions." *9th Int. Symp. on the Prevention of Occupational Accidents and Diseases in the Chemical Industry*. Lucern, Switzerland.

Van den Berg, A. C. 1985. "The Multi-Energy method—A framework for vapor cloud explosion blast prediction." *J. of Haz. Mat.* 12:1–10.

Van den Berg, A. C. 1987. "On the possibility of vapor cloud detonation." TNO Prins Maurits Laboratory report no. 1987-IN-50.

Van den Berg, A. C. 1989. "REAGAS—a code for numerical simulation of 2-D reactive gas dynamics in gas explosions." TNO Prins Maurits Laboratory report no. PML1989-IN48.

Van den Berg, A. C. 1990. BLAST—A code for numerical simulation of multi-dimensional blast effects. TNO Prins Maurits Laboratory report.

Van den Berg, A.C., Mos, A.L., "Research to improve guidance on separation distance for the multi-energy method (RIGOS)," HSE Research Report 369, prepared by TNO Prins Maurits Laboratory, ISBN 0 7176 6146 6, 2005.

van den Berg, A.C., van der Voort, M.M., Weerheijm, J., Versloot, N.H.A., BLEVE Blast by Expansion-Controlled Evaporation, Process Safety Progress 25[1], 44-51, 2006

van den Berg, A.C., van der Voort, M.M., Weerheijm, J., Versloot, N.H.A., Expansion Controlled Evaporation: A Safe Approach to BLEVE Blast, Journal of Loss Prevention in the Process Industries 17, 397-405, 2004

van den Bosch, C.J.H., R.A.P.M. Weterings, *Methods for the Calculation of Physical Effects*, Committee for the Prevention of Disasters, CPR 14E (TNO 'Yellow Book'), The Hague, The Netherlands, 1997.

van den Bosch, C.J.H., Waterings, R.A.P.M., "Methods for the Calculation of Physical Effects – Due to Releases of Hazardous Materials (liquids and gases), 'Yellow Book'," The Committee for the Prevention of Disasters by Hazardous Materials, Director-General for Social Affairs and Employment, The Hague, 2005.

Van Laar, G. F. M. 1981. "Accident with a propane tank at Enschede on 26th March 1980, Prins Maurits Laboratorium." TNO Report no. PML 1981–145.

Van Wees, R. M. M. 1989. Explosion Hazards of Storage Vessels: Estimation of Explosion Effects. TNO-Prins Maurits Laboratory Report No. PML 1989-C61. Rijswijk, The Netherlands.

van Wingerden, C. J. M. 1984. "Experimental study of the influence of obstacles and partial confinement on flame propagation." Commission of the European Communities for Nuclear Science and Technology, report no. EUR 9541 EN/II.

van Wingerden, C. J. M. 1988a. "Experimental investigation into the strength of blast waves generated by vapour cloud explosions in congested areas." *6th Int. Symp. Loss Prevention and Safety Promotion in the Process Industries.* Oslo, Norway, proceedings. 26:1–16.

van Wingerden, C. J. M. 1989b. "On the scaling of vapor cloud explosion experiments." *Chem. Eng. Res. Des.* 67:334–347.

van Wingerden, C. J. M., A. C. Van den Berg, and G. Opschoor. 1989. "Vapor cloud explosion blast prediction." *Plant/Operations Progress.* 8(4):234–238.

van Wingerden, C. J. M., and A. C. Van den Berg. 1984. "On the adequacy of numerical codes for the simulation of vapor cloud explosions." Commission of the European Communities for Nuclear Science and Technology, report no. EUR 9541 EN/I.

van Wingerden, C. J. M., and J. P. Zeeuwen. 1983. Flame propagation in the presence of repeated obstacles: influence of gas reactivity and degree of confinement." *J. of Haz. Mat.* 8:139–156.

van Wingerden, C.J.M., Experimental investigation into the strength of blast waves generated by vapour cloud explosions in congested areas, 5th Int. Symp. "Loss Prevention and Safety Promotion in the Process Industries", 1988.

van Wingerden, K., et al., *"A new explosion simulator,"* Paper presented at the ERA-Conference "Offshore Structural Design Against Extreme Loads", London, UK, 1993.

Vasilev, A. A., and Yu Nikolaev. 1978. Closed theoretical model of a detonation cell. *Acta Astronautica* 5:983–996.

Velde, B., Linke, G, Genillon, P, "KAMELION FIREEX – A Simulator for Gas Dispersion and Fires," 1998 International Gas Research Conference, http://www.computit.no/filestore/_KFX_paper_gasdispersion_fires.pdf.

Venart, J. E. S. 1990. The Anatomy of a Boiling Liquid Expanding Vapor Explosion (BLEVE). *24th Annual Loss Prevention Symposium.* New Orleans, May 1990.

Venart, J.E.S., Boiling Liquid Expanding Vapour Explosions (BLEVE), Institute of Chemical Engineers Symposium Series, Hazards XV: The Process, its Safety and the Environment , 121-137, 2000, IChemE

Venart, J.E.S., Rutledge, K., Sumathipala, K., Sollows, K., To BLEVE or not to BLEVE: Anatomy of a Boiling Liquid Expanding Vapor Explosion, Process Safety Progress 12[2], 67-70, 1993

Viera, G.A., Wadia, P.H., Ethylene Oxide Explosion at Seadrift, Part I Background and Technical Findings, Loss Prevention Symposium, March 29 – April 1, 1993.

Visser, J.G., and P.C.J. de Bruijn, *Experimental parameter study into flame propagation in diverging and non-diverging flows*, TNO Report PML-1991-C93 (1991). Data reported in: J.B.M.M. Eggen, "GAME: development of guidance for the application of the multi-energy method," TNO Prins Maurits Laboratory, publ. by HSE books, Sudbury, England, 1991

Von Neumann, J., and R. D. Richtmyer. 1950. "A method for numerical calculations of hydrodynamical shocks." *J. of Appl. Phys.* 21:232–237.

Vörös, M., and G. Honti. 1974. Explosion of a liquid CO_2 storage vessel in a carbon dioxide plant. *First International Symposium on Loss Prevention and Safety Promotion in the Process Industries.*

Walker, S. "Interpretation of experimental results from Spadeadam explosion tests," HSE Offshore Technology Report, 2001/86, ISBN 0 7 7176 2341 6, Her Majesty's Stationary Office, Norwich, UK, 2002.

Walls, W. L. 1979. The BLEVE—Part 1. *Fire Command.* May 1979: 22–24. The BLEVE—Part 2. Fire Command. June 1979: 35–37.

White, C. S., R. K. Jones, and G. E. Damon. 1971. The biodynamics of air blast. Lovelace Foundation for Medical Education and Research. Albuquerque, NM.

Whitham, G.B., "Linear and Nonlinear Waves," New York, Inter-science Publ. John Wiley and Sons, 1974

Wiederman, A. H. 1986a. Air-blast and fragment environments produced by the bursting of vessels filled with very high pressure gases. In *Advances in Impact, Blast Ballistics, and Dynamic Analysis of Structures.* ASME PVP. 106. New York: ASME.

Wiederman, A. H. 1986b. Air-blast and fragment environments produced by the bursting of pressurized vessels filled with two phase fluids. In *Advances in Impact, Blast Ballistics, and Dynamic Analysis of Structures.* ASME PVP. 106. New York: ASME.

Wiekema, B. J. 1980. "Vapor cloud explosion model." *J. of Haz. Mat.* 3:221–232.

Wilkins, M. L. 1969. "Calculation of elastic-plastic flow." Lawrence Radiation Laboratory report no. UCRL-7322 Rev. I.

Williamson, B. R., and L. R. B. Mann. 1981. Thermal hazards from propane (LPG) fireballs. *Combust. Sci. Tech.* 25:141–145.

Wilson, D. J., A. G. Robins, and J. E. Fackrell. 1982b. Predicting the spatial distribution of concentration fluctuations from a ground level source. *Atmospheric Environ.* 16(3):479–504.

Wilson, D. J., J. E. Fackrell, and A. C. Robins. 1982a. Concentration fluctuations in an elevated plume: A diffusion–dissipation approximation. *Atmospheric Environ.* 16(ll):2581–2589.

wisha-training.lni.wa.gov/training/presentations/PSMoverview1.pps

Woodward, J. L., "Estimating The Flammable Mass of a Vapor Cloud," Concept-Series, AIChE, Center for Chemical Process Safety, NY, 1998.

Woolfolk, R. W., and C. M. Ablow. 1973. "Blast waves for non-ideal explosions." Conference on the Mechanism of Explosions and Blast Waves, Naval Weapons Station. York-town, VA.

Yellow Book. 1979. Committee for the Prevention of Disasters, 1979: Methods for the calculation of physical effects of the escape of dangerous materials, P.O. Box 69, 2270 MA Voorburg, The Netherlands.

Yellow Book. 1997. Committee for the Prevention of Disasters, 1997. Methods for the calculation of physical effects of the escape of dangerous materials, 3rd ed. P.O. Box 342, 7800 AH, Apeldoorn, The Netherlands.

Yu, C.M., Venart, J.E.S., The Boiling Liquid Collapsed Bubble Explosion (BLCBE): A Preliminary Model, Journal of Hazardous Materials 46, 197-213, 1996

Zabetakis, M. G. 1965. Flammability characteristics of combustible gases and vapors. *Bureau of Mines Bulletin 627.* Pittsburgh.

Zeeuwen, J. P., C. J. M. Van Wingerden, and R. M. Dauwe. 1983. "Experimental investigation into the blast effect produced by unconfined vapor cloud explosions." *4th Int. Symp. Loss Prevention and Safety Promotion in the Process Industries.* Harrogate. UK, IChemE Symp. Series 80:D20-D29.

Zeeuwen, J.P. and Wiekema, B.J., "The Measurement of Relative Reactivities of Combustible Gases," Conference on Mechanisms of Explosions in Dispersed Energetic Materials, 1978.

APPENDIX A. VIEW FACTORS FOR SELECTED CONFIGURATIONS

In this appendix, the view factors for three configurations are given:

1. radiation from a sphere,

2. radiation from a vertical cylinder,

3. radiation from a vertical plane surface.

For other configurations, refer to Love (1968) and Buschman and Pittmann (1961). A view factor depends on the shapes of the emitter and receiver. Consider the receiver to be a small, plane surface at ground level with a given orientation with respect to the emitter. The angle between the normal to the surface and the connection between the surface and the center of the emitter (Θ) must be known.

A-1. VIEW FACTOR OF A SPHERICAL EMITTER (E.G., FIREBALL)

If the distance from the receiver to the center of the sphere is L, and Θ is the angle between the connection of the surface to the center of the sphere and the tangent to the sphere, then, for $\Theta \leq \pi/2 - \Phi$, the view factor F is given by:

$$F = \frac{r^2}{L^2} \cos(\Theta)$$

(Eq. A.1)

where:

L	=	distance between receiving surface and sphere's center	(m)
r	=	radius of sphere	(m)
Θ	=	orientation angle	(rad)

In this case, the sphere is in full sight.

When, on extension, the receiving surface intersects with the sphere ($\Theta > \pi/2 - < \Phi$), the receiver cannot "see" the total emitter. The view factor F is then given as:

$$F = \frac{1}{2} - \frac{1}{2}\sin^{-1}\left[\frac{(L_r^2 - 1)^{1/2}}{L_r}\right] + \frac{1}{\pi L_r^2}\cos\Theta\cos^{-1}\left[-(L_r^2 - 1)^{1/2}\cot\Theta\right]$$

$$- \frac{1}{\pi L_r^2}(L_r^2 - 1)^{1/2}(1 - L_r^2\cos^2\Theta)^{1/2}$$

$$(Eq.\ A.2)$$

where:

r	=	fireball radius ($r = D/2$)	(m)
D	=	fireball diameter	(m)
L	=	distance to center of sphere	(m)
Θ	=	angle between the normal to surface and connection of point to center of sphere	(rad)
2Φ	=	view angle	(rad)
L_r	=	reduced length L/r	(–)

The view factor for incomplete visibility is given in Figure A-2.

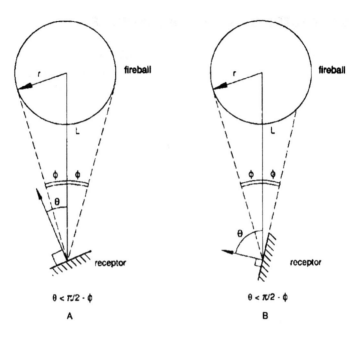

Figure A-1. View factor of a fireball.
(A) Receiver "sees" the sphere completely. (B) Receiver "sees" the sphere partially.

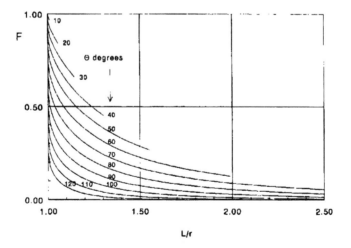

Figure A-2. View factors of a sphere as function of the dimensionless distance
(distance/radius) (incomplete view).

A-2. VIEW FACTOR OF A VERTICAL CYLINDER

A pool fire's flame can be represented (under no wind conditions) by a vertically placed cylinder with a height h and a ground surface radius r. The view factor of a plane surface at ground level whose normal lies in one vertical plane with the axis of the cylinder is given by the following equations:

$$h_r = h / r \tag{Eq. A.3}$$

$$X_r = X / r \tag{Eq. A.4}$$

$$A = (X_r + 1)^2 + h_r^2 \tag{Eq. A.5}$$

$$B = (X_r - 1)^2 + h_r^2 \tag{Eq. A.6}$$

For a horizontal surface ($\Theta = \pi/2$):

$$F_h = \frac{1}{\pi}\left[\tan^{-1}\left\{ \left(\frac{X_r -1}{X_r +1}\right)^{1/2} \right\} - \frac{X_r^2 -1+h_r^2}{\sqrt{AB}} \tan^{-1}\left\{ \left(\frac{(X_r -1)A}{(X_r +1)B}\right)^{1/2} \right\} \right]$$

$$\tag{Eq. A.7}$$

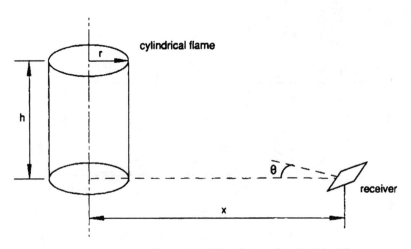

Figure A-3. View factor of a cylindrical flame.

And for a vertical surface ($\Theta = 0$):

$$F_v = \frac{1}{\pi X_r} \tan^{-1}\left\{\left(\frac{h_r^2}{X_r^2 - 1}\right)^{1/2}\right\} + \frac{h_r(A - 2X_r)}{\pi X_r \sqrt{AB}} \tan^{-1}\left\{\left(\frac{(X_r - 1)A}{(X_r + 1)B}\right)^{1/2}\right\}$$
$$- \frac{h_r}{\pi X_r} \tan^{-1}\left\{\left(\frac{X_r - 1}{X_r + 1}\right)^{1/2}\right\}$$

(Eq. A.8)

The maximum view factor is given by
$$F_{max} = (F_h^2 + F_v^2)^{0.5}$$
(Eq. A.9)

For the view factor of a tilted cylinder, refer to Raj (1977). The view factors generated by Eqs. (A.7) and (A.8) are given in Table A-1 and Figure A-4.

Table A-1. View Factors of a Vertical Cylindrical Emitter $h_r = 2L_f/d_f$; $X_r = 2X/d_f$

X_r	0.1	0.2	0.5	1.0	2.0	3.0	5.0	6.0	10.0	20.0
1. Horizontal target ($1000 \times F_h$)										
1.1	132	242	332	354	360	362	362	362	363	363
1.2	44	120	243	291	307	310	312	312	313	314
1.3	20	65	178	242	268	272	177	278	278	279
1.4	11	38	130	203	238	246	250	251	252	153
1.5	6	24	97	170	212	222	228	229	231	232
2.0	1	5	27	73	126	145	158	160	164	166
3.0			5	19	50	71	91	95	103	107
4.0			1	7	22	38	57	62	73	78
5.0				3	11	21	37	43	54	61
10.0					1	3	7	9	17	26
20.0							1	1	3	8
2. Vertical target ($1000 \times F_v$)										
1.1	330	415	449	453	454	454	454	454	454	455
1.2	196	308	397	413	416	416	416	416	416	417
1.3	130	227	344	376	383	384	384	384	384	385
1.4	94	173	296	342	354	356	356	357	357	357
1.5	71	135	253	312	329	332	333	333	333	333
2.0	28	56	126	194	236	245	248	249	249	250
3.0	9	19	47	86	132	150	161	163	165	167
4.0	5	10	24	47	80	100	115	119	123	125
5.0	3	6	15	29	53	69	86	91	97	100
10.0		1	3	6	13	19	29	32	42	48
20.0				1	3	4	7	9	14	21

(Continued next page)

Table A-1. View Factors of a Vertical Cylindrical Emitter, (*continued*)

X_r	0.1	0.2	0.5	1.0	2.0	3.0	5.0	6.0	10.0	20.0

h_r

3. Maximum view factor ($1000 \times F_{max}$)

X_r	0.1	0.2	0.5	1.0	2.0	3.0	5.0	6.0	10.0	20.0
1.1	356	481	559	575	580	581	581	581	581	581
1.2	201	331	466	505	517	519	520	521	521	521
1.3	132	236	287	448	468	472	474	474	475	475
1.4	94	177	323	398	427	433	436	436	437	437
1.5	72	138	271	355	392	400	404	404	405	406
2.0	28	56	129	208	267	285	294	296	299	300
3.0	9	19	48	88	141	166	185	189	195	197
4.0	5	10	24	47	83	106	129	134	143	147
5.0	3	6	15	29	54	73	94	100	111	117
10.0		1	3	6	13	19	30	34	45	55
20.0				1	3	4	7	9	14	22

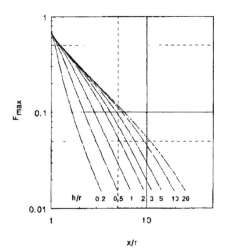

Figure A-4. Maximum view factors of a cylindrical flame as function of dimensionless distance to flame axes.

A-3. VIEW FACTOR OF A VERTICAL PLANE SURFACE

In the case of a vertical plane surface, it is assumed that the emitter and receiver are parallel to each other. The view factor is calculated from the sum of view factors from surface I and surface II (see Figure A-5). Surfaces I and II are defined as those to the left and the right of a plane through the center of the receiver and perpendicular to the intersections of the receiver with the ground.

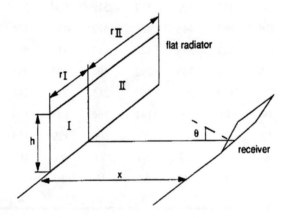

Figure A-5. View factor of a vertical plane surface.

For each of the two surfaces:

$$h_r = h/b \qquad\qquad\text{(Eq. A.10)}$$

$$X_r = X/b \qquad\qquad\text{(Eq. A.11)}$$

$$A = 1 / (h_r^2 + X_r^2)^{0.5} \qquad\qquad\text{(Eq. A.12)}$$

$$B = h_r / (1 + X_r^2)^{0.5} \qquad\qquad\text{(Eq. A.13)}$$

For a horizontal target on ground level ($\Theta = \pi/2$), the view factor is given by:

$$F_h = \frac{1}{2\pi}\left[\tan^{-1}\left(\frac{1}{X_r}\right) - AX_r \tan^{-1}(A)\right]$$

(Eq. A.14)

and, for a vertical surface ($\Theta = 0$):

$$F_v = \frac{1}{2\pi}[h_r A \tan^{-1}(A) + (B/h_r)\tan^{-1}(B)]$$

(Eq. A.15)

The maximum view factor is given by:

$$F_{max} = (F_h^2 + F_v^2)^{0.5}$$

(Eq. A.16)

It must be noted that, unless $b_I = b_{II}$, Fmax is not the maximum view factor for any distance C from the emitter.

The view factors F_h, F_v, and F_{max} can easily be found by summing the view factors calculated from the surfaces I and II. Values of the view factor F_{max} as function of X_r are given in Table A-2 and Figure A-6.

Table A-2. View Factors of a Vertical Plane Surface Emitter h_r = h/b; X_r = X/b
(See Figure A-5)

| | | | | | h_r | | | | |
X_r	0.1	0.2	0.3	0.5	1.0	1.5	2.0	3.0	5.0
				1. Horizontal target ($1000 \times F_h$)					
0.1	146	276	341	400	443	456	461	465	467
0.2	53	146	221	310	389	413	423	430	435
0.3	25	83	144	236	337	371	386	397	403
0.5	9	34	68	137	249	296	318	336	346
1.0	2	8	17	42	111	161	190	219	238
1.5	1	3	6	17	53	88	114	146	170
2.0		1	3	8	28	51	71	100	126
3.0			1	3	10	20	31	50	75
5.0				1	2	5	9	16	31
				2. Vertical target ($1000 \times F_v$)					
0.1	353	447	474	489	496	497	497	497	498
0.2	223	352	414	461	484	488	489	490	490
0.3	156	274	349	421	466	474	All	478	479
0.5	94	178	245	335	416	435	442	445	447
1.0	41	80	117	180	277	318	335	347	352
1.5	22	44	65	105	179	222	245	264	274
2.0	14	17	41	66	120	157	180	204	218
3.0	7	13	20	32	62	86	105	129	148
5.0	2	5	7	12	24	35	45	61	80

(Continued on next page)

Table A-2. View Factors of a Vertical Plane Surface Emitter, (*continued*)

				h_r					
X_r	0.1	X_r	0.1	X_r	0.1	X_r	0.1	X_r	0.1

3. Maximum view factor ($1000 \times F_{max}$)

0.1	382	525	584	632	665	674	678	681	682
0.2	229	381	469	555	621	639	647	652	655
0.3	158	286	377	483	575	602	613	622	626
0.5	94	181	255	362	484	526	544	558	565
1.0	41	80	188	185	299	356	385	410	425
1.5	22	44	66	106	187	239	270	302	322
2.0	14	27	41	67	123	165	194	227	252
3.0	7	13	20	33	62	88	109	138	165
5.0	2	5	7	12	24	36	46	63	86

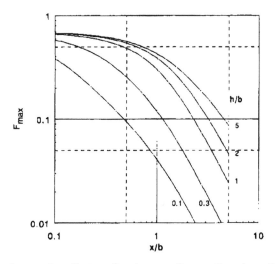

Figure A-6. Maximum view factor of a plane surface as function of dimensionless distance to emitter

APPENDIX B. TABULATION OF SOME GAS PROPERTIES IN METRIC UNITS

Table B-1. Gas Properties in Metric Units

Gas or vapor	Chemical Formula	Molecular Mass	Specific Heat Ratio	Critical Conditions	
				Abs. Press. (bar)	Abs.Temp. (K)
Acetylene	C_2H_2	26.05	1.24	62.4	309.4
Air	N_2+O_2	28.97	1.40	37.7	132.8
Ammonia	NH_3	17.03	1.31	112.8	406.1
Argon	A	39.94	1.66	48.6	151.1
Benzene	C_6H_6	78.11	1.12	49.2	562.8
n-Butane	C_4H_{10}	58.12	1.09	38.0	425.6
iso-butylene	C_4H_8	56.10	1.10	40.0	418.3
Carbon dioxide	CO_2	44.01	1.30	74.0	304.4
Carbon monoxide	CO	28.01	1.40	35.2	134.4
Chlorine	Cl_2	70.91	1.36	77.2	417.2
Ethane	C_2H_6	30.07	1.19	48.8	305.6
Ethyl chloride	C_2H_5Cl	64.52	1.19	52.7	460.6
Ethylene	C_2H_4	28.05	1.24	51.2	283.3
Helium	He	4.00	1.66	2.3	5.0
n-Heptane	C_7H_{16}	100.20	1.05	27.4	540.6
n-Hexane	C_6H_{14}	86.17	1.06	30.3	508.3
Hydrogen	H_2	2.02	1.41	13.0	33.3

Gas or vapor	Chemical Formula	Molecular Mass	Specific Heat Ratio	Critical Conditions	
				Abs. Press. (bar)	Abs.Temp. (K)
Hydrogen sulfide	H_2S	34.08	1.32	90.0	373.9
Methane	CH_4	16.04	1.31	46.4	191.1
Natural gas	—	18.82	1.27	46.5	210.6
Nitrogen	N_2	28.02	1.40	33.9	126.7
Pentylene	C_5H_{10}	70.13	1.08	40.4	474.4
Oxygen	O_2	32.00	1.40	50.3	154.4
Propane	C_3H_8	44.09	1.13	42.5	370.0
Water vapor	H_2O	18.02	1.33	221.2	647.8

APPENDIX C. CONVERSION FACTORS TO SI FOR SELECTED QUANTITIES

An asterisk before a number indicates that the conversion factor is exact, and all subsequent digits are zero.

Table C-1. Conversion Factors to SI Units

To Convert From	To	Multiply By
British thermal unit (Btu, International Table)	joule (J)	1.0550559×10^3
Btu/lb-deg F (heat capacity)	joule/kilogram-kelvin (J/kg-K)	4.1868000×10^3
Btu/hour	watt (W)	$2.93077107 \times 10^{-1}$
Btu/second	watt (W)	1.0550559×10^3
Btu/ft^2-hr-deg F (heat transfer coefficient)	joule/meter^2-second-kelvin (J/m^2-s-K)	5.6782633
Btu/ft^2-hour (heat flux)	joule/meter^2-second (J/m^2-s)	3.1545907×10^{-3}
Btu/ft-hr-deg F (thermal conductivity)	joule/meter-second-kelvin (J/m-s-K)	1.7307347
degree Fahrenheit (°F)	kelvin (K)	$t_k = (t_f + 459.67) / 1.8$
degree Rankine (°R)	kelvin (K)	$t_k = t_r / 1.8$
fluid ounce (U.S.)	meter^3 (m^3)	$*2.9573530 \times 10^{-1}$
foot	meter (m)	$*3.0480000 \times 10^{-1}$
foot (U.S. Survey)	meter (m)	3.0480061×10^{-1}
foot of water (39.2°F)	pascal (Pa)	2.98898×10^3
foot^2	meter^2 (m^2)	$*9.2903040 \times 10^{-2}$
foot/second^2	meter/second^2 (m/s^2)	$*3.0480000 \times 10^{-1}$
foot^2/hour	meter^2/second (m^2/s)	$*2.5806400 \times 10^{-5}$

To Convert From	To	Multiply By
foot-pound-force	joule (J)	1.3558179
foot^2/second	meter^2/second (m^2/s)	*9.2903040 × 10^{-2}
foot^3	meter^3 (m^3)	2.8316847 × 10^{-2}
gallon (U.S. liquid)	meter^3 (m^3)	3.7854118 × 10^{-3}
gram	kilogram (kg)	*1.0000000 × 10^{-3}
inch	meter (m)	*2.5400000 × 10^{-2}
inch of mercury (60°F)	pascal (Pa)	3.37685 × 10^3
inch of water (60°F)	pascal (Pa)	2.48843 × 10^2
inch^2	meter^2 (m^2)	*6.4516000 × 10^{-4}
inch^3	meter^3 (m^3)	*1.6387064 × 10^{-5}
kilocalorie	joule (J)	*4.1868000 × 10^3
kilogram-force (kgf)	newton (N)	*9.8066500
mile (U.S. Statute)	meter (m)	*1.6093440 × 10^3
mile/hour	meter/second (m/s)	*4.4704000 × 10^{-1}
millimeter of mercury (0°C)	pascal (Pa)	1.3332237 × 10^2
pound-force (lbf)	newton (N)	4.4482216
pound-force-second/ft^2	pascal-second (Pa-s)	4.7880258 × 10^1
pound-mass (lbm avoirdupois)	kilogram (kg)	*4.5359237 × 10^{-1}
pound-mass/foot3	kilogram/meter^3 (kg/m^3)	1.66018463 × 10^1
pound-mass/foot-second	pascal-second (Pa-s)	1.4881639
psi	pascal (Pa)	6.8947573 × 10^3
ton (long, 2240 lbm)	kilogram (kg)	1.0160469 × 10^3
ton (short, 2000 lbm)	kilogram (kg)	*9.718474 × 10^2
torr (mm Hg, 0°C)	pascal (Pa)	1.3332237 × 10^2
watt-hour	joule (J)	*3.6000000 × 10^3
yard	meter (m)	*9.1440000 × 10^{-1}

INDEX

A

accidental PVBs,
  groups of like, 294
  scenarios of, 242-244
adiabatic shock compression, VCE
  detonation and, 100
alkanes, combustion products and expansion
  ratios of, 113
alumina process, case history of PVB of, 32-
  35
American Institute of Chemical Engineers
  (AICHE), goals of, 1
American Society of Mechanical Engineers
  (ASME), PVBs and, 242
area blockage ratio (ABR), 103
  defined, 133
  flame speed and large, 137-138
  overpressures and large, 138-139
  VBR and pitch for the same, 135
atmospheric adsorption, radiation wavelength
  and, 59-60
autoignition temperature (AIT), 53-54
  shock wave and, 66
AutoReaGas as a code of CFD VCE model,
  210

B

Baker-Strehlow blast curves, PVB blast
  effects and, 261
  PES treatment, and 197
Baker-Strehlow-Tang (BST) method,
  actual side-on pressure and impulse
    computation in, 201
  analysis procedure, 199-201
  components of BST blast curves, 189
  confinement, congestion and fuel
    reactivity in, 194
  congestion assessment in, 194-196
  empirical results and, 189
  energy computation in, 193-4
  estimation of VCE energy in, 200
  flame speed correlations in, 192-193
  flame speed for, 200
    determination of maximum, 192
  fuel reactivity assessment in, 196-197
  historical account of, 188
  internal VCEs and, 194
  negative overpressure vs. distance plots in,
    191
  positive impulse vs. distance plots in, 190
  propane spill sample problem and, 233-
    235
  scaled standoff distance computation in,
    201
Baker-Tang curves, PVB blast effects and,
  261-266
'bang box' enclosures and chemical
  processing plants, 100
black body radiation, emissive power and,
  58-59
blast curve methods, 97
  Confinement Assessment Method (CAM)
    for, 175
  explosion energy and, 175-176
  family of curves vs. single, 174
  free-air vs. surface burst TNT in, 169
  initial blast strength and, 175
  one-dimensional gas charge vs. TNT
    charge in, 174-175
  PVB blast effects prediction and, 260,
    261-276
blast curves,
  negative impulse for Baker-Tang PVB,
    266
  negative overpressure for Baker-Tang
    PVB, 264
  positive impulse for Baker-Tang PVB,
    265
  positive overpressure for Baker-Tang
    PVB, 262-263
blast curves and loads, TNT equivalence and,
  166
blast effects, 97, 99
  cylindrical PVB and, 256-257
  PVBs and, 244
  VCEs, PVBs, BLEVEs and, 51

blast effects of PVB,
   blast curve methods in prediction of, 260,
      261-276
   idealized configuration in study of, 248
      computation of data for spherical
         vessel airblast, 274-277
      cylindrical PVB-specific adjustments
         for, 270-271
      elevated spherical PVB-specific
         adjustments for, 270-271
      explosion energy evaluation for, 251-
         253
      initial shock strength determination for,
         253-254
      literature experiments relating to, 248-
         249
      prediction methods for, 267-270
      review of data of numerical analysis of,
         249-250
      vessel geometry adjustments and, 270-
         274
      vessel temperature adjustments and,
         270
   non-spherical bursts and, 255-260
   plots of overpressure and impulse vs.
      energy-scaled distance, 249
   prediction methods for, 260-276
   surface bursts and, 254-255
   vessel burst in prediction of, 267
blast effects of VCE,
   1-D configuration and, 113
   2-D configurations and, 116-124
   3-D configurations and, 124-133
      with high congestion, 130-131
      with low congestion, 127-130
      with variable congestion, 131-133
      without obstacles, 124
   modeling methodologies in low
      overpressure range and, 152
   volume of vapor cloud and, 176
blast field, cylindrical PVB and pressure
   contours of, 256
blast loading,
   CFD models in estimation of, 73
   defined, 71-72
   drag force and, 73
blast overpressures,
   degree of confinement and, 118-119

vessel temperature effects in PVB and,
   267, 270
blast parameters, TNO multi-energy method
   and calculation of, 185
blast phase impulse, defined, 71
blast strength, blast curve methods and
   initial, 175
blast wave energy,
   heat of combustion and, 70
   initial vessel pressure effects in PVB and,
      267
blast wave reflection, ground surface
   reflection and, 255
blast waves, 2
   explosion process and shape of, 70, 71
   PVB and, 5, 244-245
   PVB and negative vs. positive phase of,
      245
   PVBs and strength and shape of, 248
   rarefaction waves and, 72
   side-on overpressure and, 72
blast-wave scaling, 74-75
blockage ratios,
   defined, 133
boiler, case history of PVB of, 36-40
boiling, BLEVE and thermodynamics of, 312
boiling-liquid-expanding-vapor explosions
   (BLEVEs), 1, 2
   airblast as a consequence of,
      blast effects in, 325
      blast strength and duration of, 320-323
      flashing liquid and vapor expansion in,
         320
      liquid level at failure and, 325
      MAWP as a burst pressure in, 324
      pressure relief device (PRD) set
         pressure and, 324
      ultimate tensile strength and, 325
   bubble nuclei and liquids subject to, 313
   definition, elements and example of, 5-6,
      311-312
   ductile failure vs. vessel failure in, 314
   fire and cause of, 316
   fireballs and, 6
   fire heat,
      and crack propagation in, 318-319
      and liquid wetted surfaces in, 316

flammability of liquid and, 319
fragment and debris throw as a
    consequence of,
    review of experimental work in, 342-
        343
    theoretical methods to predict, 342-343
liquid-filled vessel's response to fire heat
    in, 317
liquid flashing in, 319
liquid's response to fire heat in, 318
materials subject to, 311 method of
    predicting blast from flashing
    liquids in, 329-336
narrative of a typical, 317-320
prediction and analysis models for, 348-
    349
ranges of rocketing fragments in, 344-348
review of past experimental work on, 326-
    327
secondary events of a typical, 319-320
stress rupture of steel vessels and, 317
theoretical studies in mechanisms of, 327-
    329
thermal hazards as a consequence of,
    computation to estimate, 341-342
    fuel, fireball and, 336-339
    radiation, 339-341
thermodynamics of boiling and, 312
time to fire induced rupture estimates and,
    316
bubble nuclei,
    liquids subject to BLEVE and formation
        of, 313
    needed superheat and size of, 312-313
burning velocity,
    defined, 52
    flame speed and, 52
    laminar burning velocity and, 52
    Le Chatelier's principle and computation
        of, 196-197
    oxygen enrichment and, 52
BWTI as a code of CFD VCE model, 211

C

case histories,
    BLEVEs,

liquid CO_2 storage vessel, Proctor and
    Gamble, Worms, Germany, 40-
    41
LPG rail car derailment, Crescent City,
    Illinois, USA, 45-48
LPG railroad tank car, Kingman,
    Arizona, USA, 48-49
LPG storage facility, San Juan
    Ixhuatepec, Mexico City,
    Mexico, 41-43
propylene tank truck failure, San
    Carlos de la Rapita, Spain, 44-45
flash fires,
    olefin unit, Quantum Chemicals,
        Morris, Illinois, USA, 11-13
    propane fire, Donnellson, Iowa, USA,
        7-8
    propane fire, Lynchberg, Virginia,
        USA, 8-10
information availability and
    documentation, 7
pressure vessel burst,
    alumina process, Kaiser Aluminum,
        Gramercy, Louisiana, USA, 32-
        35
    boiler, Dana Corporation, Paris,
        Tennessee, USA, 36-40
    ethylene oxide distillation column,
        Union Carbide, Seadrift, Texas,
        USA, 35-36
vapor cloud explosion,
    chemical plant, Fixborough, UK, 13-18
    discharge from atmospheric vent, BP,
        Texas City, Texas, USA, 29-31
    hydrogen-air explosion, Jackass Flats,
        Nevada, USA, 21-22
    pipeline rupture, Ufa, West-Siberia,
        USSR, 23-25
    propane pipeline failure, Port Hudson,
        Missouri, USA, 19-21
VCE and BLEVEs,
    propylene HDPE unit, Phillips,
        Pasadena, Texas, USA, 26-29
CEBAM as a code of CFD VCE model, 211
Center for Chemical Process Safety (CCPS),
    publications on process safety by, 1
chemical energy, deflagration and detonation
    and, 100

chemical plant, case history of VCE of, 13-18

chemical processing plants and 'bang box' enclosures,100

chemical reaction, detonation wave and rate of, 153

China Lake cryogenic liquid test, flash fires and, 81

CJ-model, detonation and, 67

cloud composition, 97

cloud dispersion, Maplin Sands test and models of, 82

codes of CFD VCE model, 211

combustion, turbulence and, 65

combustion behavior,
continuous *vs.* instantaneous spills and, 82
LNG *vs.* LNP and, 82

combustion energy, actual explosive combustion and, 171

combustion mode, deflagration to detonation and, 66

combustion process, enclosure effects and, 99

combustion products, expansion ratios for alkanes and, 113

combustion rate,
high-energy ignition of unobstructed cloud and, 127
TNT equivalence and, 167,

combustion wave, VCE detonation and, 100

compressed gas, source energy computation of, 275

Computational Fluid Dynamics (CFD) model,
blast loading estimation and, 73
BST, TNO multi-energy and, 208-209
BWTI and CEBAM as codes of, 211
CFD codes and, 208
DDT modeling and, 208
description of, 208
expertise and accuracy of, 6
explosion process and, 208
EXSIM, FLACS and AutoReaGas as codes of, 210-211
flash fires and, 80
geometry representation and, 209-210
overview of, 207-208

physical sub-models of CFD codes in, 212

porosities and distributed resistances (PDR) concept and, 210

simulation of gas explosion using, 215-217

table of CFD codes, 212

three-dimensional compressible flows and, 209

validation of, 212-215

VCE prediction and, 167

confinement, 97
blast overpressure and degree of, 118-119

confinement and congestion, 97
BST method and, 194
flame acceleration and, 107
parameters of, 103
PES and zones of, 197-198
VCE and, 103

congestion,
BST method and assessment of, 194-196
1-D geometries and, 115
diameter of obstacles, ABR and pitch as components of, 194
Potential Explosion Sites (PES) as an area of, 197

Congestion Assessment Method (CAM),
adjustments in, 207
blast curve methods and, 175
congestion characteristics and, 203
data inputs for, 202
effective cloud radius determination in, 205
overview, 201
pressure decay plot for, 206
propane VCE sample problem and, 236-239
pulse duration and shape estimate in, 206-7
"Severity Index" and congestion parameters in, 203
source of overpressure determination in, 204
source overpressure and, 202

contact discontinuity. *See* contact surface

contact surface, PVB and, 244

containment,
BLEVE and failure mechanisms of, 312

BLEVE and sudden, significant loss of, 311-312

co-volume parameter, non-ideal gases and, 278

critical heat flux, liquid wetted wall and liquid's boiling, 316

critical initiation energy,
defined, 152
direct initiation of vapor cloud detonation and, 152
hydrocarbons and their, 155
some fuel-air mixtures and their, 152-153

critical tube diameter,
defined, 154
turbulent gas jet and, 156

cryogenic liquid tests, flash fires and, 81

cyclohexane-air,
flame speed variation and obstacles in explosion of, 150-151
flame speed *vs.* distance to detonation in, 132-133

cylindrical PVB,
pressure contours of blast field and, 256-257
scaled peak overpressure and positive impulse and, 258-259

D

damaging explosion, cloud ignition of high reactivity fuels and, 129

deflagration,
basic concept of, 64-66
flame arresters and, 157
ignition energy for hydrocarbon-air, 54-55
propagation mechanism of, 66

deflagration explosions, turbulent flame speeds and, 52

deflagrations and detonations,
as basic modes of combustion, 66
chemical energy and, 100
differences between, 100-101
durations of, 100
flame propagation and, 106-107
initiation energy requirements and, 101
peak overpressure and, 100

deflagration-to-detonation transition (DDT).
See also deflagration and detonation
in 1-D configurations, 156-157
effect of obstacles in flame acceleration, 157
flame propagation, 156
in 2-D configurations, 157
in 3-D configurations,
actual process plants scale and, 158
MERGE test rig and, 157-158
3-D configuration with high congestion and, 130-131
ethylene-air mixture and, 158
flame arresters and, 157
vapor cloud detonation through, 152

detonability (detonation) limits,
flammability limits and, 154
hydrocarbons and their, 155
propagation of detonation and, 156

detonability of fuels,
detonation of non-homogenous mixtures, 155-156
dynamic donation parameters, 153-155

detonation,
contiguous fuel-air mixture and, 156
continuous decay and reinitiation and, 68
detonation wave speed and overpressure in, 67
flame arresters and, 157
fuel-oxidizer mixture and cellular structure of, 68-69
gaseous fuels and direct initiation energy of, 154
ignition energy for hydrocarbon-air, 54-55
jet ignition of fuel-air mixtures and transition to, 147-148
propagation mechanism of, 66
turbulent jet and initiation conditions for, 147
vapor cloud and, 156

detonation cell size,
defined, 154
fuel, oxidizer and, 68-69, 154
of fuel-air mixtures, 70

detonation explosions, turbulent flame speeds and, 52

detonation hazard index,
defined, 154

hydrocarbons and their, 155
relative detonation sensitivity of fuels and, 154
detonation mode, VCE detonation and, 100
detonation sensitivity,
    critical initiation energy and, 152
    detonation hazard index and relative, 154
detonation wave,
    detonation parameters and hydrodynamic structure of, 153
    rate of chemical reaction and, 153
    structure of, 154
    sub-explosions and, 66
dispersion, 2
    explosive ignitable gas, 98
    Fuel Air Explosions (FAE) and, 97
dispersion models, risk analysis of flash fires and, 80
distillation column, case history of PVB of ethylene oxide, 35-36
drag and lift forces, range of free flying fragments and, 288-289
drag coefficients, 292
drag force, blast loading and, 73

E

efficiency factor. *See* trinitrotoluene (TNT) equivalence
emissive power,
    black body radiation and, 58-59
    flash fires and wide-guage radiometer measurements of surface, 90
    pool fire size and, 89
    thermal radiation and, 58
emissivity, definition and measurement of, 59
enclosure effects and combustion process, 99
equivalence ratio, defined, 157
equivalency factor. *See* trinitrotoluene (TNT) equivalence
equivalent radius, TNO multi-energy method and, 184
ethylene-air mixtures, DDT in, 158
explosion energy,
    blast curve method and, 175-176

PVBs and reduction of, 245
TNT equivalence and, 166
TNT equivalence as a conversion factor for, 169-172
vessel volume effects in PVB and, 267
explosion processes,
    blast wave and their shapes and, 71
    CFD models and, 208
explosions,
    clouds of high reactivity fuels and, 129
    flash fires and, 77
    properties of fuel-air, 53
explosion severity, TNT equivalence method and cloud's, 218
explosion source behavior, vessel pressure effects in PVB *vs.* ideal, 267
explosion source environs, secondary fragments and, 277
explosive charge,
    defined, 98
    distribution of blast parameters and weight of, 75
explosive combustion, combustion behavior and actual, 171
EXSIM as a code of CFD VCE model, 210

F

fabrication flaws, PVBs and, 243-244
failure patterns, fragments from PVB and, 277
finite-difference methods,
    Baker-Strehlow blast curves and, 261
    numerical studies of blast effects of PVB and, 249-250
fire, cause of BLEVE and, 316
fireballs,
    BLEVEs and, 6
    empirical relationship for duration and diameters of, 338
    flash fires, thermal radiation and, 51
    flash fires and, 4, 77
fire heat,
    crack propagation in a BLEVE and, 318-319
    role of liquid wetted surfaces in BLEVE and, 316

typical BLEVE and liquid-filled vessel's response to, 317

fire induced rupture, BLEVE and time estimate of, 316

fire radiation, flash fire properties and effects of, 79-80

fires and explosions, 2
  occurrences and consequences of industrial, 3
  research and insights in consequences of, 4

FLACS as a code of CFD VCE model, 210

flame, turbulence and path of, 103

flame acceleration, 97
  blockage ratios in 1-D geometries and, 115
  congestion and, 109
  in 3-D configurations without obstacles, 125
  1-D geometry and, 107, 114
  2-D geometry and, 107
  3-D geometry and, 107, 124
  3-D geometry with low congestion and, 129-130
  flame instabilities and, 107
  flame velocities of fuel-air mixtures in 2-D geometries and, 117
  fuel reactivity and, 107, 111-113, 196
  geometries and, 107-109
  Markstein-Taylor instability and, 107
  multiple obstacles and, 111, 116
  obstacle geometry *vs.* induced turbulence and, 141-142
  obstacles and images of, 110
  pitch in 1-D geometries and, 115
  positive feedback mechanism, 115
  reactivity of hydrocarbon air mixtures and, 111-113
  venting and, 115

flame arresters,
  DDT and, 157
  deflagration and, 157
  detonation and, 157

flame fronts, flash fires and, 78

flame height,
  cloud depth and, 89
  fuel-air mixtures and, 83

flame instabilities and flame acceleration, 107

flame propagation,
  apparatus to study effects of pipe racks on, 128
  comparison in TNO sets up, 149-150
  congestion in cylindrical geometry and, 121
  deflagration and detonation and, 106-107
  effect of jet ignition of fuel-air mixtures on, 147
  pitch and flame speed at the same distance of, 134

flame shape, radiation and simplifying assumptions of, 91

flame speed. *See also* flame acceleration
  arrangement of obstacles and, 140
  blockage and pitch in 2-D geometry and, 122
  burning velocity and, 52
  combustion of fuel-air cloud in congested obstacle and, 127
  confinement and effect on terminal, 120
  3-D configurations with high obstacle density and, 130-131
  degree of obstruction and, 121
  versus dimensionless distance R/P, 136-137
  versus distance for different pitches, 136
  effect of gap between congested areas on, 149
  effect of obstacle pitch on, 135
  fuel reactivity and, 111-113
  fuels in obstacles-free 3-D configurations and, 126
  gas flow velocity and, 104
  jet flame and obstacle interaction and, 127
  laminar burning velocity and, 104
  laminar flame speed in obstacles-free 3-D configurations and, 124
  large ABR and, 137-138
  of methane-air mixture, 104-105
  obstacles and acceleration of high, 132
  obstacles in cyclohexane-air explosion and variation in, 150-151
  variability with concentration and, 78
  vessel bursts induced gas cloud explosions and, 145

wind speeds and, 84-85

flame surface area, 3-D geometry and, 124

flame velocity. *See* flame speed

flammability,
   detonation limits of fuel-air mixtures and their, 154
   secondary events of a BLEVE and liquid's, 319

flammability (flammable) limits,
   aerosols and, 52
   flash fires and, 77
   oxygen enrichment and, 52
   vapor concentration and, 51

flammable aerosol, 52

flammable cloud and confinement dimensions, 103-104

flammable materials, 97
   accidental releases and hazards of, 3
   summaries of accidents involving, 3

flammable vapor cloud, 98

flash-evaporating liquid fuels, 98

flash-fire radiation model, 86

flash fires (FF), 2, 98
   CFD models and, 80
   definition and example of, 4
   dispersion models for risk analysis of, 80
   evaluation of explosions and hazards of, 1
   fire radiation effects and properties of, 79-80
   flammability limits and, 77
   Frenchman Flats cryogenic liquid tests and, 81
   HSE LPG tests and, 84
   Maplin Sands tests and, 82
   measurement of emissive power and, 90
   Musselbanks propane tests and, 83
   point of ignition and, 78
   pool-fire model and unconfined, 86-91
   radiation heat flux and, 91
   turbulent flame speed and, 52
   types and common denominators of, 77

flash point,
   flammability (flammable) limits and, 52
   flash fraction, in-cloud amount of fuels and, 173
   of fuel-air mixtures, 53

flow field divergence:

3-D geometry and degree of, 124
   turbulent flow field and, 113

fluid dynamics forces, range of free flying fragments and, 288-291

fluid dynamic studies, 97

fragmentation effects,
   PVBs and prediction of, 298
   VCE and, 99

fragment characteristics, analytical analysis for determination of, 298

fragment kinetic energy, initial fragment energy in PVBs and, 279

fragment mass,
   groups of like accidental PVBs and distribution of, 296-297

fragment ranges,
   analysis of groups of like accidental PVBs and, 294-296
   computer analysis of parameters of, 305-306
   theoretical methods and maximum, 298

fragments,
   co-volume parameter and velocity of, 278
   failure patterns and PVB, 277
   ideal-gas-filled PVBs and initial velocity of, 279-287
   primary and secondary PVB, 277
   PVB and throw distances of, 277
   PVBs and determination of number of, 300
   saturated-liquid *vs.* gas-filled PVB and hazards from, 278
   statistical analysis of accidental PVBs and, 293

free flying fragments,
   calculating range of, 304
   forces acting on and range of, 288

Frenchmen Flats cryogenic liquid test, FF and, 81

frisbee fragments, PVB and, 278

fuel-air explosion, 99

Fuel Air Explosions (FAE), 97
   dispersion, 98

fuel-air mixtures. *See also* hydrocarbon-air; cyclohexane-air; hydrogen-air; ethylene-air; methane-air
   critical initiation energy of some, 152-153

detonability limits and detonation of, 154
detonation and contiguous, 156
detonation cell size of, 70
detonation properties of, 160
equation for expansion ratio of, 105-106
explosion properties of, 53
FF and combustion of, 77
flame height and, 83
flame propagation and jet ignition of, 147
flash points of, 53
levels of congestion and occurrence of DDT in, 158
TNO multi-energy method and reactivity, cloud confinement of, 176
fuel equivalence ratio and laminar burning velocity, 104
fuel jets, inflow overpressure and inflow, 146
fuel-oxidizer mixture, cellular structure of detonation and, 68-69
fuel reactivity, 97
  BST method and, 194, 196-197
  components of, 196
  effect of obstacles on flame acceleration and, 129
  flame speeds and, 111-113
  fuels with low and high, 197
  laminar burning velocity, VCE and, 104
  laminar burning velocity and, 52
  propensity of flame acceleration and, 196
fuels,
  accidental release of, 98
  CAM fuel factor and expansion ratios for, 202
  equivalent-charge mass of TNT and in-cloud amount of, 172-173
  flash fraction and in-cloud amount of, 173
  initiation energy of detonation and gaseous, 154
  MIE of, 54-55
  oxidizer, detonation cell size and, 68-69, 154

G

gap effect in TNO sets up, 149
gas dynamic equations,
  in prediction of VCE,
    acoustic method simplification of, 167
    self-similar simplification of, 167
gas-dynamic-state variables, shock and, 71
gas explosion,
  CFD simulation example of, 215-217
  positive feedback and, 66
gas flow velocity,
  flame speed and, 104
  of methane-air mixture, 104-105
gas mixtures, blast measurements from detonating, 161
ground reflection,
  above ground explosion and, 73-74
  "hemispherical surface burst" blast curves, and 74
ground surface reflection, blast wave reflection and, 255
Guidelines for VCE, PVB, BLEVE and FF, goals of this book on, 2

H

hazards,
  case histories of, 2
  consequences of fire and explosion, 3
  explosions and FF, 1
  flammable materials' thermal and overpressure, 3
  improper prediction of building loads and its, 3
  managers overview of VCE, PVB, and BLEVE, 2
  mitigation measures and industrial, 4
  pressure vessel failure, 3
  of thermal radiation, 55
Health and Safety Executive (HSE) LPG tests, FF and, 84
heat, molecular-diffusive transport of, 66
heat of combustion,
  blast-wave energy and, 70
  thermal radiation and, 56
"hemispherical surface burst" blast curves, explosion energy and, 74
heterogeneous nucleation, boiling point of liquids and, 312
homogeneous nucleation, 313
Hopkinson scaling law, 75

hydrocarbon-air, deflagration and detonation ignition energies of, 54-55

hydrocarbons, critical initiation energies of, 155

hydrogen-air, case history of VCE of, 21-22

I

ignitable cloud, LFL and, 80

ignitable gas, dispersion of, 98

ignitable liquid, 98
   BLEVE, fireball and pressurized, 77

ignition, 2
   blasts from multiple PES and sources of, 198-199
   FF and point of, 78
   "Severity Index" and effect of, 144, 145
   sources of, 55

ignition energy, oxygen enrichment and, 52

ignition kernel, 54

impulse computation, actual side-on pressure and BST, 201

initial fragment velocity,
   comparison of methods for prediction of, 286-287
   empirical relations and calculation of, 283-285
   history of theoretical predictions and, 280
   ideal-gas-filled PVBs and, 279-287
   PVB's content and method to calculate, 301-304
   scaled pressure in PVBs and scaled, 281-283

initial trajectory angle,
   range of free flying fragments and, 304-305
   ranges for rocketing fragments and, 293

internal vapor explosion, 99

J

jet fire, FF and, 77

jet flame, flame speed, obstacle interaction and, 127

jetting release, turbulence in, 98

L

laminar burning velocity, 97
   flame speed and, 104
   of fuel-air mixtures, 53
   fuel equivalence ratio, 104
   of methane-air mixture, 104-105
   obstacles and ratio of turbulent to, 141
   reactivity of fuels and, 52
   source of values for hydrocarbons', 106
   turbulent flame speeds and, 52
   VCE, fuel reactivity and, 104

laminar flame, temperature distribution across, 64-65

laminar flame propagation, molecular diffusion and, 64

laminar flame speed:
   equation for, 106
   turbulent combustion and, 112

Le Chatelier's principle, computation of burning velocity and, 196-197

Leeds Correlation, 141

liquefied natural gas (LNG), flash fires and, 81

liquefied propane gas (LPG), flash fires and, 81

liquid CO_2, BLEVE and, 40-41

liquid contents, rocketing fragments and PVB's, 293

liquid flashes,
   rocketing fragments and PVB's, 289
   typical BLEVE and, 319

liquids,
   heterogeneous nucleation and boiling point of, 312
   liquid wetted wall and boiling critical heat flux of, 316
   secondary events of a BLEVE and flammability of, 319
   subject to BLEVE, 311

liquid wetted surfaces, fire heat in BLEVE and, 316

Lower Flammability Limit (LFL), 80

M

Mach stem,
   defined, 73
   height of explosion and, 73-74
Maplin Sands tests, FF and, 82
Markstein-Taylor instability, flame
   acceleration and, 107
maximum allowable working pressure
   (MAWP), PVB and, 242, 274
mechanical integrity, PVBs and, 242, 243
methane-air mixture, gas flow velocity of,
   104-105
minimum ignition energy (MIE), fuels and
   their, 54-55
mitigation measures, industrial hazards and,
   4
molecular diffusion, laminar flame
   propagation and, 64
Musselbanks propane tests, FF and, 83

N

non-ideal gases, co-volume parameters and,
   278
numerical simulations, 97
   PVB blast effects prediction and, 260-261

O

obstacle density. *See* congestion
obstructed regions,
   "donor" and "acceptor," 181
   TNO multi-energy method and free
     volumes of, 182
olefin unit, case history of, 11-13
one-dimensional (1-D) flame expansion, 103
one-dimensional (1-D) geometry, flame
   speed correlation and, 193
one-dimensional (1-D) numerical calculation,
   gas charge *vs.* TNT charge and, 174-
   175
"one-step" BLEVE, 318
overpressures,

computation of impulse and side-on, 274-
   277
cylindrical PVB and scaled, 258-259
3-D configurations with high obstacle
   density and, 130-131
defined, 71
deflagration and detonation and peak, 100
exploding fuel jets and inflow, 146
flash fires and, 4
large ABR and, 138-139
low-energy ignition of unobstructed fuel-
   air clouds and, 127
object shapes and blast loading, 72
partial confinement parameters and
   maximum, 146
PVB blast effects sample problem and
   calculation of, 276
PVBs and peak, 247
repeated arrays of obstructions and, 142
"Severity Index" and source, 203
sharp-edged obstacles and, 139-140
turbulent combustion in obstacle
   environment and, 142
VCE and, 5
vessel bursts induced gas cloud explosions
   and, 145
"Von Neumann spike" and detonation
   wave, 67-68
oxygen enrichment, flammability limits and,
   52

P

pipeline rupture,
   blast pressure and impulse decay in, 256
   case histories of, 19-21, 23-25
point-source model,
   thermal radiation and, 56-57
pool fire, flash fire and, 77
pool-fire model,
   flash fire dynamics and, 89
   flash-fire radiation model and, 86
   unconfined flash fire and, 86-91
positive feedback, gas explosion and, 66
positive phase characteristics, VCE
   detonations and, 162-163
positive specific impulse,

PVBs and, 247, 258-259
Potential Explosion Site (PES),
    BST method and treatment of, 197
    ignition sources and blasts from multiple,
        198-199
    zones of congestion and/or confinement
        and, 197-198
prediction methodologies, 6
    analytical method in VCE and its pros and
        cons, 167
    TNT equivalence and its limitations, 166-
        167, 168
    VCE and, 97
    VCE and blast curve method, 167-168,
        174-176
pressure at failure, PVBs and, 299
pressure relief valves (PRVs),
    BLEVE, liquid wetted surface and, 316
    MAWP and, 242
pressure vessel burst (PVB), 1, 2
    accidental, 242-244, 294
    alumina process and case history of, 32-35
    ASME code and frequency of, 242
    blast curve selection and pressure ratios
        of, 275
    blast overpressure and temperature effects
        in, 267, 270
    blast wave energy and pressure effects in,
        267
    BLEVEs and, 5-6, 241
    characteristics of, 244-245
    damage factors surrounding, 244
    explosion energy and, 244
    fabrication flaws and, 243-244
    failure patterns and fragmentation from,
        277
    fragment mass distribution and accidental,
        296-297
    fragments and statistical analysis of, 293
    frisbee fragments and, 278
    groups of like accidental, 294
    initial fragment velocity and ideal-gas-
        filled, 279-280
    MAWP and, 242, 274
    mechanical integrity and, 242, 243
    prediction of fragmentation effects and,
        298

pressure-time history of a blast wave
    from, 247
    scaled fragment velocity and scaled
        pressure in, 281-283
    scaling laws in analysis of, 246
    total energy and ideal-gas-filled, 299-300
pressure vessels,
    ASME code and PVB, 241
    fabrication materials and use of, 241
    initial fragment velocity computation and
        content of, 301-304
pressure waves, rupture of vessels and pipes
    and intensity of, 145
primary fragments,
    high explosives, PVBs and, 277
    pressure vessel, its content and, 277
process design, industry safety standards and
    practices of, 3-4
process safety, management of industrial, 3
propane fires, case histories of, 7-8, 8-10
propylene HDPE unit, case history of VCE
    and BLEVEs of, 26-29

R

radiation effects and VCE, 99
radiation hazards, FF and transient, 78
radiation heat flux, FF and, 91
radiation wavelength, atmospheric adsorption
    and, 59-60
rail car derailment, case history of BLEVEs
    of LPG, 45-48
railroad tank car, case history of BLEVEs of
    LPG, 48-49
rarefaction waves,
    blast wave and, 72
    PVB and, 245
risk management, management support and,
    4
rocketing fragments,
    BLEVE and ranges of, 344-348
    ranges of, 293, 300

S

Sachs' scaling law,
  dimensionless groups of parameters and, 75-76
  PVBs and, 246
sample problems,
  air blast from BLEVEs, 349-355
  blast effects of PVB
    failure of propane-filled cylinder and, 306-310
    failure of propane-filled cylinder during testing, 306
    spherical vessel and, 274-277
  BST method,
    propane spill, 233-235
  CAM,
    propane VCE, 236-239
  fragments from BLEVEs, 355-358
  thermal radiation from a BLEVE, 359-360
  TNO multi-energy method,
    assumed initial strength, 225-229
    initial strength per GAME, 229-233
  TNT equivalence method,
    pipe rupture VCE, 221-224
    storage site VCE, 218-221
saturation pressure, BLEVE and, 313
scaled side-on impulse,
  PVB blast effects sample problem and calculation of, 276
scaled standoff distance,
  BST method and computation of, 201
  PVB blast effects sample problem and calculation of, 275
secondary fragments, explosion source environs and, 277
"Severity Index," 143
  CAM for blast curve and use of, 203
  effect of edge ignition on, 144
  effect of jet ignition on, 145
  effect of length-to-width ratio on, 144
  influence of spacing between congested volumes on, 148
  plot of scaled source overpressure as a function of scaled, 205
  source overpressures and, 203

TNO severity levels and, 203
sharp-edged obstacles, turbulence and, 139-140
Shell Congestion Assessment Method (CAM) correlation, 143
shock front, blast wave and, 72
shock initiation, 98
  numerical studies of blast effects of PVB and insight into, 249
shock pressure, uneven ruptures and, 255
shock waves,
  AIT and, 66
  autoignition temperature and, 66
  Mach stem and coalescing of, 73
  PVB of ellipsoid and, 255
solid-flame model, thermal radiation and, 57-64
source energy, of compressed gas, 275
source overpressure, CAM and, 202
spherical PVB,
  pressure contours of blast field and elevated, 257-258
  sample problem of, 274-277
spills,
  combustion behavior of continuous *vs.* instantaneous, 82
  Musselbanks propane tests and propane, 83
  sample FF calculations for massive propane, 92-96
  sample problem involving propane, 233-235
statistical methods, fragment characteristics and use of, 298
steel vessels,
  BLEVE and stress rupture of, 317
  temperature of wall and strength of, 316
storage facility, case history of BLEVEs of LPG, 41-43
storage vessel, case history of BLEVEs of liquid CO_2, 40-41
sub-explosion, detonation wave and, 66
submicron nucleation, boiling of liquids, 312
superheat, size of bubble nuclei and needed, 312
"superheat limit temperature," 312-313

T

tank truck, case history of BLEVEs involving
    propylene, 44-45
tetryl, 153
The Port Huron explosion, 156
thermal radiation, 2
    emissive power and, 58
    FF *vs.* jet or pool fires and, 78
    fireballs, FF and, 51
    heat of combustion and, 56
    measurement of the effects of, 55-56
    source and hazards of, 55
    value of point-source model and, 57
    view factors and, 57-58
three-dimensional (3-D) flame expansion,
    101
    geometry of, 102
TNO GAME Correlation, 142
TNO multi-energy method,
    basic geometrical shapes and, 180
    blast curves and VCE prediction in, 176-
    177
    blast history construction, MERGE data
        set and, 187
    blast parameters calculation in, 185
    combined energies of blast sources and,
        184
    Critical Separation Distance and, 181-182,
        186
    deflagration *vs.* detonation consideration
        in, 179
    "donor" and "acceptor" obstructed regions
        and, 181
    equivalent radius and, 184
    flame propagation direction and, 180
    free volume of obstructed region and, 182
    GAMES project and, 182-183
    in-cloud flammable mass of vapor and,
        179
    initial blast strength, gas-explosion blast
        behavior and, 179
    overpressure and duration curves in, 178
    plant geometry and VCE hazard in, 179
    potential blast sources and, 180
    sample problem using assumed initial
        strength, 225-229

sample problem using initial strength per
    GAME, 229-233
source strength estimation and, 183-184
unobstructed part of vapor cloud and, 184
volume of unobstructed part of vapor
    cloud and, 182
TNO Yellow Book, Baker-Strehlow blast
    curves and, 261
total amplitude of characteristics, VCE
    detonations and, 162
total variation diminishing (TVD) scheme,
    blast effects of PVB and, 250
transition to detonation, jet ignition of fuel-
    air mixtures and, 147, 148
transmissivity, definition and measurement
    of, 59-61
transmissivity curves, adsorption, scattering
    and, 90
trinitrotoluene (TNT),
    blast curve method and free-air vs.surface
        burst, 169
    combustion rate and equivalent-charge
        mass of, 167
    in-cloud amount of fuel and equivalent-
        charge mass of, 172-173
    side-on blast peak overpressure and
        equivalent-charge mass of, 173-174
trinitrotoluene (TNT) equivalence,
    as a conversion factor for explosion
        energy, 169-172
    explosion energy and, 166
trinitrotoluene (TNT) equivalence method,
    97, 218
    accidental VCEs and, 171
    "average major incident conditions" and,
        172
    cloud's explosion severity and, 218
    dissipation rate of energy in detonating
        TNT charge and, 171
    energy or explosion of TNT and, 171
    ground burst data and, 171
    heat of combustion of the amount of fuel
        and, 172
    portions of fuel and calculation of, 171
    PVB blast effect prediction and, 260
    safe and conservative value of, 171

sample problem involving pipe rupture VCE and, 221-224

sample problem of storage site VCE hazard assessment and, 218-221

trinitrotoluene (TNT) hemispherical surface bursts, parameters for, 169-170

turbulence,
    combustion and, 65
    measurement of, 53-54
    pre- and post-ignition vapor cloud properties and, 53
    sharp-edged obstacles and, 139-140
    VCE and the role of, 65

turbulent flame, flash fire and two-dimensional, 86

turbulent flame speed,
    FF and, 52
    pool-fire model for FF and, 86

turbulent flow field, flow field divergence and, 113

turbulent gas jet,
    detonation of fuel-air mixture and, 156
    initiation of detonation and, 147

two-dimensional (2-D) flame expansion, 102, 103

"two-step" BLEVE, 319

U

ultimate failure pressure, PVBs and, 243

unburned gas, combustion products and, 113

unobstructed cloud, combustion rate and ignition of, 127

V

vapor cloud detonations, 97
    blast effects by,
        accidental detonations and benefits of evaluating, 159
        blast wave parameters, 165
        detonation parameters determination in, 159
        from heterogeneous mixtures vs. incomplete combustion, 165
        measurements of blast effects from, 161

motor fuel-air experiments, 165
    positive impulse vs. distance from heterogeneous detonations, 166
    positive overpressure and impulse for, 163-164
    positive overpressure vs. distance from heterogeneous detonations, 165
    from spherical, 163
    critical initiation energy and direct initiation of, 152
    direct initiation and, 152-153
    non-homogenous mixture and, 155

vapor cloud dispersion, atmospheric, 51

vapor cloud explosion (VCE), 1, 2
    air blast effects and, 98
    BST method and internal, 194
    confined/congested area and, 98
    congested zone and combustion rate of, 141
    congestion and, 103
    definition and example of, 5, 99
    dispersal of materials in, 98
    fireball and, 98
    flammable cloud layout and, 101
    flash fires and, 78
    ignitable mixture and, 98
    ignition source and, 98
    mode of accidental, 159
    necessary conditions for, 98
    overpressure and, 5
    phenomena, 97
    prediction methodologies for, 97, 166-168, 174-177
    prediction of fragmentation effects and, 99
    releases, 98
    role of confinement in, 101
    secondary fire and, 98
    turbulence and combustion interaction and, 65
    unconfined VCE and, 99

vapor cloud explosion (VCE) deflagrations, 97, 99
    burning velocity and, 99
    chemical reaction propagation and, 100
    2-D experiments in topside structures and, 124
    effect of congestion,
        obstacle blockage ratio and. 137-139

obstacle pitch and, 134-137
obstacle shape and arrangement and,
    139-140
effect of other factors,
    scale and, 140-141
    unburned mixture displacement and,
        140
flame speed and, 99, 101
fuel and combustion products and, 99
test results in 2-D configuration and, 116-
    117, 123
theory and research in, 104
tube-like apparatus experiments, 114
under unconfined and low congestion
    conditions, 129
VCE and, 99
vapor cloud explosion (VCE) detonation, 100
vapor cloud explosion (VCE) energy, BST
    method in estimation of, 200
vapor cloud ignition, material properties and,
    51
vapor clouds, 97
    blast effects of VCE and volume of, 176
    concentration profile of, 51
    detonation and, 156
vapor concentration,
    flammable limits and, 51
    flash point and, 52
vent discharge, case history of VCE of
    atmospheric, 29-31
venting,
    1-D geometries and, 115
    2-D geometries and, 119
vessel,
    size of bubble nuclei and clean smooth-
        walled, 312
    typical BLEVE and rated working
        pressure of, 319
vessel failure,
    BLEVE, PVB and mechanism of, 313
    BLEVE and possible fates of, 314-315
    critical crack length and catastrophic, 314
    hazards of, 3
    speed of sound and, 314
vessel material, BLEVE and conditions for
    failure of, 313
view factors,

cylindrical and plane vertical transmitters,
    90
measurement of, 61-64
thermal radiation and, 57-58
volume blockage ratio (VBR), 103
    defined, 133
    large ABR and, 138
"Von Neumann spike," detonation wave
    overpressure and, 67

W

wave speed, VCE detonation and, 100
weapon systems, development of, 97
wind speed, flame speed and, 84-85

Y

yield factor. *See* trinitrotoluene (TNT)
    equivalence

Z

Zel'dovich-Von Neumann-Döring (ZND)
    model, detonation and, 67-68